CRITICAL TERMS *for* ANIMAL STUDIES

T0138477

CRITICAL
TERMS
for
ANIMAL
STUDIES

Edited by
LORI GRUEN

THE UNIVERSITY OF CHICAGO PRESS
Chicago and London

The University of Chicago Press, Chicago 60637
The University of Chicago Press, Ltd., London
© 2018 by The University of Chicago
Chapter 2, "Activism," © Jeff Sebo and Peter Singer

Published 2018
Printed in the United States of America

27 26 25 24 23 22 21 20 19 18   1 2 3 4 5

ISBN-13: 978-0-226-35539-9 (cloth)
ISBN-13: 978-0-226-35542-9 (paper)
ISBN-13: 978-0-226-35556-6 (e-book)
DOI: https://doi.org/10.7208/chicago/9780226355566.001.0001

Library of Congress Cataloging-in-Publication Data

Names: Gruen, Lori, editor.
Title: Critical terms for animal studies / edited by Lori Gruen.
Description: Chicago ; London : The University of Chicago Press, 2018. | Includes
    bibliographical references and index.
Identifiers: LCCN 2018008776 | ISBN 9780226355399 (cloth : alk. paper) | ISBN 9780226355429
    (pbk. : alk. paper) | ISBN 9780226355566 (e-book)
Subjects: LCSH: Human-animal relationships. | Animals—Social aspects. | Animals
    (Philosophy). | Human-animal relationships—Moral and ethical aspects. | Animal rights.
    | Cognition in animals.
Classification: LCC QL85 .C75 2018 | DDC 591.5—dc23
LC record available at https://lccn.loc.gov/2018008776

♾ This paper meets the requirements of ANSI/NISO Z39.48–1992
(Permanence of Paper).

Contents

INTRODUCTION

Lori Gruen

Animal Studies is almost always described as a new, emerging, and growing field. A short while ago some Animal Studies scholars suggested that it "has a way to go before it can clearly see itself as an academic field" (Gorman 2012). Other scholars suggest that the "discipline" is a couple of decades old (DeMello 2012). Within cultural studies as well as the social sciences, there have been multiple attempts to locate the beginning of Animal Studies in the 1990s, and each proposed origin story is accompanied by specific aspirations for the field. The various hopes evoked by Animal Studies are part of what makes the field so exciting and at times contentious.

Histories

In one of the first journals dedicated to Animal Studies, *Society and Animals*, editor Ken Shapiro wrote in 1993 that "the main purpose" of the journal "is to foster within the social sciences, a substantive subfield, animal studies." And he described this subfield as primarily concerned with providing a "better understanding of ourselves"; "through animal studies we wish to understand our varied relations to them, and to assess the costs—economic, ethical, and most broadly, cultural—of these relations" (Shapiro 1993, 1). Social scientists also assessed the benefits of these relations, insofar as they existed, for both humans and other animals.

Around the same time that *Society and Animals* was launched, in literary studies, according to Robert McKay (2014), "very few scholars were concerned with the near omnipresence of nonhuman animals in literary texts or how they formed part of a much longer story about creatural life that the humanities, in dialogue with other disciplines, could document and interpret"(637). This paucity of attention may have been the result of a discomfort that emerges when, as Susan McHugh (2006) describes it, "a systematic approach to reading animals in literature necessarily involves coming to terms with a discipline that in many ways appears organized by the studied avoidance of just such questioning." But that avoidance was beginning to dissipate by the late 1990s, when we find "the peculiar correlation that gave birth to Animal Studies at that time: the commitment to developing both scholarly knowledge of an as yet unthought subject of inquiry (always a serious business) and also the responsibility needed to show the proper respect for, to take seriously as subjects of experience, the animals whose lives are represented in cultural texts" (McKay 2014, 637). Cary Wolfe (2009) reflects further:

> One would think animal studies would be more invested than any other kind of "studies" in fundamentally rethinking the question of what knowledge is, how it is limited by the over determinations and partialities of our "species-being" (to use Marx's famous phrase); in excavating and examining our assumptions about who the knowing subject can be; and in embodying that confrontation in its own disciplinary practices and protocols (so that, for example, the place of literature is radically reframed in a larger universe of communication, response, and exchange, which now includes manifold other species). (Wolfe 2009, 571)

Within the broad rubric of cultural studies, there was a different focus than that of the social scientists, and different types of questions were being asked. And even within cultural studies we can see tensions. Is the project of Animal Studies to take animal representations seriously within literature or to take animals seriously as subjects or to come to new understandings by recognizing the difficulties and possibilities of moving beyond the human as the only subjects of cultural knowledge?

Though these questions were being asked by a growing number of literary scholars, animals were not quite "unthought subjects of inquiry" in other disciplines. Important books had already been published: Donna Haraway's *Primate Visions* (1989), Harriet Ritvo's *The Animal Estate* (1987), Carol Adams' *The Sexual Politics of Meat* (1990), and Donald Griffin's *Animal Minds* (1992) are just a few, representing quite different perspectives on "the question of the animal." The 1990s marks an important moment in the growth of work in Animal Studies to be sure, but I hesitate to call it the origin. Thinking about and with animals has been a central concern across a number of academic disciplines going back a very long time.

Within philosophy, for example, two of the most well-known scholars thinking about ethical and political obligations to other animals, Peter Singer and Tom Regan, published before the 1990s (Singer's *Animal Liberation* first appeared in 1975 and Regan's *The Case for Animal Rights* appeared in 1983), but animals as subjects of philosophical inquiry can be found all the way back to antiquity. Henry Salt, in his 1892 book *Animal Rights: Considered in Relation to Social Progress*, draws readers to his "immediate question"—"if men have rights, have animals their rights also?"—and notes,

> From the earliest times there have been thinkers who, directly or indirectly, answered this question with an affirmative. The Buddhist and Pythagorean canons, dominated perhaps by the creed of reincarnation, included the maxim "not to kill or injure any innocent animal." The humanitarian philosophers of the Roman empire, among whom Seneca and Plutarch and Porphyry were the most conspicuous, took still higher ground in preaching humanity on the broadest principle of universal benevolence. (Salt 1892, 2–3)

While philosophers were interested in what sort of ethical claims animals made on us, work in the sciences provided some reasons as to why we might owe other animals our attention and concern, why they are worthy subjects of study, and how they may be subjects in their own right.

In the mid-late 1800s, Charles Darwin's work radically altered the view of other animals and our relationships to them. Humans and other animals were not separable by kind, he suggested, only by degree. He argued that like us, animals express emotion, can experience their worlds in vivid ways, and he suggested that they can even reason:

> Only a few persons now dispute that animals possess some power of reasoning. Animals may constantly be seen to pause, deliberate, and resolve. It is a significant fact, that the more the habits of any particular animal are studied by a naturalist, the more he attributes to reason and the less to unlearnt instincts. (Darwin [1874] 1998, 77)

Though questions of animal emotion and reason were and are topics for debate (see chaps. 8, 20), Darwin's observations led to rich interdisciplinary explorations of animal intelligence (George Romanes 1882) and their *Umwelts* (Jakob von Uexküll 1934), and new fields of inquiry, including comparative psychobiology (Robert Mearns Yerkes 1925), gestalt psychology (Wolfgang Köhler 1947), ethology (Konrad Lorenz 1961 and Niko Tinbergen 1963), and eventually cognitive ethology (Donald Griffin 1976; Dale Jamieson and Marc Bekoff 1992). Like all scholarly investigations, these various areas of inquiry were shaped by the accepted theories of the times as well as particular social and cultural anxieties. The insights that emerged from these investigations led to important developments that couldn't help but inform what we now call Animal Studies. Central to these earlier explorations was a commitment to understanding other animals as subjects and often, although not always explicitly, understanding ourselves in relation to them.

Given this long history of inquiry, and I have only mentioned here a very small fraction of it, I find it odd that the novelty of Animal Studies is so often remarked on. Animal Studies seems to have had an extended developmental period akin to what is referred to in evolutionary biology as *neoteny*. *Neoteny*, coming from the Greek words *neos*, as in juvenile, and *teinein*, meaning extended, is thought to be especially advantageous for our species, *Homo sapiens*. By having an extended childhood, we come to develop our individual wit and charm, and perhaps more importantly, better abilities to cope with the complexities of our environments. Neoteny is one of the explanations for why we humans are still around when the estimated twenty-seven or so other hominid species perished (Walter 2014). Perhaps the lengthy time Animal Studies has been thought to be "developing" will similarly insure its success as a mature, interdisciplinary field.

If the intensity of scholarly attention to Animal Studies is any indication, the signs of successful maturity are good. There are conferences and workshops across a wide range of topics in Animal Studies around the globe occurring almost weekly. There are at least ten book series, a dozen or more dedicated journals, and a growing number of academic programs, some offering undergraduate and graduate degrees in response to demand from students seeking to pursue focused work in Animal Studies. And there are a large number of highly respected senior scholars working in the area, many of whom have written the chapters that follow.

While it is not necessarily a bad thing to remain in perpetual development, there is a time when focus on whether Animal Studies is yet a field can be redirected toward more interesting topics. My hope is that the publication of this volume and the quality of the discussions contained in it are indications of the field's maturity. Of course, maturity as a field doesn't mean that the state of inquiry is static or that there is consensus about what counts as the proper objects of study or best methods of inquiry.

Most "mature" disciplines have rich, often transformative debates about these issues, and this is particularly true in interdisciplinary fields.

Contestations

Activism and the "Real World"

Interdisciplinary fields such as Women's Studies, African American Studies, and more recently Environmental Studies, Queer Studies, and Disability Studies emerged as scholarly tentacles of political movements. Though the connections to activism can vary considerably depending on the experiences of scholars and teachers doing the academic work, there is a general sense that a scholar working in any one of these areas is committed to some of the goals of the political movements to which they are, or should be, accountable.

The scholarly connection to activism in these cases is not just to opinions or arguments or texts, nor is it only to the study of the social movement in question (although there is important scholarship along these lines), but to a shared normative commitment, as I call it, that motivates social movement. Normative commitments in all of these interdisciplinary fields and the movements they are connected to are ethical/political aspirations about eliminating the conditions that subjugate, erase, deny, violate, or destroy the subjects of study. And when a scholar in these fields appears indifferent to these goals or seems not to share the aspirations, it is especially noticeable. Consider an environmental scientist who discovers dangerous levels of pesticides in a particular river who, rather than reporting it to the local environmental protection department or letting the parents of the children swimming in the river know, keeps the data quiet to compare with more data that will be collected two years or five years hence. This scientist should not be surprised when challenged by environmental studies colleagues or environmental activists if and when they learn of this.

Of course, the normative commitments that scholars have, even within the same area of study, will vary as they often do within the movements with which such study is connected. Debates in women's studies about who is a "woman" were going on as women's studies programs were starting and continue to this day. Different, sometimes contradictory, conceptions and waves of feminism animate much activism and scholarship. The meaning and politics of intersections between gender, race, sexuality, class, physical ability, gender expression, and other dimensions of power and privilege generate complex disagreements and move theory and practice in new directions. To a large extent, work in women's studies and ethnic studies in the 1970s and 1980s provided important space for discussions about the ways that academic inquiry is always imbued with normative commitments, and that in turn empowered students politically. Connections to political movements have taken a variety of forms within both teaching and scholarship, and these connections can often be a source of contention, but scholars within these interdisciplinary areas are rarely completely detached from the political goals of the movements.

For example, as the recent Black Lives Matter protests occurred on the streets, African American studies programs as well as ethnic studies and gender studies programs sponsored events and offered courses addressing the issues raised by the movement. Political syllabi were made available online for those teaching courses as well as people, both within and outside of the academy, interested in more study. There have, of course, been protests on campuses, too, and this has often led to changes within universities as well as stronger university-community partnerships. Links between academics and activists have generated important scholarly collaborations that promise to reshape curriculum and research.

Although there is contention about the texture, depth, and content of the various normative commitments within these interdisciplinary areas, that there are ethical and political aspirations that accompany scholarship is not particularly controversial. But within Animal Studies, embracing normative commitments and being accountable, in some way, to the animal protection movement, also known as the animal rights movement, seems more vexed.

I think part of the reluctance to acknowledge one's ethical or political views stems from a fear of criticism from various corners. In one corner, there is that part of the animal rights movement that is loud and unforgiving. When one is attempting to explore new topics that colleagues question as being connected to inquiry in their particular field, there may also be a worry about being targeted by activists. In a different corner, there are activists, in the animal protection movement as well as other social movements, who find "theory" too far removed from "the real world" and can be critical or, more often than not, dismissive. This sort of detachment was what made "academic feminism" bad words. There are certainly animal activists who ignore Animal Studies scholarship, finding it too far removed from the lives and deaths of real animals. This, too, may serve as a disincentive to make one's work accountable to a movement that isn't particularly receptive. In yet another corner are scholars within Animal Studies who are disciplined in more historical or textual or scientific methodologies and don't see a clear connection to contemporary advocacy. Some of them think that scholars shouldn't dirty their hands with "activism." I'll say more about the anxiety about advocacy below.

Another source of reluctance to make one's political commitments known undoubtedly has to do with the depth and breadth of anthropocentrism (see chap. 3). Animal Studies provides insights into the ideologies and frameworks according to which some forms of life are enabled to thrive while others are oppressed and destroyed. Using animals in various ways is not just part of the structures that shape our lives and to which much work in Animal Studies is directed; it is also part of our daily practices. Questions about our own use of other animals certainly heighten discomfort. In human-centered scholarship, animals are relegated to the background. Animal Studies, in bringing other animals to the fore as sentient subjects who can have meaningful lives and relationships, presents challenges to our own ways of living. These challenges can be difficult to acknowledge in the classroom and at faculty meetings as well as in our personal lives. The discomfort that these challenges elicit can lead to a desire to disconnect theory from practice, scholarship from advocacy.

Institutionalization

Another contested issue has to do with the institutionalization of Animal Studies. A look at the history of women's studies programs is again instructive here. When faculty and students came together on college campuses in consciousness-raising sessions in the 1970s to protest ubiquitous sexism on campus and off, discussions began about building common curriculum to combat the silencing of and violence against women. Scholars working in many different disciplines convened and debates emerged about whether to build centralized, interdisciplinary programs, or to push for integrating feminist scholarship into more courses within disciplines. It quickly became clear that one could do both, create new women's history courses or feminist ethnography courses, for example, that could be cross-listed courses in women's studies. Feminist faculty, together with their students, began developing interdisciplinary methods for teaching and scholarship, and hundreds of women's studies programs emerged.

While there are now a great many courses offered on topics in Animal

Studies around the globe, there is one distinct difference between the creation of institutional homes for women's studies and other interdisciplinary fields and more centralized programs for Animal Studies, and that is that the subjects of study are not organizing curriculum, mobilizing faculty, or agitating for inclusion. More precisely, one of the central areas of scholarly concern in Animal Studies involves representing animals (see chap. 21) not only as symbols or metaphors for human interests and projects but as subjects themselves. Animal Studies has been at the forefront of efforts to foster new epistemological paradigms for recognizing and articulating the agency of other animals, but "speaking for" others is always tricky, especially so when the subjects don't speak human languages. Within women's studies classrooms—where important interventions about the exclusion of the experiences of black women, women of color, queer women, transwomen, and gender nonconforming people continue to occur—the excluded subjects' perspective can be articulated, usually by the subjects themselves. Feminist scholarship on just how to respectfully attend to the perspective of the other has deeply informed feminist practice over the years. This is not so easy with other animals, where not only language but entire ways of living are vastly different (see, e.g., chaps. 4, 7). The very category "animal" is so vast and includes such diverse beings as orangutans and coral, butterflies and cows, parrots and sharks, it is hard to identify a commonality other than that they are "not human."

Their status as not human has institutional ramifications as well. While sexism, racism, and other forms of prejudice exist in institutions of higher learning, whether overt or implicit, animals are there as objects of use. At large research institutions there may be laboratories containing dogs, cats, cows, pigs, and monkeys. Even at smaller institutions, rats, mice, fish, birds, and frogs are being used in the sciences. Those who use animals may complain about the idea that there is a field of study at their very institution that questions the legitimacy of their work. And this has generated tensions about institutionalizing Animal Studies. Of course, scholarly disagreements are at the heart of intellectual exploration, and questions about legitimacy themselves are centrally important for opening new avenues of inquiry. Any field, whether biology, psychology, sociology, or history, becomes static when it resists challenges.

And such challenges often come from within a discipline. I mentioned earlier that even when women's studies programs were getting started, there were questions about who it is that women's studies studies along with questions about whether women's studies is good for women. At the turn of this century, many programs began reflecting on whether their success at institutionalization had become a liability and whether the intellectual and political excitement of the field was becoming dulled as programs worked to "secure their boundaries, define an exclusive terrain of inquiry, and fix their object of study" (Brown 2005, 122). Institutionalization comes with costs. In response to these and other challenges, women's studies programs began changing their names to better reflect not just the diversity of women and women's issues differentially experienced racially, sexually, ethnically, religiously, in terms of class, ability, and gender expression but also questions about the different ways of understanding how these complex, often intersectional social positions influenced affective orientations and social institutions. Many women's studies programs became gender studies programs, others became feminist and gender studies programs, others became gender and sexuality studies programs, and there are other naming combinations as well.

What's in a Name?

These efforts to rename women's studies programs were, to a large extent, designed to more accurately represent the objects of study, but there is also a normative (in the sense I described earlier) dimension of naming. Politics and perception play a role in the naming contestations that have occurred in some interdisciplinary fields, and this is certainly true in Animal Studies.

When scholars first began describing their work as Animal Studies, there was occasionally confusion—some people, including many scientists, thought that meant scholars were working directly with animals, for example, in laboratories or in the wild. This led some scholars to adopt the name Human-Animal Studies (HAS) and emphasize the relationships that the field was devoted to examining, understanding, and critically evaluating. But this, too, led to further confusion, particularly about the meaning of *human*.

Posthumanism, for example, works toward developing new frameworks that don't center the human, often urging recognition of claims for other animals to flourish on their own terms and not in reference to categories and characteristics that are tied to human flourishing. Posthumanism challenges the assumptions, desires, and imperatives of humanism, the very theoretical framework that is often used to extend rights to other animals (see chap. 22), and takes the distinction between human and animal as a site for theorizing.

The posthumanist branch of Animal Studies is not alone in challenging the human-animal binary—those working in feminist Animal Studies have long challenged it, and theorists and activists in the developing area of scholarship on race and animals pointedly remind us that the "human" in human-animal studies is a social construction steeped in racist history (see chap. 1). Independent scholar and activist Syl Ko writes,

> In her 1994 open letter to her colleagues, cultural theorist Sylvia Wynter noted, "You may have heard a radio news report which aired briefly during the days after the jury's acquittal of the policemen in the Rodney King beating case. The report stated that public officials of the judicial system of Los Angeles routinely used the acronym N.H.I. to refer to any case involving a breach of the rights of young Black males who belong to the jobless category of the inner city ghettos. N.H.I. means 'no humans involved.'" . . .
>
> It's no wonder that one way we have historically sought and continue to seek social visibility is by asserting our "humanity."
>
> I used to be that kind of black activist. You know: *"We're human, too!"* But now, I question this strategy. . . .
>
> The domain of the "human" or "humanity" is not just about whether or not one belongs to the species homo sapiens. Rather, "human" means a certain way of being, especially exemplified by how one looks or behaves, what practices are associated with one's community, and so on. So, the "human" or what "humanity" is just is a *conceptual way to mark the province of European whiteness as the ideal way of being homo sapiens.* . . . This means that the conceptions of "humanity/human" and "animality/animal" have been constructed along *racial* lines. (Ko 2017, 20–23)

The racial and gendered social history of both the human and the animal are important areas of theoretical work. And the relationships among the various beings that are seen to fall into one or the other category, both as groups and as individuals, as well as the conceptual roles these relationships play in social, cultural, practical, and theoretical knowledge, are the objects of Animal Studies.

There is another group of scholars who take up the name Critical Animal Studies, in part as a reaction to the Human-Animal Studies nomenclature and its claims. HAS scholar Margot DeMello (2012), for example, notes in her text *Animals and Society: An Introduction to Human-Animal Studies* that "there is nothing in the field of HAS that demands that researchers, instructors, or students take an advocacy or political position of any kind" (17). Of course, not taking an advocacy position is itself political. "What defines critical animal studies," notes Claire Jean Kim (2013), "is that it is fiercely, unapologetically political. Critical animal studies scholars aim to end animal exploitation and suffering and have little patience for work that just happens to be about animals" (464). So Critical Animal Studies scholars reject the name and the claims of Human-Animal Studies.

But there are other scholars who argue that Human-Animal Studies does include a commitment to respecting and acting on the behalf of other animals. For example, Samantha Hurn (2010) writes about her fieldwork in Ceredigion, in which she observed Hindu monks campaigning for an individual animal's right to life in the face of opposition from the farming community, that "lent itself more to the approach of what is referred to as 'human-animal studies' (HAS). HAS differs, in my opinion, from anthro-zoological research through the process of 'bringing in' the animal. In other words, the hyphen in 'human-animal studies' places all of the research sub-jects on a level playing field, recognizing the interconnectedness between humans and our fellow living beings" (27).

And then there is *Anthrozoology*, a term that prioritizes the human in scholarship and tends to be more focused on the scientific aspects of human-animal relations. One anthrozoology program suggests, "At its core, the field of anthrozoology is about helping people live better lives. . . . Anthrozoology is about embracing the bond between humans and animals, and touching lives" (Carroll College, n.d.). There is a clear normative com-mitment noted here, a type of advocacy, but it is not the same sort of advocacy that one sees in Critical Animal Studies, for example.

Importantly, there are individual scholars who may identify their work with any one of these names but have a different set of political and prac-tical commitments. I view Animal Studies as an expansive field of study that encompasses aspects of all of these positions. Animal Studies uses a variety of methodologies to explore relationships of various kinds to help us understand the ways in which animals figure in each other's lives, in our lives, and we in theirs. Some of this variety is represented in the chapters that follow. Like other interdisciplinary fields, Animal Studies will con-tinue to be shaped by lively debates about normative commitments and disciplinary frameworks as well as changes in our understanding of our various relationships and, inevitably, by the prerogatives of institutions, both social and academic.

Critical Terms

One of those prerogatives shapes this book, and that is the constraint on the number of pages that limited the critical terms that are included in the volume. There are many terms that don't appear here, and my choices require some explanation. Given that the two most prominent objects of Animal Studies are the animals themselves and our relationships to them, one might expect to see chapters on chimpanzees or chihuahuas or cheetahs and chapters that specifically address our most common relationships with animals—as companions, as scientific models, as entertainment, or as food. They don't appear because these aren't really "critical terms." Critical terms might be thought of as tools to help solve the conceptual problems that are raised within Animals Studies, they provide a framework for helping us think more methodically about animals as subjects, and they are resources for analyzing our relationships with other animals. Fortunately, given the growth of Animal Studies, there are many places to find books on other animals. For example, the Reaktion series Animal, edited by Jonathan Burt, starts with albatross, ant, and ape and ends with whale, wild boar, and wolf, with seventy-six books to date, each devoted to a particular animal in between. And there are a growing number of collections in Animal Studies, some organized by discipline and others that are more interdisciplinary, focused on particular kinds of relationships with animals (e.g., as research subjects or as food).

The critical terms in this volume are centrally important for Animal Studies, and each term demands, and often elicits, varying interpretations. The authors were encouraged to bring their own distinctive voices and perspectives to "their terms." In some cases this means that the normative commitments that I mentioned above are front and center; some discussions are more descriptive, some more analytical, some significantly political. Since the authors are well-respected experts, they were not asked to provide standard descriptions of their terms or simply review various ways the term is employed within particular disciplines. Rather, they were invited to explore what they thought was most exciting about the term, and each of the chapters identifies the term's conceptual developments and theorizes in ways that help readers rethink the term's role for Animal Studies. In some chapters, the traditional or expected understanding of the term is being stretched and challenged, and this will undoubtedly raise debates and perhaps raise blood pressure, all with the hope of eliciting future engagement.

Of course, not every conceptual issue is addressed. There were practical decisions that I made about what critical terms, of the many that could have been included in a two- or three-volume work, would ultimately appear here. Fortunately, many of the terms that could have been their own chapter are discussed in other chapters. For example, *agency* is explored in the chapters on behavior, mind, personhood, rationality, and sociality; *analogy* is explored in the chapters on difference, law, and sentience; *domestication* is explored in the chapters on captivity and sanctuary; *consciousness* comes up in chapters on pain and sentience; *race* is analyzed in chapters on abolition, biopolitics, empathy, and postcolonial. But there are nonetheless gaps; no book of this sort can be comprehensive.

My hope is that *Critical Terms for Animal Studies* provides readers who are already engaged in Animal Studies as well as those who are curious about it with opportunities for thinking deeply and differently about our relationships with other animals, our conceptions of what it means to be a human animal, how we might engage practically and intellectually with other animals, and how our attitudes and actions might more positively affect the more than human world.

References

Adams, Carol. 1990. *The Sexual Politics of Meat*. New York: Continuum.

Brown, Wendy. 2005. *Edgework*. Princeton, NJ: Princeton University Press.

Carroll College. N.d. https://www.carroll.edu/academic-programs/anthrozoology.

Darwin, Charles. (1874) 1998. *The Descent of Man*. London: J. Murray. Reprint, New York: Prometheus Books.

DeMello, Margo. 2012. *Animals and Society: An Introduction to Human-Animal Studies*. New York: Columbia University Press.

Fraiman, Susan. 2012. "Pussy Panic versus Liking Animals: Tracking Gender in Animal Studies." *Critical Inquiry* 29: 89–115.

Gorman, James. 2012. "Animal Studies Cross Campus to Lecture Hall." *New York Times*, January 2. http://www.nytimes.com/2012/01/03/science/animal-studies-move -from-the-lab-to-the-lecture-hall.html.

Griffin, Donald. 1976. *The Question of Animal Awareness*. New York: Rockefeller University Press.

———. 1992. *Animal Minds*. Chicago: University of Chicago Press.

Haraway, Donna. 1989. *Primate Visions*. New York: Routledge.

Hurn, Samantha. 2010. "What's in a Name? Anthrozoology, Human-Animal Studies, Animal Studies, or . . . ?" *Anthropology Today* 26 (3): 27–28.

Jamieson, Dale, and Marc Bekoff. 1992. "On Aims and Methods of Cognitive Ethology." *Philosophy of Science* 1992 (2): 110–24.

Kim, Claire Jean. 2013. "Introduction: A Dialogue" *American Quarterly* 65 (3): 461–79.

Ko, Aph and Syl. 2017 *Aphro-ism: Essays on Pop Culture, Feminism, and Black Veganism from Two Sisters*. New York: Lantern Books.

Köhler, Wolfgang. 1925. *The Mentality of Apes*. London: K. Paul, Trench, Trubner.

———. 1947. *Gestalt Psychology*. New York: Liveright.

Lorenz, Konrad. 1961. *King Solomon's Ring*. Translated by Marjorie Wilson. London: Methuen.

McHugh, Susan. 2006. "One or Several Literary Animal Studies?" H-Animal. https://net works.h-net.org/node/16560/pages/32231/one-or-several-literary-animal-studies -susan-mchugh.

McKay, Robert. 2014. "What Kind of Literary Animal Studies Do We Want or Need?" *Modern Fiction Studies* 60 (3): 636–44.

Regan, Tom. 1983. *The Case for Animal Rights*. Berkeley: University of California Press.

Ritvo, Harriet. 1987. *The Animal Estate*. Cambridge, MA. Harvard University Press.

Romanes, George. 1882. *Animal Intelligence*. London: K. Paul, Trench.

Salt, Henry. 1894. *Animal Rights: Considered in Relation to Social Progress*. New York: Macmillan.

Shapiro, Kenneth. 1993. "Editor's Introduction to Society and Animals." *Society and Animals* 1: 1–4.

———. 2002. The state of Human-Animal Studies: Solid at the margin! *Society and Animals* 10 (4): 331–37.

Singer, Peter. 1975. *Animal Liberation*. New York: Harper Collins.

Tinbergen, Niko. 1963. "On Aims and Methods of Ethology." *Zeitschrift für Tierpsychologie* 20 (1963): 410–33.

Uexküll, Jakob von. 1957. "A Stroll Through the Worlds of Animals and Men." In *Instinctive Behavior: The Development of a Modern Concept*, translated and edited by C. H. Schiller, 5–80. New York: International Universities Press.

Walter, Chip. 2014. *The Last Ape Standing*. New York: Bloomsbury.

Wolfe, Cary. 2009. "Human, All Too Human: 'Animal Studies' and the Humanities" *PMLA* 124 (2): 564–75.

Yerkes, Robert. 1925. *Almost Human*. New York: Century.

1 ABOLITION

Claire Jean Kim

Abolition is the interminable radicalization of every
radical movement. JARED SEXTON

The giant exhibition panel, titled "Enslaved," featured two photographs
side by side: one of a black person's naked leg chained around the ankle,
and the other of an elephant's leg, similarly shackled. This panel, along
with others that extended the comparison between racial slavery and animal exploitation ("Branded" and "Sold Off") and still others that pulled
the comparison forward into the era of lynching and Jim Crow ("Hanging,"
"Experimented On," and "Beaten"), made up People for the Ethical Treatment of Animals' (PETA's) exhibit "We Are All Animals" that toured the
United States in 2005. When civil rights groups expressed concern, PETA
apologized, inserted new panels featuring other nonwhites and white
women and children (presumably with the thought that diluting the focus
on blackness would dilute the outrage of black viewers), and eventually
canceled the tour. But they were not quite done. After moving the exhibit
online for a time, in 2012 they launched a new traveling exhibit ("Glass
Walls") that resurrected the slavery analogy. At the same time, PETA sued
Sea World, charging that the Thirteenth Amendment to the US Constitution, ratified at the close of the Civil War to formalize the end of slavery,
prohibited the theme park from keeping orcas. It was as if there were a
truth that PETA could not *not* speak: *Animals are the new slaves.*

My thanks to the participants and speakers at the Race and Animals Institute held at Wesleyan
University in June 2016 and to my coorganizers, Lori Gruen and Timothy Pachirat. Conversations held at the institute enriched this chapter in numerous ways.

Granting PETA's penchant for spectacularizing animal suffering through shocking publicity stunts, in this instance the organization merely sought to amplify a theme that has become increasingly central to the project of animal liberation in the United States in recent years. If we think of the modern US animal movement as composed mostly of "welfarists," or those seeking to reform institutions in order to improve animal welfare, the radical remainder are "liberationists," who embrace an uncompromising view and call for the wholesale dismantling of animal exploiting institutions. It is the political imaginary of the latter that has come to rest upon the conceptual linchpin of "animal slavery," with many liberationists describing themselves as "abolitionists" and self-consciously locating themselves in the tradition of the nineteenth-century abolitionists who took on the institution of racial slavery.[1]

The allure of this analogy stems from its unique resonance in Western political culture and its singular capacity to endow the contemporary animal rights movement with an aura of world-historical significance, moral urgency, and historical possibility. In liberationists' long-range struggle against powerful industries (including those involved in "animal agriculture" and biomedical and pharmaceutical research) and the machinery of repression deployed by a state beholden to those industries—in their struggle, that is, against neoliberal forces on neoliberal terrain—there is obvious value in a symbolic frame that distinguishes good from evil, elevates the dignity of life over profit accumulation, and provides a moral justification for lawbreaking to boot. As corporate and state interests escalate their rhetorical and legal war against animal liberationists, whom they now officially designate "terrorists," the latter can counter that they are, rather, "abolitionists" or "freedom fighters" (Best and Nocella 2004).

The Absent Presence in Animal Abolition

Animal abolition, however, proliferates a certain kind of danger. The briefest acquaintance with antebellum public discourse in the United States reveals that likening black people to animals—to apes in the jungles of Africa, to "livestock" animals such as oxen and horses, to savage "brutes"—was a central, perhaps even indispensable, ideological practice for enacting and stabilizing the institution of slavery. For proslavery ideologues, drawn

1. Sinha (2016) traces the roots of US abolition back to the 1600s, although it is customary to speak of abolitionism primarily as a nineteenth-century phenomenon.

to the stark distinctions of polygenesis under the pressure of abolitionism, the black was more animal than man and thus properly enslaved (Jordan 1968; Frederickson 1971). For abolitionists, the very crime and sin of racial slavery was that it treated the black person, a human made in God's image, as a beast (Sinha 2016). If we consider that defenders of slavery sought always to narrow the distance between slave and animal while abolitionists sought always to enlarge it, the dangers of PETA's exhibit—whose title is "We Are All Animals" but whose content says "black people are animals"— are thrown into relief. With its insistence on blurring the line between "human" and "animal" through the specific mechanism of the historically freighted slave-animal analogy, PETA can escape the charge of reproducing the logic of slavery only to the extent that black people have been entirely reincorporated into the "human" and are thus able to serve as nondescript exemplars of this category—a conclusion that animal abolitionists assert but that is, to say the least, contestable.

No wonder, then, that some animal abolitionists proffer their arguments with trepidation. Consider the title of Marjorie Spiegel's book, *The Dreaded Comparison: Human and Animal Slavery* (1988).[2] Spiegel wants us to know that she recognizes the taboo against invoking the slave-animal analogy and that she does not violate it lightly. She hesitates, as decency requires, but in the end she *must* take the risk and articulate the comparison. Juxtaposing slave and animal images, as PETA will do nearly two decades later, Spiegel provides very few captions or explanatory comments to go along with the images, as if we need only see the images in close proximity to understand what they mean—that these are analogous oppressions, two variations on the same phenomenon. The female slave in the scold's bridle and the muzzled dog; the male slave in a spike collar and the rabbit immobilized for cosmetics testing; the pilloried slave and the monkey in a laboratory restraint.

Like all visual images, however, Spiegel's images are unruly, jumping the interpretive frame and subverting the "message." Slaves were put in scold's bridles, spike collars, and pillories not as a routine part of extracting their labor but specifically as punishment for daring to defy the master or overseer's authority—for speaking out or running away or disobeying— and what was intended here, in addition to physical immobilization, was

---

2. Consider, as another example, the conference entitled "Inadmissible Comparisons," held at the New York University Law School in 2007. This event brought together activists from the LGBT, labor, peace, animal liberation, black liberation, and feminist movements.

public humiliation, the shaming and debasement of the slave, as well as the instillation of terror in the hearts of other slaves who observed the punishment. These images of slave punishment, in other words, even as they highlight the similarities in technologies of control between racial slavery and animal exploitation, also highlight the slaves' distinctively human forms of transgression, their distinctively human vulnerabilities to certain types of psychic as well as physical sanction, and their distinctively human potential for challenging their conditions and participating in their own liberation.[3] Slaves knew themselves to be human and indeed declared their freedom through the revolutionary language of humanism. And even as slaveholders derided the revolution in Saint Domingue as the rampaging of wild beasts, they grasped, if only through the fog of negrophobic anxiety, the possibility that their own captives would follow the Haitian example. Indeed, a good portion of the slaveholder's psychic energy was devoted at all times to preventing slaves from gathering, sharing information, aiding each other's escapes, and plotting rebellion.

As a project that analogizes the animal to the slave, then, animal abolition represses (and is therefore haunted by) a series of disjunctures and contradictions between the status of the slave and that of the animal. How the difference between slave and animal was imagined and understood, say, in the nineteenth century, how recognition of this difference was built into law and practice, and how this shaped the forms of violence and coercion inflicted on slaves and animals, respectively—these are the questions Spiegel's juxtaposed images bring to mind even as her analogical frame forbids them. All we need to know about the slave from the vantage point of animal abolition is that she was treated like an animal or a thing.

And that she is no longer a slave. To keep animals center stage, animal abolition relentlessly displaces the issue of black oppression, deflecting attention from the specificity of the slave's status then and mystifying the question of the black person's status now. According to animal abolition's narrative of racial temporality, black people at some point (variously, emancipation, Reconstruction, the civil rights movement) moved demonstrably from slavery to freedom, from the outside in, from abjection to inclusion. They were resutured into the "human." Animals, by contrast, remain in their original state of abjection. Spiegel (1988) writes,

3. This is not to deny the varied and rich psychic life of animals or the fact that our understanding of it remains underdeveloped but only to emphasize some of the ways in which humans may experience and respond to captivity differently than other animals.

Most members of our society have reached the conclusion that it was and is wrong to treat blacks "like animals." But with regard to the animals themselves, most still feel that it is acceptable to treat them, to some degree or another, in exactly the same manner. . . . A line was arbitrarily drawn between white people and black people, a division which has since been rejected. But what of the line which has been drawn between human and non-human animals? (Spiegel 1988, 19–20)

Here Spiegel is borrowing from Jeremy Bentham, whose famous passage she quotes a bit later:

[Slaves] have been treated by the law upon the same footing as in England, for example, the . . . animals are still. . . . [Some] have already discovered that the blackness of the skin is no reason why a human being should be abandoned without redress to the caprice of a tormentor. It may come one day to be recognized, that the number of legs, the villosity of the skin, or the termination of the *os sacrum*, are reasons equally insufficient for abandoning a sensitive being to the same fate. (Bentham [1789] 2007, 311n, quoted in Spiegel 1996, 32).

The angel of history moves with extensionist purpose down the rungs of the Great Chain of Being—or outward, if one prefers a different spatial metaphor, toward the farthest rings of the concentric circles of moral considerability. But crucial to the success of what she is about to do (liberate animals) is what she has already done (banish racial slavery). Animals can be recognized as the "new slaves" only if the "old slaves" have vacated their position.

Animal abolition takes as an operating premise the putative resolution of racial slavery as a problem of history. This can be seen in the writings of Gary Francione and Steven Best, respectively, the two scholars most closely associated with animal abolition as a specific political stance. Francione, unrestrained by Spiegel's diffidence, has branded his approach in several books (2009, 1996, 1995, Francione and Charlton 2015, Francione and Garner 2010) and a website (www.abolitionistapproach.com), forthrightly claiming the mantle of nineteenth-century abolitionism. The title of his book *Rain without Thunder: The Ideology of the Animal Rights Movement* (1996), for instance, references a famous speech in which escaped slave and abolitionist Frederick Douglass expounded on the necessity of social agitation and unrest in the antislavery cause. In his most recent book, *Animal Rights: The Abolitionist Approach* (2015), Francione and coauthor Anna Charlton condemn welfarism, which, the authors explain, undermines the animals' cause by sanctioning their property status, touting "humane" reforms, and telling the public that they can have their beef and eat it, too. Abolitionism, for Francione and Charlton (2015), is the refusal to succumb to these lies, the insistence that immoral institutions must be opposed utterly: "We would all agree that beating one's slaves less is better than beating one's slaves more, but the institution of slavery is still morally wrong. . . . No one would promote the 'humane' treatment of slaves as something that would eradicate the injustice of the institution of slavery" (24).

The front cover of *Animal Rights: The Abolitionist Approach* (2015) features an image designed by artist Sue Coe. Several enchained figures encircle the earth, on which rests a banner proclaiming "ABOLITION" in large red letters (the only color in an otherwise black and white image). The figures—which include a cow, a lamb, a duck, a turkey, a rooster, a goose, a pig, and fishes—are all animals bred and raised for human consumption, except for the two humans at the bottom: a black woman slave holding a small banner featuring a picture of an enchained black fist, and a kneeling male slave with shackled wrists outstretched.[4] The latter is immediately recognizable as a replication of the engraving by potter Josiah Wedgwood that became the iconic image of eighteenth-century abolitionism but with one difference. Coe reproduces the image of the enchained, kneeling slave but not the caption that always accompanied it: "Am I not a man and a brother?"

The absent presence of the caption reminds us of what Coe, Francione, and Charlton would have us forget: that those fighting racial slavery in the eighteenth and nineteenth centuries, including slaves themselves, framed their appeals in the humanist discourses available to them and that these discourses specifically sought to resuture the "human" against the "animal." Abolitionists' insistence that black people not be treated like animals, in other words, functioned ideologically to reinscribe animal abjection. This is not to deny that abolitionists of that period recognized and discussed commonalities between the slave's plight and the animal's (Douglass 2013; Gossett 2015; Keralis 2012; Quallen 2016) or that many abolitionists were involved in welfarist advocacy for animals (Beers 2006;

---

4. I surmise that the black woman is a slave (although she is not in chains) because of her mode of dress, the banner she is carrying (picturing an enchained black fist), and the fact that all of the other figures in the image are depicted as enslaved.

Li 2000) but only to suggest that animal abolition draws our attention to the very things it depends on repressing—namely, the nonfungibility of slave and animal and the commitment of abolitionists to bringing slaves back into the "human" fold.[5]

Consider Sue Coe's artwork on the back cover of *Animal Rights: The Abolitionist Approach*. The front cover, as described above, depicts the misery of a world afflicted by slavery. The back cover, on the other hand, depicts the joy of a liberated world. Unchained animals encircle and embrace the earth, and the banner that used to say "ABOLITION" now says "Vegan." The sun rises red and strong in the upper right corner, signaling the start of a new day, and the lit sky is strewn with stars. The two black slaves, however, have disappeared. In a new world where liberation is defined as universal veganism, they cannot be depicted. To show them as before would highlight the fact that veganism did not secure their liberation. To show them standing free and unbowed would raise the unanswerable question of how veganism secured their liberation. Once again, an absent presence reminds us of what is not accounted for. In response, we might ask: What is animal abolition if it does not include liberated black people in its emancipatory tableau? And what could animal abolition be if it did?

Like Spiegel, Francione and Charlton (2015) are less interested in exploring the character of racial slavery than in pronouncing it dead and naming animal slavery as its successor:

We now accept that every human being—whatever their level of intelligence, talent, beauty, etc.—holds a pre-legal moral right not to be treated as property. They have the right to be a moral person and not a thing. Yet, we have not extended this right to animals. The principle of equal consideration says that we must treat similar interests similarly. Both humans and nonhumans have an interest in not being treated exclusively as resources. . . . This different treatment is speciesist. (Francione and Charlton, 2015, 29)

Given that people were not forced into slavery in the US South because of a deficit of "intelligence, talent, beauty," but rather because of their blackness, we can perhaps conclude that the authors are being coyly abstract

---

5. It is interesting that many leading abolitionists were involved in animal advocacy in the 1800s but in a capacity that one would today call "welfarist." The British abolitionist William Wilberforce, for example, was one of the founders of the Royal Society for the Prevention of Cruelty to Animals. Animal abolition only emerged as a distinct political project in the late 1900s.

here, but the rhetorical dodge is symptomatic of animal abolition's overall difficulty with race. Francione and Charlton (2015) do not go so far as to claim that black people have achieved full equality, only that current discrimination is an irrational trace of what is now safely past:

> The right not to be a slave is different from the right not to be the victim of discrimination. We may discriminate against humans in all sorts of ways and that is morally wrong but, unfortunately, there is a great deal of dispute about what constitutes discrimination. There is, however, *no* dispute about using humans as replaceable resources. When we do think of some sort of treatment as discriminatory, we condemn it. But enslaving humans is qualitatively worse than just about anything we can do to them. That is because when we are talking about discrimination, we are talking about discrimination against persons, or members of the moral community who should not be victims of discrimination. . . . Discrimination is certainly not a good thing. But when we are talking about enslaving humans or otherwise treating them as replaceable resources, we have removed those humans *entirely* from the class of persons. We have placed them in the category of things. (Francione and Charlton 2015, 16, italics in original)

While the "fifth principle" of the Francione-Charlton "manifesto" calls for the "reject[ion of] all forms of human discrimination" (2015, 9), then, we can see that the very term *discrimination*, so defined, operates to contain and displace racial oppression (backwards in time). Yet certain questions assert themselves: At what point in time did black people become "members of the moral community?" Is there really "no dispute about using [them] as replaceable resources?" How, then, do we make sense of the Movement for Black Lives emerging in force at this historical juncture, a movement dedicated to naming and challenging the pervasive, fundamental devaluation of black life?

In his article "The New Abolitionism: Capitalism, Slavery and Animal Liberation" (n.d.), Steven Best, too, seeks to displace racial injustice as the moral dilemma of our time:

> The horrors of slavery were the burning ethical and political issue of modern capitalism. Over a century after the liberation of blacks in the 1880s [*sic*], however, slavery has again emerged as a focal point of debate and struggle, as society shifts from considering human to animal slaves and a new abolitionist movement seeking animal liberation emerges as a flashpoint for moral evolution and social transformation.

Not until the civil rights struggles of the 1950s and 1960s and the Civil Rights Act of 1964 did brutality diminish, the walls of apartheid come down, and significant social progress become possible. As black Americans and anti-racists continue to struggle for justice and equality, the moral and political spotlight is shifting to a far more ancient, pervasive, intensive, and violent form of slavery that confines, tortures, and kills animals by the billions in an ongoing global holocaust.

Moral advance today involves sending human supremacy to the same refuse bin that society earlier discarded much male supremacy and white supremacy. . . . Animal liberation is the culmination of a vast historical learning process whereby human beings gradually realize that arguments justifying hierarchy, inequality, and discrimination of any kind are arbitrary, baseless, and fallacious. . . . Having recognized the illogical and unjustifiable rationales used to oppress blacks, women, and other disadvantaged groups, society is beginning to grasp that speciesism is another unsubstantiated form of oppression and discrimination.[6]

Best's teleological language suggests that moral extensionism is an immanent force in history or that history itself is the unfolding of collective reason and enlightenment. Black liberation took place, the walls of apartheid came down, social progress became possible, and racism has been recognized as a moral error. Now the "far more" troubling phenomenon of animal slavery demands our attention. Animal abolition gestures toward the moral centrality of black oppression but only after it safely incarcerates it in the past.[7]

6. See http://www.drstevebest.org/essays.htm and Best and Nocella (2004) for further elaboration of Best's abolitionist views.

7. Best's principal argument with Francione and Charlton concerns the latter's categorical insistence on nonviolence. Animal abolition, Best contends, should properly embrace a wide range of tactical orientations. However, Best quickly moves away from this ecumenical posture and back toward the viewpoint he is challenging (that violence is presumptively wrong), emphasizing that the Animal Liberation Front (ALF, which Best supports and Francione and Charlton condemn) is actually "nonviolent" because it destroys property and not human life and contrasting the ALF with various organizations that do endorse assassination. But if nineteenth-century abolitionists believed that violence was justified in the fight against racial slavery—and more and more of them came to believe this during the antebellum period, as evidenced, for example, by the open rebellions mounted to resist fugitive slave renditions after 1850 (Sinha 2016; Lubet 2010)—then why should the ALF disavow "violence" in the fight against "animal slavery"? Are animal activists not justified in using force—even lethal force—to effect animal liberation? If not, why not?

Radicalizing Animal Abolition

Animal abolition's insistence upon displacing black subjugation so that we can get on with the business of saving animals creates a boomerang effect: black subjugation is sent back into the past only to return to the present with greater force. The more animal abolition denies the nonequivalence of slave and animal, conceals the slave's recourse to humanistic claims, and hedges on the continuation of black subjugation, the more these phenomena assert themselves and demand consideration. In this sense, animal abolition performs a service by inadvertently leading us to ask, first, how we might understand the historically productive articulations and disarticulations of blackness and animality that first emerged during the period of formal racial slavery and, second and relatedly, how we might emplot the narrative of black oppression. In the remainder of this chapter, I offer some preliminary thoughts on these lines of inquiry.

We might begin with recent scholarship in black studies (and related fields) that fundamentally unsettles conventional understandings of what racial slavery is and what time frame it inhabits.[8] Racial slavery, Frank Wilderson argues in *Red, White and Black: Cinema and the Structure of U.S. Antagonisms* (2010), is not just an exploitative labor system but rather "a condition of ontology" (18). Emerging in the late Middle Ages, racial slavery inaugurated an ontological rupture between Africans and all others by designating the former as uniquely and categorically eligible for enslavement because of their physical features, thus breaking with the historical practice of enslaving groups only as a result of contingent events (such as losing a war or committing a crime). The gratuitous violence that came to be inflicted on the African/slave as a matter of course "destroy[ed] the possibility of ontology" (38) by "reconfigur[ing] the African body into Black flesh" (18) and rendering the black "an object made available . . . for any subject" (38). The slave, split off from the human, stripped of all constitutive relations, and dispossessed of being, lived in a state of "social death."[9] Rendered thus, the slave performed a vital function in the symbolic and libidinal economies of the human, serving as "the very antithesis of a Human subject" (9), the counterpoint against which the human could gain

8. The works I discuss below prompted an evolution in my thinking on the slave-animal analogy compared with my earlier writing on the subject.

9. Wilderson borrows this concept from Patterson (1982).

self-knowledge and coherence: "The Human was born, but not before it murdered the Black, forging a symbiosis between the political ontology of Humanity and the social death of Blacks" (21).

Slaveness, Wilderson writes, "stuck to the African like Velcro" (2010, 18), remaining indissociable from blackness long after the formal institution of racial slavery was dismantled. Here Wilderson joins Saidiya Hartman in disturbing the temporality of racial subjection asserted in the traditional historiography of slavery and Reconstruction. Challenging the prevailing notions that slavery ended after the Civil War and that Reconstruction was derailed by contingent developments, in *Scenes of Subjection: Terror, Slavery, and Self-Making in Nineteenth-Century America* (1997), Hartman shows that the "freedom" whites were willing to grant to freedpersons (and free black persons) was always only a new form of "subjection" in disguise, a fact that ensured the continuation of slavery and the devastating collapse of Reconstruction.

Under slave law, Hartman (1997) writes, the slave was "legally recognized as human only to the degree that he [wa]s criminally culpable" (24). In order that he might feel the full disciplinary power of the law, in other words, the slave was acknowledged to possess a limited sentience, but "the degree of sentience had to be cautiously calibrated in order to avoid intensifying the antagonisms of the social order" (9). This "savage quantification of life and person" (95) signaled the refusal of the black person's humanity under slavery rather than its recognition. Views of the freedman after the war reflected a similar racial calculus. The "freedom" extended to the freedman during Reconstruction did not resemble a white man's freedom—it was not a maximally robust notion that included freedom of motion, freedom of residence, freedom to attend school, freedom from discrimination, freedom from violence, freedom from gratuitous state interference, and so on. Rather, the freedman's "freedom," as imagined by the Freedmen's Bureau, for example, was the "gift" of not being technically enslaved, a gift given in a dubious spirit and accompanied by exhortations that the recipient strive continually to prove himself worthy.

After emancipation, the Freedmen's Bureau distributed practical manuals such as "Advice to Freedmen," which informed former slaves that they must repay the "gift" of their freedom by staying on the plantation and "bend[ing their] back[s] joyfully and hopefully to the burden" (Hartman 1997, 135) of labor. Thus, Hartman notes, the only way to earn one's "freedom," to prove oneself worthy of it, was to acquiesce in one's virtual reenslavement right down to fidelity to one's former master and his plan-tation.[10] Lest these counsels be ignored by freedpersons seeking a mean-ingful change from enslavement, Southern states hurried to pass a slew of laws—vagrancy laws; breach of contract laws; antienticement laws; laws against open hunting, fishing, and grazing, among others—that forced most freedpersons to remain on the plantation and delivered would-be escapees to the clutches of the convict lease system, an institution that created a new slave economy under the cover of penal administration (Blackmon 2008). Rather than rescuing the slave from slavery, "the lan-guage of freedom" became "the site of the re-elaboration of that condition" (Hartman 2003, 185).

"Emancipation," Hartman (1997) concludes, "appears less the grand event of liberation than a point of transition between modes of servitude and racial subjection" (6). Rather than speaking of slavery's definitive end, then, she speaks of the "afterlife of slavery" (Hartman 2007) to capture its ongoingness across epochal boundaries. It is not that slavery died and left a diminishing trace (e.g., in the form of "discrimination") but that slavery was pronounced dead for official purposes and *has since taken on new life*. Racial dynamics unfold through the modality of historical time, and spe-cific regimes of racial domination morph and adapt in the flow of historical contingency, but black subjection goes on and on and on. Reconstruction's collapse is, on one level, a story about how Northern and Southern eco-nomic and political elites reached an unholy agreement to maintain black subjugation as the enduring foundation of the nation's life, and, on a more abstract level, a story about racial power's dissimulation and resilience, its habit of announcing its own demise at the very moment of its reincarna-tion, and its uncanny ability to augment itself through the very terms that appear to diminish it.

The "afterlife of slavery" is instantiated in a racial order that is essen-tially and enduringly antiblack. In *Her Majesty's Other Children: Sketches of Racism from a Neocolonial Age* (1997), Lewis Gordon (1997) identifies the two principles structuring this order as "(1) be white, but above all, (2) don't be black" (63). The Negrophobia of the slaveholding republic is borne aloft into the present, where it finds expression in environmental racism, resi-dential and educational segregation and degradation, the school-to-prison

10. See Fick (2007) for a striking parallel in how "freedom" for former slaves had been sim-ilarly redefined during and after the revolution in Saint Domingue also with the imperative of maintaining the productivity of the plantation economy.

pipeline, mass incarceration, police violence, and other institutionalized mechanisms for continuing the "savage quantification of life and person" of which Hartman wrote (Alexander 2010; Wacquant 2002; James 2007; Davis 2005). White supremacy, the valorization of whiteness, plays a role here, but as Gordon's ordering of principles suggests, antiblackness, the aversive or phobic reaction to blackness, is the more powerful and foundational force of the two. In Frantz Fanon's (2016) unforgettable words, "Simple enough one has only not to be a nigger" (18). Although much of the ideological machinery of society is committed to disavowing the existence of this racial order and instead proclaiming a triumphalist-mythical story of racial "progress"—in this sense, animal abolition simply borrows its narrative of racial temporality from the mainstream—the Movement for Black Lives is now speaking back: "We take as a departure point the reality that by every metric—from the hue of its prison population to its investment choices—the U.S. is a country that does not support, protect or preserve Black life."[11]

This, then, is the historical backdrop against which slavery has become what Jared Sexton (2010a) calls "the grounding metaphor of social misery" (47).[12] The frequent recourse to slavery analogies today by disenfranchised groups of various kinds, Sexton argues (2010a and in a number of other articles), is neither random nor innocent but rather symptomatic of an antiblack racial order that depends on denying the singularity of racial slavery and confining it in the distant past. Responding to those who compare slavery to exile, colonization, land dispossession, exclusion from privileges, forced migration, and forced labor, he returns our focus to the slave's social death, the destruction (of being) that exceeds all other forms of destruction: "Slavery is not a loss that the self experiences—of language, lineage, land, or labor—but rather the loss of any self that could experience such loss" (Sexton 2016, 591). To generalize from the status of black people to that of other groups—for example, through the concept "people of color"—is therefore to "misunderstan[d] the specificity of antiblackness and presum[e] or insis[t] upon the monolithic character of victimization under white supremacy—thinking (the afterlife of) slavery as a form of exploitation or colonization or a species of racial oppression

---

11. "About Us." https://policy.m4bl.org/about/.

12. Slavery analogies are as old as the republic—American revolutionaries frequently described themselves as enslaved by British tyrants in the late 1700s (Furstenberg 2003), and white workers complained that "wage slavery" was worse than racial slavery in the 1800s (Sinha 2016)—and have become especially popular in the contemporary era.

among others" (Sexton 2010a, 48). Gestures of identification and solidarity by other groups, to the extent that they rest on this type of conceptual elision, become gestures of misrecognition and disavowal, insidiously extending the half-life of antiblackness (Sexton 2010b).

If we (re)center black existence in our analytic frame rather than displacing it through analogical substitution, the architecture of the racial order in its entirety comes into view, which is to say that we understand the situation of other groups correctly only if we also understand the situation of black people. Sexton (2010a) writes,

> Every analysis that attempts to understand the complexities of racial rule and the machinations of the racial state without accounting for black existence within its framework—which does not mean simply listing it among a chain of equivalents or returning to it as an afterthought—is doomed to miss what is essential about the situation. Black existence does not represent the total reality of the racial formation—it is not the beginning and the end of the story—but it does relate to the totality; it indicates the (repressed) truth of the political and economic system. That is to say, the whole range of positions within the racial formation is most fully understood from this vantage point. (Sexton 2010a, 48)

The slave, it turns out, must be reckoned with after all. What is essential, then, is "denatur[ing] the comparative instinct altogether in favor of a *relational* analysis" (47, italics in original). Instead of comparing the refugee, colonial subject, or "person of color" to the slave in a manner that effaces the latter, we might consider how the refugee, colonial subject, or "person of color" stands *in relation to* the slave, or how those dispossessed of homeland, self-rule, territory, and rights are positioned relative to those dispossessed of being itself. In this way, black existence stops serving as an evacuated historical prop dragged on- and offstage in someone else's drama and is illuminated as the fulcrum of the modern machinery of power.

How to think relationally about the animal and the slave? We might begin with the archive of racial slavery in the United States, from Senate debates to plantation bills of sale, which reminds us of the institution's continuous and intimate dependence on the symbolic (as well as material) figure of the animal. For many slaveholders, the black person's humanity was a technicality to be ritually counteracted through myriad practices of slave-animal conflation, including branding slaves, selling slaves at auction, and feeding slaves at the trough with pigs (Bay 2000). Slaves understood this all too well. Reflecting on the ex-slave narratives collected by the Work Progress Administration, Mia Bay (2000) recounts that "ex-slaves remembered being fed like pigs, bred like hogs, sold like horses, driven like cattle, worked like dogs, and beaten like mules" (119). That is, they understood that the slaveholder saw his relationship with them as, in many ways, a human-animal relationship.

One possibility, then, is to think of the ontological schema of slavery as essentially triadic, with the key terms being *human*, *slave*, and *animal*. The *human*, the primary term, is produced through the *simultaneous abjection of slaveness/blackness and animality*. *Slave* and *animal*, in other words, serve as dual ontological counterpoints to the *human*. At the same time, *slave* and *animal* are also defined in relation to each other, completing the triangular dynamic. To put it differently, the human-animal dyad, the human-slave dyad, and the slave-animal dyad are all subtended by the more fundamental *human-slave-animal triad*. It bears emphasizing that bringing black people into the analytic frame of animal liberation should not be read as an act of anthropocentric restitution at the expense of the animal but, on the contrary, as a bringing together of the two ethical issues, a *putting into relation*. Indeed, the ontological schema of slavery also bring animals into the analytic frame of black liberation.

This kind of relational analysis has the potential to appreciably shift the politics of animal abolition. It might lead, for instance, to animal abolition dropping the "animal" qualifier and folding itself into the larger abolition movement, understood as the historic effort to dismantle the antiblack racial order and secure meaningful freedom for black people for the first time. Animal abolition's radical instincts are entirely correct: it is right in its unforgiving critique of welfarism and right in its call for revolutionary change. The problem is not, as is sometimes argued, that it is too extreme, but rather that it is not extreme enough in its conceptualization of the problem, its sense of what needs to be done, and its vision of the future. By delinking their cause from black liberation, substantively if not rhetorically, animal abolitionists have to this point been "doomed to miss what is essential about the situation." By relinking their cause to black liberation, animal liberationists can not only achieve a clearer understanding of the structures of power they are struggling against and the world they hope to create but in turn can radicalize abolition by questioning its continuing humanist assumptions. It is not only black people who should not be treated "like animals." It is not only animal liberationists who should be concerned with the status of the "animal." The war against the "human" is most effectively, and perhaps necessarily, waged on these two fronts at once, from both sites of ontological opposition.

Suggestions for Further Reading

Francione, Gary, and Anna Charlton. 2015. *Animal Rights: The Abolitionist Approach.* N.p.: Exempla Press.

Kim, Claire Jean. 2015. *Dangerous Crossings: Race, Species, and Nature in a Multicultural Age.* Cambridge: Cambridge University Press.

———. 2017. "Murder and Mattering in Harambe's House." *Politics and Animals* 2 (1): 37–51.

Sexton, Jared. 2010. "People-of-Color-Blindness: Notes on the Afterlife of Slavery." *Social Text* 28 (2): 31–56.

Wilderson, Frank, III. 2010. *Red, White and Black: Cinema and the Structure of U.S. Antagonisms.* Durham, NC: Duke University Press.

References

Alexander, Michelle. 2010. *The New Jim Crow: Mass Incarceration in the Age of Colorblindness.* New York: New Press.

Bay, Mia. 2000. *The White Image in the Black Mind: African-American Ideas about White People, 1830–1925.* Oxford: Oxford University Press.

Beers, Diane. 2006. *For the Prevention of Cruelty: The History and Legacy of Animal Rights Activism in the United States.* Athens, OH: Swallow.

Bentham, Jeremy. (1789) 2007. *An Introduction to the Principles of Morals and Legislation.* Mineola, NY: Dover.

Best, Steven. N.d. "The New Abolitionism: Capitalism, Slavery and Animal Liberation." http://www.drstevebest.org/TheNewAbolitionism.htm.

Best, Steven, and Anthony Nocella II, eds. 2004. *Terrorists or Freedom Fighters? Reflections on the Liberation of Animals.* New York: Lantern Books.

Blackmon, Douglas. 2008. *Slavery by Another Name: The Re-Enslavement of Black Americans from the Civil War to World War II.* New York: Anchor Books.

Davis, Angela. 2005. *Abolition Democracy: Beyond Empire, Prisons, and Torture.* New York: Seven Stories.

Douglass, Frederick. 2013. *Narrative of the Life of Frederick Douglass, An American Slave.* Hollywood, FL: Simon and Brown.

Fanon, Frantz. 2016. "The Fact of Blackness." In *Postcolonial Studies: An Anthology,* edited by Pramod Nayer, 15–22. Malden, MA: Wiley / Blackwell.

Fick, Carolyn. 2007. "The Haitian Revolution and the Limits of Freedom: Defining Citizenship in the Revolutionary Era." *Social History* 32 (4): 394–414.

Francione, Gary. 1995. *Animal, Property, and the Law.* Philadelphia: Temple University Press.

———. 1996. *Rain without Thunder: The Ideology of the Animal Rights Movement.* Philadelphia: Temple University Press.

———. 2009. *Animals as Persons: Essays on the Abolition of Animal Exploitation.* New York: Columbia University Press.

Francione, Gary, and Anna Charlton. 2015. *Animal Rights: The Abolitionist Approach.* N.p.: Exempla Press.

Francione, Gary, and Robert Garner. 2010. *The Animal Rights Debate: Abolition or Regulation?* New York: Columbia University Press.

Fredrickson, George. 1971. *The Black Image in the White Mind: The Debate on Afro-American Character and Destiny, 1817–1914.* Middletown, CT: Wesleyan University Press.

Furstenberg, François. 2003. "Beyond Freedom and Slavery: Autonomy, Virtue, and Resistance in Early American Political Discourse." *Journal of American History* 89 (4): 1295–1330.

Gordon, Lewis. 1997. *Her Majesty's Other Children: Sketches of Racism from a Neocolonial Age.* Lanham, MD: Rowman and Littlefield.

Gossett, Che. 2015. "Blackness, Animality, and the Unsovereign." VersoBooks.com, September 8. http://www.versobooks.com/blogs/2228-che-gossett-blackness-animality-and-the-unsovereign.

Hartman, Saidiya. 1997. *Scenes of Subjection: Terror, Slavery, and Self-Making in Nineteenth-Century America.* New York: Oxford University Press.

———. 2007. *Lose Your Mother: A Journey Along the Atlantic Slave Route.* New York: Farrar, Straus and Giroux.

Hartman, Saidiya and Frank Wilderson III. 2003. "The Position of the Unthought." *Qui Parle* 13 (2): 183–201.

James, Joy, ed. 2007. *Warfare in the American Homeland: Policing and Prison in a Penal Democracy.* Durham, NC: Duke University Press.

Jordan, Winthrop. 1968. *White over Black: American Attitudes toward the Negro, 1550–1812.* Chapel Hill, NC: University of North Carolina Press.

Keralis, Spencer. 2012. "Feeling Animal: Pet-Making and Mastery in the *Slave's Friend.*" *American Periodicals* 22 (2): 121–38.

Li, Chien-hui. 2000. "A Union of Christianity, Humanity, and Philanthropy: The Christian Tradition and the Prevention of Cruelty to Animals in Nineteenth-Century England." *Society and Animals* 8 (3): 265–83.

Lubet, Steven. 2010. *Fugitive Justice: Runaways, Rescuers, and Slavery on Trial.* Cambridge, MA: Belknap.

Patterson, Orlando. 1982. *Slavery and Social Death: A Comparative Study.* Cambridge, MA: Harvard University Press.

Quallen, Matthew. 2016. "Making Animals, Making Slaves: Animalization and Slavery in the Antebellum United States." Honors thesis, Georgetown University.

Sexton, Jared. 2010a. "People-of-Color-Blindness: Notes on the Afterlife of Slavery." *Social Text* 28 (2): 31–56.

———. 2010b. "Proprieties of Coalition: Blacks, Asians, and the Politics of Policing." *Critical Sociology* 36 (1): 87–108.

————. 2016. "The Veil of Slavery: Tracking the Figure of the Unsovereign." *Critical Sociology* 42 (4/5): 583–97.

Sinha, Manisha. 2016. *The Slave's Cause: A History of Abolition*. New Haven, CT: Yale University Press.

Spiegel, Marjorie. 1996. *The Dreaded Comparison: Human and Animal Slavery*. Rev. and expanded ed. New York: Mirror.

Wacquant, Loïc. 2002. "From Slavery to Mass Incarceration: Rethinking the 'Race Question' in the US." *New Left Review* 13 (January/February): 41–60.

Wilderson, Frank, III. 2010. *Red, White and Black: Cinema and the Structure of U.S. Antagonisms*. Durham, NC: Duke University Press.

2 ACTIVISM

Jeff Sebo and Peter Singer

The history of animal activism is rife with controversy about tactics. Some controversies involve pragmatic disagreement about which tactics are most effective. Other controversies involve principled disagreement about which tactics are morally right or wrong independently of whether or not they are effective. Our focus here will be on pragmatic disagreements in animal activism as well as on the relationship between animal activism and Animal Studies.

Some pragmatic disagreements concern what issues activists should address, what goals they should aim for with respect to these issues, and what paths they should take to achieve these goals. For example, can activists do more good overall by focusing on animals in food, research, education, or entertainment? Either way, what is the best outcome that activists can realistically achieve? For instance, in the case of animals used for food, should activists aim to regulate this practice to ensure that people treat animals as well as possible, or should they aim to abolish this practice to ensure that people do not harm animals at all? Finally, no matter what goal activists select, what is the best way to achieve this goal? For example, should activists aim to bring about their ideal food system through a series of incremental reforms to the current system or through an effort to dismantle the current system and then build a new one in its place?

Other pragmatic disagreements concern how activists should engage with the public as well as with other social movements. For example, can activists do more good overall by advocating primarily for individual behavioral change (e.g., through vegan advocacy) or by advocating primarily for structural social, political, and economic change (e.g., through advocacy for legal personhood for animals or research and development of cultured meat)? Also, what kind of tone should activists take in their advocacy? Should they take a conciliatory approach, asking people to reduce consumption of animal products? Or should they take a confrontational approach, protesting in spaces where people consume animal products? And should activists focus exclusively on nonhuman animals in their advocacy, or should they aspire to stand in solidarity with other social movements by working to uproot racism, sexism, ableism, and other forms of oppression, too?

Still other pragmatic disagreements concern what the limits of effective animal activism are. Should activists always follow the law, or should they sometimes break the law by performing open rescues or undercover investigations even when states have passed laws against that? Moreover, if activists do sometimes break the law, should they always do so civilly, limiting any illegal activity to protest, liberation, and investigation, or should they sometimes do so "uncivilly," perhaps by engaging in property destruction as well? Finally, if activists do sometimes break the law "uncivilly," and if (as the vast majority of activists believe) they should never engage in violence, how can they draw the line between nonviolent and violent direct action in practice? Here activists have to consider not only what counts as violence but also what political opponents can frame as violence for purposes of representing activists as terrorists.

Academics and activists have been asking these questions about effective animal activism for decades without much progress. Part of the reason for this lack of progress has been lack of evidence about what would in fact be most effective. In the absence of evidence that would decide the matter, it is easy for each side of a debate to construct plausible, albeit speculative, arguments about why their approach is right and other approaches are wrong.[1]

Some academics and activists are now attempting to solve this problem by finding and promoting evidence about the effectiveness of different approaches to animal activism. However, evidence-based activism raises concerns as well. One concern is that a commitment to evidence-based activism could lead to a bias in favor of certain types of activism (e.g., types whose benefits are relatively measurable and/or whose costs are relatively nonmeasurable) and against certain other types of activism (e.g., types whose benefits are relatively nonmeasurable and/or whose costs are

---

1. For a history of the animal advocacy movement, see Phelps (2007). And for general discussion about the ethics of many of the issues mentioned above, see Gruen (2011) and Schlottmann and Sebo (2018).

relatively measurable). The challenge, then, is to find and promote evidence about effective animal activism while mitigating this risk of measurability bias as much as possible.

Many Animal Studies scholars find themselves in a dual role concerning this topic because they are activists as well as academics, and therefore they have a personal as well as professional stake in this discussion.[2] This dual role has benefits and costs. On one hand, it can motivate one to answer these questions accurately (since one might hope to learn how to do activism effectively) and to draw from personal as well as professional experience while doing so. On the other hand, it can also motivate one to answer questions about effective activism self-servingly (since one might hope to learn that one is already, in fact, doing activism effectively) and draw from personal experience more than one should while doing so. We will explore further connections between Animal Studies and effective animal activism below.

Effective Animal Activism

Responding to the need for evidence-based activism, a growing number of animal activists are joining forces with effective altruism (EA), a social movement that attempts to use evidence and reason to do the most good possible. The hope is that if animal activists take an evidence-based approach to their work, as EAs do, they can resolve some of the disagreements and dissolve some of the tensions that currently stand in the way of progress in the animal rights movement.

In one sense, EA is a very old idea. A commitment to using evidence and

2. The authors of this chapter are no exception. We both work in animal activism and Animal Studies, and we also both work in the effective altruism movement that we will be discussing here. For example, Jeff Sebo has been involved in animal activism since he was a college student, and he is currently a board member at Animal Charity Evaluators, a board member at Minding Animals International, executive committee member at the Animals & Society Institute, and director of the Animal Studies MA program at New York University. Similarly, Peter Singer has a history of activism going back to 1971 when, as a graduate student at the University of Oxford, he organized the first Oxford protest against battery cages and veal crates. He was for many years president of Animal Liberation (Victoria), cofounder and president of the Australian Federation of Animal Societies (now Animals Australia), and cofounder of The Great Ape Project. He was a close advisor to Henry Spira and, after Spira's death, president of Spira's organization, Animal Rights International. He continues to speak at events for animals and is a board or advisory board member of several animal organizations. This work provides us with valuable experience, but it also positions us in this discussion in ways that we need to be mindful of.

reason to do the most good possible has been implicit in the work of many academics and activists, including those who work on animal issues. But in another sense, EA is a very new idea. In particular, the term *effective altruism*, as well as the growing network of organizations explicitly committed to this ideal, is an exclusively twenty-first-century phenomenon.

How do EAs in general, and effective animal activists (EAAs) in particular, attempt to use evidence and reason to do the most good possible? William MacAskill has developed an influential model that invites people to ask several questions (MacAskill 2015).[3] On this model, an EAA should first ask about the *scale* of a problem: How much harm does this problem cause relative to other problems? Second, an EAA should ask about *neglectedness*: How much attention are people currently paying to this problem relative to other problems? Third, an EAA should ask about *tractability*: How much of a difference, if any, are people likely to make on this problem relative to other problems? Finally, an EAA should ask about *personal fit*: What are their personal talents, interests, and background, and how well-suited are they for certain kinds of work? The more harmful, neglected, and tractable a problem is, and the better suited an EAA is for working on this problem, the more an EAA should prioritize working on this problem.

EAAs believe that this framework supports animal activism as a high-priority area in general, since animal suffering is a relatively massive, neglected, and tractable issue that many people are well suited to address. (Other high-priority areas include global poverty, climate change, and reducing the risk of the extinction of intelligent life on Earth.)

EAAs also believe that this framework supports some areas in animal activism more than others. For example, Animal Charity Evaluators (ACE) estimates that more than 99 percent of all domesticated animals used and killed by humans are farmed animals, yet farmed animal advocacy organizations receive less than one percent of all donations for domesticated animals used and killed by humans (Animal Charity Evaluators 2016d). They also estimate that efforts to save or spare companion animals cost several hundred dollars per animal, whereas efforts to save or spare farmed animals cost less than ten cents per animal (Animal Charity Evaluators 2016a). If these estimates are even roughly accurate, it follows that farmed animal suffering is a massive, neglected, and tractable issue relative to companion animal suffering.

---

3. For more on the ethics of EA, see Singer (2015). And for case studies of people committed to EA, see MacFarquhar (2015).

For a more complicated example, consider wild animal suffering. On one hand, wild animal suffering is both massive and neglected, making it potentially high priority as well. On the other hand, wild animal suffering is not, at present, as tractable as domesticated animal suffering. Experts do know how to relieve wild animal suffering somewhat, for example, by assisting animals with migration, but they do not know how to address the most pressing challenges that wild animals face, such as drought, famine, or predation. Moreover, even modest initiatives such as assisted migration risk interfering in delicately balanced ecosystems in ways that are difficult to predict. As a result, although some EAAs support the idea of researching interventions in wild animal suffering, few people think that they are in a position to responsibly carry out such interventions now.

What about the many other views within animal activism about the most effective strategies to pursue, for example, about what kind of tone to strike in animal advocacy or about what kind of education and outreach to engage in? EAAs attempt to resolve disagreements by seeking evidence about their efficacy. One approach involves applying existing evidence about advocacy in general to questions about animal advocacy in particular. Another involves working to gather new evidence about advocacy in general and about animal advocacy in particular. For example, several animal charities are currently funding, conducting, and/or promoting original research on whether and to what degree tactics such as leaflets, online ads, and corporate outreach are an effective tool for persuading people to reduce or eliminate consumption of factory farmed animal products.

We can be confident that if animal activists adopt strategies supported by this research, they can expect to save or spare many animals from suffering.[4] The more activists engage in vegan advocacy, the more consumers will select vegan options. And, the more activists engage in corporate outreach, the more vegan options consumers will have to select. But do the benefits of these strategies outweigh the costs? That depends on their long-term effects, and these effects are, of course, harder to measure.

Measurability Bias

Critics of EAA worry that evidence-based activism has important limits. In particular, they worry that a focus on measurable evidence will bias people

---

4. For information about recent victories animal activists have achieved through strategies supported by this research, see Pitney (2016).

in favor of approaches whose benefits are relatively easy to measure and/or whose costs are relatively hard to measure and against approaches whose benefits are relatively hard to measure and/or whose costs are relatively easy to measure, with negative consequences for the overall composition and impact of the animal rights movement.[5]

For example, consider the contrast between conciliatory and confrontational approaches to animal activism. On the conciliatory side, we can grant that, when the Reducetarian Foundation persuades people to reduce their consumption of factory farmed animal products, they do real measurable good in the short term, since they persuade many people to make positive behavioral changes right away. What are the long-term effects? This is harder to say. It might be that the Reducetarian Foundation does real, albeit less measurable, good in the long run as well by motivating people to gradually change their beliefs, values, and practices over time so that they eventually become less inclined to consume any animal products at all. Or it might be that they do real, albeit less measurable, harm in the long run by motivating people to think that reducing consumption of factory farmed animals is all that activists need to do in order to treat animals well. Either way, the long-term effects are hard to measure, and so the short-term benefits will be especially salient for activists looking to measurable evidence for guidance.

Meanwhile, on the confrontational side, we can grant that, when Direct Action Everywhere (DxE) activists protest production and consumption of "humane" animal products by chanting "it's not food, it's violence" at Chipotle or Whole Foods, they do real measurable harm in the short term, since they alienate potential allies and risk leading potential Chipotle and Whole Foods customers to shop at places with lower welfare standards. What about the long-term effects? Once again, this is harder to say. It might be that DxE does real, albeit less measurable, harm in the long run as well by motivating people to become even more entrenched in their resistance to animal rights. Or it might be that they do real, albeit less measurable, good in the long run, since the kinds of disruptive tactics that alienate potential allies in the short term also challenge oppressive ideologies, shift the center of debate, and pave the way for radical change in the

5. For discussion of measurability bias in EAA, see Singer and Sullivan (2015). For discussions of measurability bias in EA, see Herzog (2016), Todd, Farquhar, and Mills (2012), Rubenstein (2015), and Srinivasan (2015). And for discussion of measurability bias in nonprofit organizations, see INCITE! (2007).

long run.[6] Either way, as before, the long-term effects are hard to measure, and so the short-term costs will be especially salient for activists looking to measurable evidence for guidance.

For another example (which sits at the intersection of animal activism and Animal Studies), consider the contrast between direct and indirect approaches to education and outreach about animal issues. On the direct side, if an EAA wants to know how much of a difference they can make through, say, financial support of leafleting or online advertising, they can calculate the expected impact of this work roughly as follows: They can estimate how many leaflets or online ads they can make possible with each donation, how many people are likely to see these messages, how many of the people who see these messages are likely to go vegan, how many of the people who go vegan are likely to stay vegan, how many animals a vegan is likely to save or spare in a given period of time, and so on. They can then combine these values to calculate how many animals they are likely to save or spare by taking this approach. Granted, there are uncertainties at each stage in this analysis. But even if EAAs make conservative estimates about these impacts, the expected value of leafleting and online ads is likely to be high. This is part of why, for example, ACE evaluates the Humane League and Mercy for Animals so highly: They estimate that both organizations engage in campaigns that likely save or spare multiple animals per dollar spent (Animal Charity Evaluators 2016b, 2016c).

Meanwhile, on the indirect side, if an EAA wants to know how much of a difference they can make through, say, financial support of Animal Studies in higher education (which is itself a scholarly rather than an activist enterprise), they cannot calculate the expected impact of their work nearly as easily. Sure, they can calculate how much research, teaching, and programming they can make possible with each donation, how many people are likely to be exposed to this activity, and so on. But they cannot easily calculate the degree to which this activity will contribute to bringing about the kind of long-term ideological change that they might be hoping for. After all, when activists work to promote Animal Studies in colleges and universities, they are trying to do more than promote information and arguments about the value of veganism. They are also trying to create space for sustained conversations about the root causes of domination and exploitation of human and nonhuman animals and what, if anything, people can do to bring about a more just, decent society for human and

6. For discussion of this issue, see Young (2001).

nonhuman animals. And of course any benefits that flow from this deeper level of engagement will necessarily be long-term, indirect, and diffuse. Thus, any attempt to estimate the impacts of working to promote Animal Studies in this kind of way is likely to be very difficult.

The general worry, then, is that a focus on measurable evidence risks distorting evaluations about effective animal activism in two related ways: It risks leading people to place too much emphasis on the direct, individual impacts of particular actions, and it risks leading people to place not enough emphasis on the indirect, structural impacts of sets of actions. As a result, it risks biasing assessments in favor of approaches that aim for relatively direct benefits and away from approaches that aim for relatively indirect benefits, and it also risks alienating potential allies who take these latter approaches.

This critique of EAA is an internal critique. It is not an argument against reducing animal suffering as much as possible or against using evidence and reason in the course of doing so. It is rather an argument that, if one aspires to reduce animal suffering as much as possible and to use evidence and reason in the course of doing so, then one should take care not to focus disproportionately on the direct, individual impacts of particular interventions since doing so will lead to an incomplete, and probably incorrect, analysis of what animal activists should be doing individually and collectively.

The Future of Effective Animal Activism

How can EAAs find, promote, and implement evidence about effective animal activism while mitigating the risk of measurability bias? We have two suggestions. First, EAAs can *improve* their risk-benefit analyses by expanding their methods of assessment, expanding their scope of assessment, and correcting for biases in their assessments. Second, EAAs can *supplement* their risk-benefit analyses with other approaches to reducing animal suffering (approaches which risk-benefit analysis supports at a higher level). We will briefly discuss each of these suggestions in turn.

Ways in Which EAAs Can Improve Their Risk-Benefit Analyses

EAAs Can Expand Their Methods of Assessment

First, EAAs can consult a wider range of fields than they sometimes do, including history and social, political, and economic theory. This will allow

EAAs to see the role that particular tactics can play in a broader division of labor within and across social movements. For example, historical studies of the civil rights movement, the feminist movement, the LGBTQ+ movement, and more allow one to see how conciliatory and confrontational approaches to activism can interact in ways that make the whole different from and greater than the sum of its parts. Historical studies also allow us to see how the emergence of new academic fields can correspond to increased engagement with the topics that they address even if the causal pathways are difficult to detect. Of course, ideally, EAAs would be able to combine assessments of individual approaches and assessments of sets of approaches into a single, unified judgment about what activists should do to help animals as effectively as possible. It might be difficult to do this work well, but it is not impossible to do so. In the meantime, making both kinds of assessments available so that people can make up their own minds is a good first step.

EAAs Can Expand Their Scope of Assessment

Second, EAAs can apply these expanded methods of assessment to a wider range of questions than they sometimes do, including questions about approaches whose benefits are difficult to measure. For example, even if one might not currently be able to reliably compare conciliation with confrontation or direct with indirect education and outreach, one might at least be able to reliably compare interventions *within* each category. For instance, one can look to the history of confrontational activism (or conduct experiments in confrontational activism) for guidance about how to preserve the expected long-term benefits of this approach while mitigating the risk of alienating people in the short term. Similarly, one can look to the history of other interdisciplinary fields (or look at the limited evidence currently available about Animal Studies as an interdisciplinary field) for guidance about how to develop Animal Studies in an intrinsically and instrumentally valuable way. Of course, these kinds of studies might not tell particular individuals whether to engage in these approaches in the first place. But they might at least provide them with partial guidance about how to engage in these approaches insofar as they choose to do so.

EAAs Can Correct for Biases in Their Assessments

Third, as Jennifer Rubenstein and Amia Srinivasan have suggested, EAAs can consider how the history and demographics of the animal rights and EA movements might be limiting their perspective (Rubenstein 2015;

Srinivasan 2015). For example, many EAAs have relatively privileged identities and backgrounds. Does that make them more trusting of current social, political, and economic systems than they should be? If so, then they have reason to expand their demographics and discount their intuitions in favor of systems from which their current demographics benefit accordingly. Relatedly, EAAs can consider whether and how observational biases might be shaping their collection and interpretation of evidence. For instance, are they unduly influenced by the streetlight effect, that is, the tendency to look for answers where they are more apparent; the availability heuristic, that is, the tendency to rely on familiar examples; and/or selection bias, that is, a bias that results from selecting unrepresentative samples? If so, then they have reason to question their reliance on leafleting, online ads, and other tactics whose benefits are relatively familiar and measurable. Granted, these corrections might not eliminate measurability bias in effective animal activism. But they can at least reduce this bias, and if they do, that will be a step in the right direction.

Ways in Which EAAs Can Supplement Their Risk-Benefit Analyses

As Jasmin Singer and Mariann Sullivan have suggested, EAAs can supplement their risk-benefit analyses with other approaches to reducing animal suffering (approaches which risk-benefit analysis supports at a higher level; Singer and Sullivan 2015). Compare John Stuart Mill, whose impartial, benevolent, pragmatic moral theory partly inspired the EAA movement and who argued that if people want to promote good outcomes as individuals, then they should not attempt to promote good outcomes with each and every action they perform. Instead, they should often proceed indirectly, by thinking about which rules and character traits will lead them to promote good outcomes overall and then following those rules and cultivating those character traits for the most part in everyday life (Mill [1861] 2002). Similarly, he argued that if people want to make use of the state to promote good outcomes, then they should not attempt to have the state impose a particular way of life on citizens. Instead, they should proceed indirectly, by thinking about what kind of political society will lead citizens to live well overall and then attempting to have the state bring about that kind of political society (Mill [1859] 1978). This is why Mill advocated, on utilitarian grounds, a liberal, pluralistic state. Yes, in a liberal, pluralistic state, people will disagree about what to do, and some people will make bad decisions. But they can also experiment with different ways of living, learning from each other and making more progress over time than they would if the state imposed a single, unified vision of how to live on everybody.

We believe that EAA would benefit from a similarly indirect approach. In particular, if animal activists want to reduce animal suffering as much as possible, then they should not attempt to save or spare the most animals possible with each and every action they perform. Instead, they should often proceed indirectly, by thinking about what kinds of approaches will lead people to save or spare the most animals overall and then engaging in these approaches for the most part in everyday life and promoting these approaches for the most part within the animal movement.

What would the animal movement look like if people took this more indirect approach? In our view, it would be an informed, rational movement in which activists engage in direct risk-benefit analysis much more than they do now. And we think EAAs are correct to focus on the value of this approach in the short term. But it would also be a liberal, pluralistic movement in which activists take other approaches, too, within certain limits. And this is why we think, it would be a mistake for EAAs to focus exclusively on the value of direct risk-benefit analysis in the long run.

With that said, we must qualify our support for pluralism in two ways. First, it does not extend to the use of violence on behalf of animals. This is a position for which one of us (Peter Singer) has consistently argued since the 1970s. Despite support for this view from many leaders of the early animal rights movement, including Henry Spira, in the 1980s the movement suffered a significant setback when a handful of incidents carried out by a tiny minority made it possible for political opponents to brand all animal activists as "terrorists."[7] The accusation was absurd but nevertheless damaging. In the present climate, violence would be even more disastrous.[8] Individuals or organizations sanctioning violence as a practical strategy for achieving animal liberation cannot be tolerated within a movement that is

---

7. Perhaps the most dramatic of these actions occurred in November 1982, when an organization calling itself the Animal Rights Militia mailed letter bombs to UK prime minister Margaret Thatcher, the government minister responsible for overseeing animal experimentation, and to the leaders of the three main opposition parties. No one was injured by these devices. Two years later, the same organization placed explosive devices under the cars of two scientists who used animals in experiments. Both vehicles were damaged beyond repair. For further incidents, see the Wikipedia entry on the Animal Rights Militia. And for the use of the label "terrorism," see Miller (2006).

8. For more, see Potter (2011).

seeking to persuade the general public that living an ethical life is incompatible with the violence of eating animals or of other kinds of animal use.[9]

The second point is that participants in a pluralist movement should draw a clear line between those who are defending the continuing abuse of animals—essentially, factory farmers, breeders and users of laboratory animals, users of captive animals for education and entertainment, and so on—and those who are on the side of the animals. Being part of a pluralist movement means, of course, that people can disagree about strategy with other members of the same movement. Animal activists can and should air these disagreements respectfully at the appropriate occasions. They should not, however, become so consumed by these differences that instead of focusing most of their limited resources on the exploiters of animals, they focus most of these resources instead attacking those who are, from a broader perspective, allies in the struggle for animal liberation.

In any case, as we indicated above, the need for moderate pluralism within EAA is not especially pressing at the moment since EAAs still represent a minority of animal activists. As a result, any measurability bias that exists within EAA has not yet led to a similar measurability bias in the broader animal rights movement. Indeed, if anything, the broader animal rights movement still has a bias *against* measurability, since, for example, people still donate much more to companion animal organizations than to farmed animal organizations in spite of the fact that farmed animals still experience much more suffering than companion animals do. But as EAA grows in prominence, it will become more pressing for people in this branch of the animal rights movement to strike an internal balance between direct and indirect approaches to animal activism—and since this will be difficult to do, academics and activists should start thinking about these issues now.

A challenge for Animal Studies scholars interested in animal activism, then, is to think about how academic research, teaching, and service regarding animals can contribute directly and indirectly to efforts to help animals as effectively as possible. As observers of the movement, Animal Studies scholars can engage by evaluating different strategies for helping animals and examining possible biases in these evaluations and examinations. Meanwhile, as participants in the movement, Animal Studies scholars may or may not think that academic work is the most effective means that they have for helping animals (and they may or may not think that

9. For discussion about the nature and ethics of violence, see Regan (2004).

they should be evaluating their work primarily against this standard in the first place). Either way, if our discussion here is correct, then we can at least say this much: It would be unreasonable for EAAs to dismiss Animal Studies simply on the grounds that the value of this work is difficult to measure, and it would also be unreasonable for Animal Studies scholars to dismiss EAA simply on the grounds that evidence-based activism risks measurability bias. Instead, academics and activists should work together to arrive at an informed, rational, and balanced perspective about how to do the most good for animals possible.

Suggestions for Further Reading

Gruen, Lori. 2011. *Ethics and Animals: An Introduction*. Cambridge: Cambridge University Press.

INCITE! Women of Color Against Violence, ed. 2007. *The Revolution Will Not Be Funded: Beyond the Non-profit Industrial Complex*. Cambridge, MA: South End.

MacAskill, William. 2015. *Doing Good Better: How Effective Altruism Can Help You Make a Difference*. New York: Penguin.

Phelps, Norm. 2007. *The Longest Struggle: Animal Advocacy from Pythagoras to PETA*. New York: Lantern Books.

Schlottmann, Christopher, and Jeff Sebo. 2018. *Food, Animals, and the Environment: An Ethical Approach*. London: Routledge.

Singer, Peter. 2015. *The Most Good You Can Do: How Effective Altruism Is Changing Ideas about Living Ethically*. New Haven, CT: Yale University Press.

Srinivasan, Amia. 2015. "Stop the Robot Apocalypse: The New Utilitarians," *London Review of Books* 37 (18): 3–6.

References

Animal Charity Evaluators. 2016a. "Donation Impact." Last modified December 2016. https://animalcharityevaluators.org/research/donation-impact/.

———. 2016b. "The Humane League." Last modified November 2016. https://animalcharityevaluators.org/research/charity-review/the-humane-league/

———. 2016c. "Mercy for Animals." Last modified November 2016. https://animalcharityevaluators.org/research/charity-review/mercy-for-animals/

———. 2016d. "Why Farmed Animals?" Last modified November 2016. https://animalcharityevaluators.org/donation-advice/why-farmed-animals/

Gruen, Lori. 2011. *Ethics and Animals: An Introduction*. Cambridge: Cambridge University Press.

Herzog, Lisa. 2016. "Can 'Effective Altruism' Really Change the World?" *Transformation*, February 22.

INCITE! Women of Color Against Violence, ed. 2007. *The Revolution Will Not Be Funded: Beyond the Non-profit Industrial Complex.* Cambridge, MA: South End.

MacAskill, William. 2015. *Doing Good Better: How Effective Altruism Can Help You Make a Difference.* New York: Penguin.

MacFarquhar, Larissa. 2015. *Strangers Drowning: Grappling with Impossible Idealism, Drastic Choices, and the Overpowering Urge to Help.* New York: Penguin.

Mill, John Stuart. (1859) 1978. *On Liberty.* Indianapolis, IN: Hackett.

———. (1861) 2002. *Utilitarianism.* Indianapolis, IN: Hackett.

Miller, John J. 2006. "In the Name of the Animals: America Faces a New Kind of Terrorism." *National Review*, July.

Phelps, Norm. 2007. *The Longest Struggle: Animal Advocacy from Pythagoras to PETA.* New York: Lantern Books.

Pitney, Nico. 2016. "Revolution on the Animal Farm." *Huffington Post*, July 6.

Potter, Will. 2011. *Green Is the New Red: An Insider's Account of a Social Movement under Siege.* San Francisco: City Lights.

Regan, Tom. 2004. "How to Justify Violence." In *Terrorists or Freedom Fighters? Reflections on the Liberation of Animals*, edited by Steven Best and Anthony John Nocella III, 213–16. New York: Lantern Books.

Rubenstein, Jennifer. 2015. "Effective Altruism Is a Movement That Excludes Poor People." *Boston Review*, July 1.

Schlottmann, Christopher, and Jeff Sebo. 2018. *Food, Animals, and the Environment: An Ethical Approach.* London: Routledge.

Singer, Jasmin, and Mariann Sullivan. 2015. "Effective Altruism as It Relates to Animal Rights: An Open Ended Approach to Advocacy." *Our Hen House*, October 28.

Singer, Peter. 2015. *The Most Good You Can Do: How Effective Altruism Is Changing Ideas about Living Ethically.* New Haven, CT: Yale University Press.

Srinivasan, Amia. 2015. "Stop the Robot Apocalypse: The New Utilitarians," *London Review of Books* 37 (18): 3–6.

Todd, Ben, Sebastian Farquhar, and Pete Mills. 2012. "The Ethical Careers Debate." *Oxford Left Review* 7: 4–9.

Wikipedia. 2016. "Animal Rights Militia." Last modified December 11, 2016. https://en.wikipedia.org/wiki/Animal_Rights_Militia.

Young, Iris Marion. 2001. "Activist Challenges to Deliberative Democracy," *Political Theory* 29 (5): 670–90.

3 ANTHROPOCENTRISM

Fiona Probyn-Rapsey

Anthropocentrism refers to a form of human centeredness that places humans not only at the center of everything but makes "us" the most important measure of all things. As a core problem in Animal Studies, anthropocentrism is at once everywhere and nowhere, meaning that it is difficult to pin down precisely. To understand how anthropocentrism can be both everywhere *and* nowhere, we might each think about the space in which we are currently located. How is it structured with *humans* in mind rather than nonhuman animals? Take my home office as an example. My office at home is adjacent to an area of the Australian bush, part of Gundungurra country, one hour south of Sydney. From my office I can see and hear many species of birds (including chickens), and inside Alice and Billy (canine companions) often come in and sleep when I open the door for them. My home is designed and built on a human scale, with spaces and structures for ease of human use, just like the town and the city more generally, where encounters with animals are also conditional upon them fitting into structures and places that are not designed with them in mind. These structures are not just architectural in the strict sense, they are also manifestations of cultural beliefs about our place in relation to the other animals, that they fit in with us and not the other way around. These ideas, like the buildings we live in, might appear to have been there all along, preceding us and to a certain extent structuring the relations with animals that we find ourselves in now. It is the very intimacy and *homeliness* of anthropocentrism, the way it takes up residence as "our" residence, as "us," that makes displacing or challenging anthropocentrism such a tricky and vital project. Just as you think you might have located anthropocentrism somewhere specific, it can go "dorsal" (Wills 2008); it's right behind you.

In the same way that feminists have long argued that sexism is not reducible to sexist speech or sexist behavior but is present in institutions and knowledge systems, so, too, with anthropocentrism. Anthropocentrism is expressed by individuals in particular acts or statements that indicate a chauvinist attitude to animals, such as "animals are mindless," but it also informs our epistemologies or what we think we know about animal "mindlessness," for example. All three interacting levels of life (the personal, the cultural, and the epistemological) inform anthropocentrism, leaving us with a generalized sense of complicity that limits what we can claim to be outside of but also indicates a deep and abiding sense of responsibility to a complex problem that takes a huge toll on nonhuman animals.

When it comes to anthropocentrism, Animal Studies is united by a core objective, which is to avoid the worst of anthropocentrism's dominance over nonhuman life. Critical responses to the problem of anthropocentrism vary. They include an acceptance and even embrace of anthropocentrism as an unavoidable limit or standpoint from which to begin; an attempt to frame anthropocentrism as a political problem rather than only a problem of being human; a call to question the "anthropos" embedded in the concept, and lastly, a claim to have escaped or exceeded anthropocentrism in some way. All of these strategies need to be measured by the extent to which they benefit nonhuman animals and how they make space for the possibility that animals themselves also have a view on the matter. The idea that nonhuman animals also have a stake in challenging our anthropocentrism is imperative for Animal Studies, and it is one that appears, to varying degrees, within the strategies I'll discuss in this chapter.

Inevitable Anthropocentrism

The argument that anthropocentrism is *inevitable* exists across the field but to different degrees. Australian ecofeminist philosopher Val Plumwood characterizes anthropocentrism as linked to a largely benign human centeredness that can also lead to a pernicious form of human self-centeredness. We are "inevitably rooted in human experience in the world, and humans experience the world differently from other species" (Plumwood 2002, 132). An example of this might be that without machines we cannot breathe underwater for very long, see certain things like ultraviolet light, and that by comparison with dogs we have a very poor sense of smell, sight, and inadequate fur to keep us warm. These bodily and sensory differences root us in a "human experience" that we cannot avoid and that frames our perceptions. But as we'll see later, Plumwood does not see this as a pernicious or problematic form of anthropocentrism, it is not chauvinism or "arrogance" (to use Gruen's word), it's simply locatedness. Rob Boddice also argues in favor of accepting the inevitability of anthropocentrism. Tired of hearing anthropocentrism being used as an accusation ("You're anthropocentric!"), Boddice suggests that these sorts of accusations are best met with a sort of "so are you" response. He writes, "when accusations of anthropocentrism are leveled, it is not often that we are convinced that the prosecutor has a less anthropocentric perspective, although we may assume it to be benevolent" (Boddice 2011, 6). This is a useful reminder of the impossibility of any "pure" or innocent position, because if we take anthropocentrism to be inevitable, then it is impossible for anyone to accuse anyone else of being anthropocentric from a position of noncomplicity.

But just what we do with that noncomplicity remains an important political question. Boddice distinguishes between the "who" and the "what" of anthropocentrism by suggesting that we should critique an "anthropocentrist world view" but that our starting point is "based in the anthropocentric" (Boddice 2011, 13). For Boddice, this "base" is not defined in relation to species capabilities or human perspectives but in the constitution of politics and ethics itself. We cannot actually reject anthropocentrism, Boddice argues, because to do so would be an ethical and political act, and it is only humans who are capable of acting politically and ethically. Boddice writes that "we humans—either have to accept that our ethics make us the exceptional being, or we have to call time on ethics." And because we cannot "call time on ethics," anthropocentrism should be rethought as "not the great evil to be denounced and eliminated, but the great problem to be embraced and directed" (12).

Boddice's approach to embracing anthropocentrism stems, in part, from the epistemological limits of the "human" as understood by the two theorists, Bruno Latour and Giorgio Agamben, who guide his thinking but who were not interested in nonhuman animals per se. They are more interested in how nonhuman animals figure as a mostly negative foil through which humans can define themselves. Animals are not interesting in and of themselves, for what they might tell us about themselves, but they are useful for tracing the limits of how we might imagine ourselves. Such a position on nonhuman animals can operate in a number of ways. It might operate as an ethical limit on human knowledge (we cannot know them, finally), or it can also operate more anthropocentrically as a refusal or a decision made in advance that the nonhuman animal is not worth consid-ering in its own right. The claim made by Latour and taken up by Boddice, that ethics and politics are uniquely "human" functions not only as a state-ment of fact about "all" nonhuman animals but also functions as a claim for human uniqueness that forms an anthropocentric base (or bias). Its anthropocentrism lies in the fact that it decides in advance that nonhuman sociality cannot count as political or involving ethics–something which numerous Animal Studies scholars would dispute or at least complicate. The social lives of animals also include care, attention, territoriality, altru-ism, play, and negotiated interactions,[1] none of which should necessarily be excluded from any definition of politics and ethics (see chap. 25). This makes the claim that humans are the only ones with politics and ethics an anthropocentric one. The type of anthropocentrism that such a decision evinces might be *inevitable*, as Boddice points out, but it also risks being *arrogant*, a "type of human chauvinism" that places humans at "the cen-ter of everything" and "elevates the human perspective above all others" (Gruen 2015, 24). The line between inevitable and arrogant anthropocen-trism is often blurry and based on what we think we know about animal difference or whether we care to know about animal difference at all.

Agamben's work is often relied on to think the "human" in relation to anthropocentrism in that he analyzes the way the human is precariously realized through what he calls the "anthropological machine" or an "optical machine": "a series of mirrors in which man, looking at himself, sees his own image always already deformed in the features of an ape" (2004, 26). The anthropological machine that produces "the human" can only see the animal as a repudiated self out of which the human is projected. In her insightful reading of Agamben, Kelly Oliver makes the point that Agamben does not consider the possibility that animals could be seen as someone other than a denigrated human. His attention is not drawn to nonhuman animals as such; rather, his thinking is focused on the precariously defined "human," "an empty ideal produced through the continual disavowal of the failure of homo sapiens to escape their animality" (Oliver 2007, 8; see also Oliver 2009). When the "animal" is positioned as that which needs to be escaped or transcended, then the "animal" in all her specificity and difference has already been erased.

It seems to me that an embrace of anthropocentrism in the way that

---

1. See, for example, the work of Bekoff and Pierce (2010), Safina (2015), Massumi (2010), Gruen (2015), and Oliver (2009) for their discussion of the moral and political lives of animals.

Boddice is proposing resigns itself to a sort of disengaged perspective on animals themselves. Often it is assumed that nonhuman animals are either not worthy of being thought about or are impossible to think about as thinking entities. But as Jacques Derrida observed in relation to his cat, not only do we look at them, but they look back at us with their own perspectives. These perspectives, however impossible to reach, serve as a useful reminder that if we write off the possibility of displacing anthropocentrism, then we also write off the possibility of nonhuman animal (in)difference to us. Given that Animal Studies is in part motivated to appreciate, recognize, and theorize nonhuman animal difference, it is vital that nonhuman animals are situated as more than merely negative foils for thinking the human (see chap. 7). After all, they probably have more important things to think about than merely being our negative foils. They might be busy resisting all the other uses to which we have put them (Hribal 2010).

Animal Critics

Unless nonhuman voices on the matter of anthropocentrism are considered, our insights remain one sided and, well, anthropocentric. Animals are perhaps among the most vocal and active critics of anthropocentrism; slaughterhouses are designed to cope with animal resistance, pet owners with forms of restraint and "training," farmers with electrified fences, prods, traps, poisons. It is this toll of violence that needs to be recognized and understood as being unleashed by and also authorized by anthropocentrism. In his account of how our relations with animals are best understood as a "war" whose violence is sublimated by the civil political sphere, Dinesh Wadiwel (2015) argues that challenging the "epistemic violence of the war against animals necessarily implies de-centring a human perspective" (35). Wadiwel argues that it is not only that the war against animals is conducted at the level of the "intersubjective, institutional and epistemic" but that all three are tied up with the notion of human sovereignty that underwrite them; we have made ourselves sovereign *through* our continual violent domination of animals. Barbara Noske also describes animal agriculture as part of an "animal industrial complex," a concept derived from the *military*-industrial complex. She uses "animal industrial complex" to describe the meshing of animal life with global capital, a shift from the small farm to the vertically integrated intensive factory farm that controls all aspects of an animal's life: "control over life supporting activities has passed from the animals themselves to machines and managers," and an "animal's life-time has truly been converted into . . . round the clock production" (1997, 17). The shift involves many actors, including corporations ("meat" producers, pharmaceutical companies, biotechnology; see also Twine 2012), government agencies (responsible for animal welfare, terms of trade, regulation of quota systems, funding for research) and universities (investing in agricultural knowledge production as well as the biotechnological and pharmaceutical products to maintain the animals in confinement). The animal-industrial complex also relies on the colonization of land belonging to indigenous people (Belcourt 2015), the cultivation of pastoral myths and legends (Boyde 2013), the normalization of meat consumption in everyday life, and the invisibility of the process (Pachirat 2011; Vialles, 1994) of converting the animal body into "meat" (Adams 1995; Shukin 2009).

To describe this violent, organized, and pervasive domination of nonhuman animals as anthropocentric may sound rather weak. Indeed Wadiwel's refiguring of human domination as "war" and Noske's account of the "animal industrial complex" suggest that, as Erika Cudworth (2008) astutely observes, anthropocentrism is perhaps "too weak a term politically to capture some of the severity of violence and exploitation involved" (34). Cudworth suggests that *anthroparchy*, human domination, is a more apt term. The difference is important in a strategic sense: how might Animal Studies critics who accept the inevitability of anthropocentrism accept being *anthroparchal*? The term draws attention to the ethical and political nature of the "war" against animals and not just the locatedness and limits of human perspectives. In other words, anthropocentrism may not be a useful term in Animal Studies if it fails to indicate the domination, force, war, and violence that are involved in human domination of animals. It would have to retain a sense of cultural politics, it would have to refer not only to epistemic location but also to political and cultural practices of domination that are embedded at the level of knowledge, culture, and self.

Anthropocentrism: Akin to Racism and Sexism

Anthropocentrism is most useful to Animal Studies when it is framed as a political problem that can be addressed rather than as something that is in any *simple* way inevitable. The two are linked but function quite differently when put to work in the name of challenging human domination of animal life. This is why Val Plumwood makes a distinction between human-centeredness and human *self*-centeredness, the latter being anthropocentrism of the sort that requires political redress, and Lori Gruen makes a useful distinction between *"inevitable* anthropocentrism" and *"arrogant* anthropocentrism." Inevitable anthropocentrism is related to the fact that one is human, with human perspectives that do not cancel out the possibility that we can also learn to appreciate and understand the perspectives of others (Gruen 2015, 24). Our locatedness, as she calls it, should not preclude an openness to others. If it does, then we have settled into a form of arrogant anthropocentrism that allows us to resign from the problem or claim some superhuman detachment from it. If we name all that "we" are (our locatedness) as inevitably anthropocentric, then we might imagine that critique is hopeless. Or we might imagine that only some form of critical distance or even transcendence (or what Plumwood dubs "superhuman detachment" [2002, 138]) from "humanness" will allow us to challenge it. And because superhuman detachment is neither possible nor desirable (not desirable because it removes us from the problem we are attempting to challenge), then we might just give up on the project to displace anthropocentrism entirely. While I think this is an understand-able set of reactions—to want to be superhuman (*reigning over*) or to curl up and disappear (*resigned* to it)—neither of these tactics is useful. The resignation tactic is particularly incongruous when we make the compari-son to other social justice movements, as Plumwood (2002) does: "it is no more necessary for humans to be human-centred than it is for males to be male-centred, or for whites to be Eurocentric or racist in their outlook" (134). Put that way, the idea of embracing anthropocentrism or accepting it as merely inevitable is tantamount to accepting a major political failing. By making the comparison between anthropocentrism and other forms of centrism (such as sexism and racism), Plumwood makes it clear how anthropocentrism's "moral and political failing" (133) links individuals to broad cultural and institutional structures. As is the case with sexism and racism, anthropocentrism manifests at multiple sites: at the personal level through individual beliefs and actions, at the level of identity, and through institutions and systems of knowledge.

While it is not *necessary* for humans to be human centered or males to be male centred or whites to be racist, the very fact that we are able to call them male, whites, or human already suggests that something more than a neutral locatedness has occurred. The male, the white person, and the human are already racialized, gendered, and situated as a member of a privileged species; separated from the rest of the animals by virtue of complex, fraught, culturally specific histories that also determine who and who does not count as "human." I imagine that Plumwood herself would likely raise a suspicious eyebrow if, for example, a white male coloniser claimed "epistemic locatedness" but denied that this included the privilege of having had "human experiences" defined in their image. In other words, while we are all epistemologically located, we are not all located in the same way, with the same ability to stand in for or critique a totalized "humanity" or "human experience," however generalized. Just who "counts" as the *anthropos* to which *anthropocentrism* refers is *inevitably* a gendered, racialized, and ableist affair already politically framed.

Who Is the "Anthropos" in Anthropocentrism?

The epistemic location of the "human" as a species is complicated even further when we consider the various claims made on the term as it is deployed in rights discourses, delineated in biology, and disputed in political and cultural spheres. As many feminists and postcolonial scholars have pointed out, "women" have not been at the center of any conception of the human (Adams and Donovan 1995, Plumwood 1997), nor have *all* women been at the center of any definition of *woman* either (Deckha 2012; Wynter 1990). The distance from any central term such as *human* or *woman* very much depends on histories and structures of power that determine who gets to claim to be universal, to speak on behalf of all. Colonized people are also historically and culturally cast at the margins of human communities and also on the margins of Eurocentric definitions of the human. In Australia, Aboriginal people were not counted as part of the human population of the commonwealth until a referendum to change the constitution was successful in 1967, and some Aboriginal people were governed by state bureaucracies that also managed flora, fauna, and wildlife. This failure to "count" as part of the human population of the commonwealth and their association with flora and fauna management gave rise to a tacit sense that Aboriginal people in Australia were not quite fully human. Similarly, racialized minorities in other colonies (Mbembe 2001), settler countries and multicultural states (Kim 2015), as well as the differently abled (Taylor 2017) are also figured as outside of, exceptional to, the norm of the "human" when figured in universal terms. Given this, the burden of responsibility for anthropocentrism, for actualizing "human" chauvinism on a large scale, is not *evenly* distributed. This point will become more salient as we unpack how anthropocentrism works to privilege not only some human perspectives over others but also how intrahuman conflict (who gets to be at the center?) also affects nonhuman animals. Anthropocentrism can operate very differently depending on the "anthropos" it is referring to, or indeed, bringing into relation with nonanthropos.

The reach of anthropocentrism and its myriad forms also suggests that "anthropos" might be better re-formed rather than displaced, embraced, or erased. David Kidner has an interesting take on anthropocentrism in which he suggests that it might be possible to reshape "anthropos" in progressive ecological terms, making anthropocentrism less of a problem for the planet. He points out that ultimately, the ecological crises of the Anthropocene are not anthropocentric in that they are not in "our" favor; rather, the crisis is born from industrialism, which he characterises as a parasite on human life. Kidner suggests that if we accuse ourselves of anthropocentrism, then we conceal the larger, nonhuman problem of industrialism behind a myth of choice and personal moral failing. The "anthropos" that we should aspire to be, according to Kidner (2014), is a "passionate, feeling, congruent, relational member of the biotic community" (480), and if anthropocentrism were based around those characteristics, then who would object to the organization of life in *their* name?

Kidner (2014) is right to suggest that the "anthropos" that Animal Studies (and environmental studies) often critiques is one that conforms to the "fundamental tenets of industrialism, including a ravenous appetite for consumption, the expectation of an ever-growing material standard of living, and a belief that all other forms of life exist to serve us" (472). This anthropos is rather like Plumwood's "masters of the Universe" or "Empire of Men" (Plumwood 2002, 236), detached from ecologies, disembodied and hyperseparated from nature (Plumwood 1997). Alternative anthropos are possible, already in existence, and political projects to think about these alternatives not surprisingly come out of places of alterity, difference, and marginality, and they include projects within the feminist care tradition (Adams 1990; Adams and Donovan 1995), ecofeminism (Gruen 2015; Plumwood 2002; Kheel 1990), decolonization and postcolonial projects (TallBear 2015; Belcourt 2015), as well as critical disability projects (Taylor 2017) and antiblack racism (Yarborough 2012; Harper 2010; Ko and Ko 2017) and queer studies (Grubbs 2012; jones 2005). Kidner notes, with a sense of irony, that we tend to describe the "passionate, feeling, congruent, relational member[s] of the biotic community" as being "non-anthropocentric" whereas it would be ideal perhaps to install these characteristics as *properly* anthropos rather than not at all related to *anthropos*.

Resisting Anthropocentrism

Many scholars look to indigenous knowledges for an alternative view that situates the human not as "hyper-separated" (to use Plumwood's word) from nature but connected and interdependent with it. Kim TallBear (2015) makes the point that when Western scholars "discover" alternatives to instrumentalist relationships with nonhumans, they need to be aware that "indigenous people have never forgotten that nonhumans are agential beings engaged in social relations that profoundly shape human lives" (234). Not only are "new" anthropos possible, they have always already been there, though subordinated by coexisting systems of domination. Billy Ray Belcourt (2015) argues that anthropocentrism is connected to other axes of oppression and that it functions as the "anchor of speciesism, capitalism *and* settler colonialism" offering "a decolonial ethic that accounts for animal bodies as resurgent bodies" (3–4).

In the Australian context, anthropologist and multispecies studies scholar Deborah Bird Rose suggests that "Indigenous philosophical ecology" offers a way of "resituating" the human apart from "hegemonic anthropocentrism" because of four important differences. First, indigenous philosophical ecology does not make sentience and agency "solely a human prerogative but is located throughout other species and perhaps throughout country itself" (Rose 2005, 302). Second, "life processes . . . do not prioritise human needs and desires"; humans are "one species among many others, both giving and receiving benefit" (303). Third, a structure of "kinship with nature . . . ensure[s] that non-humans and humans are part of the same moral domain" (303). Finally, humans are not situated as the masters of ecological systems that need them to act but are "called into action by the world" (303) in a recursive system of mutual benefits.

In later work, Rose describes indigenous philosophical ecologies as "non-anthropocentric" (Rose 2014, 431) because they live beyond the "west's nature-culture/mind-matter binaries" (432), where humans are situated "within a great family of participants" (433). Such a knowledge system represents a significant difference to the rapacious, extractive, and detached anthropos that Plumwood describes as the "masters of the Universe" who position nature and nonhumans as a mere resource. But, as Kidner suggests, we may well want to describe this alternative way of life not as "non-anthropocentric" but as "anthropocentric" on the basis that it seems much better suited to ecological survival of the human species and others.

Yet any claim to be "non-anthropocentric" or a *good* "anthropocentric" does not necessarily allow for finer distinctions between modes of being that are good for humans and ones that are good for animals. This is important because interconnectedness of, for example, indigenous people with other species does not always mean equal consideration. Animals might find themselves on the side of a *refuted* social obligation, as a gift/sacrifice: "the one little wallaby I shot cannot begin to repay this social debt, although it is a start" (Povinelli 1995, 511). The wallaby is an item of exchange, not a creature with his *own* set of social relations that might exclude use by humans. In his discussion of indigenous hunting practices, Craig Womack questions whether or not animals "would agree that they have signed" the agreement that makes them edible kin who agree to give themselves to others. He suggests that animals, too, have kin and that "we would do well to imagine them if we want to take into account all—not some—of our relations" (Womack 2013, 24; see chap. 12). Womack's work suggests that indigenous practices are not necessarily "non-anthropocentric" when it comes to animals: "there is a compelling case to be made for trying to imagine their [animals'] perspectives, no matter how fraught the process" (20). Prioritizing an imagined animal perspective as an ultimate test of the claim to be nonanthropocentric is characteristic of Animal Studies scholarship.

Claire Kim's analysis of the Makah whaling case in the United States also highlights the ways that conflicts between indigenous and settler states over animal practices are often conducted anthropocentrically as a dispute between human groups that both make different claims on non-human animals. Kim points out that in the debates that occurred between different groups over the rights of the Makah to resume hunting the gray whale, that much of the discussion was anthropocentric, pitting human groups (ontologies and worldviews) against each other, and had the effect of displacing any consideration of the whale's interests. Asking whether or not one can critique Makah whaling while also supporting their claims to sovereignty—Kim suggests that a "multi-optic" approach that takes into account the varied stakes, claims, and perspectives involved (human and nonhuman) can enable a critique of practices that impinge on animal life as well as being supportive of indigenous sovereignty. In her reading, the Makah should take seriously "what whales deserve and want" and "question whether their own cultural understandings, too, might bear traces of domination (and self-rationalization)" (Kim 2015, 250). She argues that we should act as if the gray whales wish to live; otherwise, "we humans, Native and non-native, run the risk of imposing our own systems of meaning on those who lack the power to contradict us" (245). In other words, it is not a question of one being anthropocentric and the other not but of both being anthropocentric in different ways and with different opportunities to impose that view. Kim's account of the gray whales own culture and systems of kinship is largely adduced from Western science and recent histories of nonhunting interaction with the whales rather than from indigenous cosmologies. Neither is free of imposing a system of belief on whales, but they do so from very different positions with different potential for actualization. And because of this, I would suggest that anthropocentrism remains a useful concept and tool not because we agree on who "anthropos" is (any more than we use *sexism* because we agree on what "sex" is or *racism* because we agree on what "race" is) but because we need a name for a kind of domination that structurally privileges the perspectives of humans over nonhuman animals. The term's usefulness thus intersects with and gets caught up with any presumed or contextual "anthropos." This explains why we can find it in unexpected places, including places that either claim to have transcended it or paradigms that claim to be including animals in a moral community, that is, within Animal Studies itself.

Anthropocentrism of Animal Studies

The anthropocentrism of Animal Studies comes in different forms but frequently is linked to theorists who draw a line, extend a circle, or widen the hierarchy (to use some calculating metaphors) to include all sentient creatures in a moral community (Singer 1975; Regan 1983; Beckoff and Pierce 2010). This is often interpreted by critics as not only exclusive but also ultimately anthropocentric because it values what resembles "us." Much of the work of resemblance making comes from a shift in thinking about animal minds that would have been "unthinkable" up until fairly recently, and now accounts of animals doing "human" things abound in popular culture. As Virginia Morrell points out, it seems that every week new findings announce newly recognized animal capabilities: "Whales Have Accents and Regional Dialects," "Fish Use Tools," "Squirrels Adopt Orphans," "Honeybees Make Plans," "Sheep Don't Forget a Face," 'Rats Feel Each Other's Pain," "Elephants See Themselves in Mirrors," "Crows Able to Invent Tools," and (for me as a dog lover, a favorite) "Dog Has Vocabulary of 1,022 Words" (Morrell 2013, 2). Plan-making honeybees and tool-inventing crows not only put into question the exceptional human but also profoundly rewrite the "animal" in all their multiplying differences. Cary Wolfe (2003) refers to the "veritable explosion" of work in multiple discipline areas that "have called into question our ability to use the old saws of anthropocentrism (language, tool use, the inheritance of cultural behaviours, and so on) to separate once and for all from animals" (xi). This interest in animals is key to understanding how "generic" rather than exceptional *anthropos* might be and how we are different in degree rather than kind (as Charles Darwin concluded). But a question remains—does the fact that we now value these stories relate to a fundamental shift in our old anthropocentric ways, or is it just a continuation in another form?

Environmental philosophers, multispecies studies, and posthuman scholars are also critical of Animal Studies' anthropocentrism, specifically the focus on nonhuman animals to the exclusion of other biota and abiota that may also express complex intentionality and preferences for life over death. Plumwood (2002), for instance, finds Singer's "indifference to plant lives . . . deeply shocking" (258), while multispecies studies finds the exclusion of abiota an "unjustifiable bias" (Kirksey and van Dooren 2016). Animal studies does indeed focus on primarily sentient nonhuman animals, but this should not make it any *more* anthropocentric than studies of plants or microbiota that are justified because of their subject's "intentionality," "liveliness," or "agency," all of which might reasonably be described as anthropocentric, too—recalling Boddice's "throwing stones in glass houses" response to the accusation of anthropocentrism. One of the reasons that Animal Studies focuses on sentient creatures relates to the fact that it is sentient creatures (chickens, cattle, pigs, sheep) that make up the bulk of the world's agricultural animals, those designated "livestock," whose status as valued animals outside of being commodities, such as being part of our "environmental" world or "nature" is ambivalent. While environmental or multispecies studies privilege places and critters in encounter, Animal Studies looks largely into places of confinement, incarceration, industrialized death, and the backstories of domestic animal breeding (Tuan 2007). In this case, the focus on sentience, on who does or does not feel pain, fits the terrain in which we find the bulk of the world's animals. After all, as Barbara Noske points out, it was the sentience and sociability of livestock and "domesticated" breeds of animal that made them able to be domesticated (Noske 1997, 18). While Animal Studies approaches are understandably criticized for drawing a line at sentience, environmental or multispecies approaches are also at risk of drawing a line at the "animal gulag" (the phrase is Plumwood's), avoiding the extent of the violence that exists in "relationality" and "mutuality," too (see Weisberg 2009; Wadiwel 2015). This is why examining anthropocentrism from multiple perspectives is so vital, not necessarily because it settles the differences by agreement once and for all but because it shows just how energetic and slippery the concept of anthropocentrism can be. This is especially the case when it comes to claims to have escaped it, to have become nonanthropocentric.

It is not uncommon within the literature on anthropocentrism to find critics desiring an escape from anthropocentrism. Again, this is completely understandable given the violence and domination that we associate with it. There are different examples of this rhetoric of escape in work cited in this chapter. For example, Kidner (2014) refers to a sense of "unfreedom" within anthropocentrism. Kirksey and van Dooren (2016) refer to multispecies work that has "escaped the tunnel vision of anthropos" (14), Boddice (2011) refers to our "captivity" within anthropocentrism while Braidotti (2013) calls anthropocentrism our "cage" (80). But where things get tricky is where this desire to be free of anthropocentrism morphs into a claim to have actually escaped it. Pausing over the rhetoric for a moment, it is worth noting that critics employ a rhetoric of captivity and liberation that is applicable to the lives of animals who are in *actual* cages from birth to death and in prisons, factory farms, enclosures, houses, towns, suburbs, and epistemologies all of *our* making. As such, the rhetorical gesture that sees "anthropos" (rather than "animal") as the captive of anthropocentrism can risk centering the self all over again as if an equivalence is being implied between the experiences of those animals victimized and imprisoned by anthropocentric domination and the critics of that domination. Their suffering is not the same. This is where the liberation rhetoric rings especially hollow also because just as the claim to have escaped sexism and racism can actually diminish its pervasive and mobile reach, so, too, can the claim to be outside of anthropocentrism indicate a diminished and noncomplicit perspective on its ubiquity.

The human-centeredness that defines anthropocentrism is rarely perceived in neutral terms, as something that Animal Studies shouldn't be challenging in some form or another. Indeed, the belief that we are perhaps *imprisoned* by anthropocentrism is itself a restriction, not allowing us to register the ways that the nonhuman animal has always been at the heart of our thinking (as well as our biologies—we are, after all, animals ourselves). Anthropocentrism describes a turning away from the relationships that we are *already in* as well as a refusal to acknowledge how we are already changed by them as anthropomorphic beings ourselves, where our "projections" can also be openings for communication and knowledge (Taylor 2011). The critic of anthropocentrism needs multiple lenses, optics, but also mirrors, a feeling for blind spots, and an appreciation of what it might mean to be under the gaze of the nonhuman animal who is (in)different to us.

Suggestions for Further Reading

Belcourt, Billy Ray. 2015. "Animal Bodies, Colonial Subjects: (Re)locating Animality in Decolonial Thought." *Societies* 5: 1–11.

Boddice, Rob. 2011. "The End of Anthropocentrism." In *Anthropocentrism: Humans, Animals, Environments*, edited by Rob Boddice, 1–18. Leiden: Brill.

Oliver, Kelly. 2007. "Stopping the Anthropological Machine." *Phaenex* 2: 1–23.

Plumwood, Val. 2002. "The Blindspots of Centrism and Human Self Enclosure." In *Environmental Culture: The Ecological Crisis of Reason*. London: Routledge.

Taylor, Nik. 2011. "Anthropomorphism and the Animal Subject." In *Anthropocentrism: Humans, Animals, Environments*, edited by Rob Boddice, 265–79. Leiden: Brill.

References

Adams, Carol. 1990. *The Sexual Politics of Meat*. New York: Continuum.

Adams, Carol, and Josephine Donovan, eds. 1995. *Animals and Women: Feminist Theoretical Explorations*. Durham, NC: Duke University Press.

Adams, Carol, and Lori Gruen, eds. 2014. *Ecofeminism: Feminist Intersections with Other Animals and the Earth*. London: Bloomsbury.

Agamben, Giorgio. 2004. *The Open: Man and Animal*. Translated by Kevin Attell. Stanford, CA: Stanford University Press.

Bekoff, Marc, and Carron A. Meaney. 1998. *Encyclopaedia of Animal Rights and Animal Welfare*. London: Routledge.

Bekoff, Marc, and Jessica Pierce. 2010. *Wild Justice: The Emotional Lives of Animals*. Chicago: University of Chicago Press.

Belcourt, Billy Ray. 2015. "Animal Bodies, Colonial Subjects: (Re)locating Animality in Decolonial Thought." *Societies* 5: 1–11.

Boddice, Rob. 2011. "The End of Anthropocentrism." In *Anthropocentrism: Humans, Animals, Environments*, edited by Rob Boddice, 1–18. Leiden: Brill.

Boyde, Melissa. 2013. "'Mrs Boss! We gotta get those fat cheeky bullocks into that big bloody metal ship!': Live Export as Romantic Backdrop in Baz Luhrmann's *Australia*." In *Captured: The Animal within Culture*, edited by M. J. Boyde, 60–74. New York: Palgrave Macmillan.

Braidotti, Rosi. 2013. *The Posthuman*. London: Polity.

Cavalieri, Paola, and Peter Singer, eds. 1993. *The Great Apes Project: Equality beyond Humanity*. New York: St. Martin's Griffin.

Cudworth, Erika. 2008. "'Most Farmers Prefer Blondes': The Dynamics of Anthroparchy in Animals' Becoming Meat." *Journal for Critical Animal Studies* 5 (1): 32–45.

Deckha, Maneesha. 2012. "Toward a Postcolonial, Posthumanist Feminist Theory: Centralizing Race and Culture in Feminist Work on Nonhuman Animals." *Hypatia* 27 (2): 527–45.

Derrida, Jacques, 2002. "The Animal That Therefore I Am (More to Follow)." *Critical Inquiry* 28 (2): 369–418.

Grubbs, Jennifer, ed. 2012. "Inquiries and Intersections: Queer Theory and Anti-Speciesist Praxis." Special issue, *Journal of Critical Animal Studies* 10 (3).

Gruen, Lori. 2015. *Entangled Empathy: An Alternative Ethic for Our Relationships with Animals*. New York: Lantern Books.

Harper, Breeze A., ed. 2010. *Sister Vegan: Black Female Vegans Speak on Food, Identity, Health and Society*, New York: Lantern Books.

Hribal, Jason. 2010. *Fear of an Animal Planet*. California: CounterPunch.

jones, pattrice. 2005. "Of Brides and Bridges: Linking Feminist, Queer, and Animal Liberation movements." *Satya*. http://www.satyamag.com/jun05/jones_bridges.html.

Kheel, Marti. 1990. "Ecofeminism and Deep Ecology". In *Reweaving the World: The Emergence of Ecofeminism*, edited by Irene Diamond and Gloria Feman Orenstein, 128–37. San Francisco: Sierra Club Books.

Kidner, David W. 2014 "Why 'Anthropocentrism' Is Not Anthropocentric." *Dialectical Anthropology* 38: 465–80.

Kim, Claire Jean. 2015. *Dangerous Crossings: Race, Species, and Nature in a Multicultural Age*. Cambridge: Cambridge University Press.

Kirksey, Eben, and Thom van Dooren. 2016. "Multispecies Studies: Cultivating Arts of Attentiveness." *Environmental Humanities* 8: 1–23.

Ko, Aph, and Syl Ko. 2017. *Aphro-ism: Essays on Pop Culture, Feminism, and Black Veganism from Two Sisters*. New York: Lantern Books.

Massumi, Brian. 2010. *What Animals Teach Us about Politics*. Durham, NC: Duke University Press.

Morrell, Virginia. 2013. *Animal Wise*. London: Random House.

Mbembe, Achille. 2001. *On the Postcolony*. Berkeley: University of California Press.

———. 2003. "Necropolitics." *Public Culture* 15(1): 11–40.

Noske, Barbara. 1997. *Beyond Boundaries: Humans and Animals*. Montreal: Black Rose Books.

Oliver, Kelly. 2007."Stopping the Anthropological Machine" *Phaenex* 2: 1–23.

———. 2009. *Animal Lessons: How They Teach Us to Be Human*. New York: Columbia University Press.

Pachirat, Timothy. 2011. *Every Twelve Seconds: Industrialized Slaughter and the Politics of Sight*. New Haven, CT: Yale University Press.

Plumwood, Val. 1997. *Feminism and the Mastery of Nature*. London: Routledge.

———. 2002. *Environmental Culture: The Ecological Crisis of Reason*. London: Routledge.

Povinelli, Elizabeth.1995. "Do Rocks Listen: The Cultural Politics of Apprehending Australian Aboriginal Labor."*American Anthropologist* 97 (3): 505–18.

Regan, Tom. 1983. *The Case for Animal Rights*. Berkeley: University of California Press.

Rose, Deborah. 2005 "An Indigenous Philosophical Ecology: Situating the Human." *Australian Journal of Anthropology* 16 (3): 294–305.

———. 2014. "Arts of Flow: Poetics of "Fit" in Aboriginal Australia." *Dialectical Anthropology* 38: 431–45.

Safina, Carl, 2015. *Beyond Words: What Animals Think and Feel*. New York: Henry Holt.

Shukin, Nicole. 2009. *Animal Capital: Rendering Life in Biopolitical Times*, Minneapolis: University of Minnesota Press.

Singer, Peter. 1975. *Animal Liberation*. New York: Harper Collins.

TallBear, Kim. 2015. "An Indigenous Reflection on Working beyond the Human/Not Human." *GLQ: A Journal of Lesbian and Gay Studies*. 21 (2/3): 230–35.

Taylor, Nik. 2011. "Anthropomorphism and the Animal Subject." In *Anthropocentrism: Humans, Animals, Environments*, edited by Rob Boddice, 265–79. Leiden: Brill.

Taylor, Sunaura. 2017. *Beasts of Burden: Animal and Disability Liberation*. New York: New Press.

Tuan, Yi-Fu 2007. "Animal Pets: Cruelty and Affection." In *The Animals Reader: The Essential Classic and Contemporary Writings*, edited by Linda Kalof and Amy Fitzgerald, 141–53 London: Berg.

Twine, Richard. 2012. "Revealing the 'Animal-Industrial Complex": A Concept and Method for Critical Animal Studies?" *Journal for Critical Animal Studies* 10 (1): 13–28.

Vialles, Noelie. 1994. *Animal to Edible*. Cambridge: Cambridge University Press.

Yarborough, Anastasia. 2012. "Afro-Animal: Our Shared Struggle." March 25. https://animalvisions.wordpress.com/category/afro-animal/.

Wadiwel, Dinesh. 2015. *The War against Animals*. Leiden: Brill.

Weisberg, Zipporah. 2009. "The Broken Promises of Monsters: Haraway, Animals and the Humanist Legacy." *Journal of Critical Animal Studies* 7: 22–59.

Wills, David. 2008. *Dorsality: Thinking Back through Technology and Politics*. Minneapolis: University of Minnesota Press.

Wolfe, Cary. 2003. *Animal Rites: American Culture, the Discourse of Species, and Posthumanist Theory*. Minneapolis: University of Minnesota Press.

Womack, Craig. 2013. "There Is No Respectful Way to Kill an Animal." *Studies in American Indian Literatures* 25 (4): 11–27.

Wynter, Sylvia. 1990 "Afterword: Beyond Miranda's Meanings: Un/silencing the 'Demonic ground' of Caliban's 'Woman.'" In *Out of the Kumbla: Carribean Women and Literature*, edited by Carole Boyce Davies and Elaine Savour Fido, 355–72. New Jersey: Africa World Press.

4 BEHAVIOR

Alexandra Horowitz

How can we know anything about nonlinguistic others? One answer is that we can learn about these others by studying their *behavior*. Ethologists study animal behavior in natural settings; comparative psychologists note resemblance (or lack of resemblance) of mental function of different species based on their observed behavior; Animal Studies often takes as its subject the interaction, or result of interaction, between two or more species.

In each case, *behavior* is the empirical anchor: the fields revolve around the identification and analysis of independently verifiable acts of organisms. Thus, any meaning discovered therein must emerge from an understanding of what, exactly, this phenomenon called "behavior" is. Yet the term is often not defined, even in textbooks of "animal behavior" and "human behavior." The opening pages of the *Encyclopedia of Human Behavior* (Ramachandran 2012) introduce behavior only as "complex, quixotic, and difficult to fathom" (xxix). The influential behaviorist and psychologist B. F. Skinner did not bother to define behavior at any point in his *Science and Human Behavior* (1953). He wrote only that researchers must "observe human behavior carefully from an objective point of view" (5), which certainly raises the question of what, exactly, it is to be observed and whether the particular "point of view" might get in the way of observing it, if we leave the definition to the observer/subject.

Animal behavior texts often do one better. One suggests that behavior "includes a wide range of responses, from simple movements of a limb to complex social interaction" (Barnard 2003, 31). Those movements may range, another specifies, from a "series of muscle contractions, perhaps performed in clear response to a specific stimulus" to "enormously complex activities, such as birds migrating across the world, continuously assessing their directions and positions with the help of various cues from stars, landmarks and geomagneticism" (Jensen 2009). Behavior, with this lens, is some kind of movement, either reflexive and simple or complex. While these definitions seem straightforward, each introduces an implied precursor: the movement is done as a *response*—presumably to a stimulus provided by the behaver's environment. In that way, a behavior is prompted by, in some cases, another behavior: an egg caused by a chicken caused by an egg.

While even those behavior textbooks that define the term soon leave the topic of definition and continue on to analysis of examples, in this chapter I will linger on the alleged meanings and use of the term in an attempt to explicate it. By considering the issues, especially methodological, raised in the study of behavior, the complexities of the concept and the slippery use of the word are highlighted.

Studying Behaviors

The name for the study of behavior is not *behaviorism*. Behavior is the studied datum in fields as varied as economics and anthropology. But the field of behaviorism, which dominated psychology and much study of animals in the early twentieth century, is a good place to begin a consideration of behavior studies. As advocated by psychologists John Watson and, later, B. F. Skinner, behaviorism was driven by a rejection of "mind" or "mental states" as the topic of psychology—especially where nonhuman animals[1] were concerned. As late nineteenth-century books on animal intelligence (George Romanes) and animal emotions (Charles Darwin) gained audiences, the study of animal "mind" was given credibility (see chap. 16). Watson and Skinner saw these usages as overattributions. Given that *states of mind* as such are not externally verifiable, Skinner, for instance, described them as unfit subjects for scientific inquiry. Further, he suggested, given the seeming success in behavioral explanations, alleged properties of the mind of human and nonhuman animals could be simply "fiction": a kind of convenient but inaccurate explanatory story for seen actions. As a field, behaviorism demoted the subjective and promoted an (ostensibly) objective understanding of a subject.

---

1. Herein also "animals."

While "mind" remains, to this day, more seemingly complex than "behavior,"[2] the use of behavior as a description for "what animals do" only came into currency in psychological literature around the turn of the twentieth century (Barnard 2003, 2). The term, glossed as the "manner in which a thing acts under specified conditions or circumstances, or in relation to other things"[3] was uncommonly used of any animal, including human animals. Surely "behavior" of animals had been observed and at least adventitiously studied ever since there were humans to observe animals and animals to observe each other. For behavior begets other behaviors, and thus, observing behavior allows for prediction of future behaviors—such as behaviors of a predator that may threaten a prey animal's life, or behaviors of a prey animal that would allow a predator (such as a human) to capture them.

But behaviorism coincided with, and arguably contributed to, a change in that usage. Behaviorists argued that animals *only* exhibited behavior and that there was not necessarily a corresponding mental state behind any behavior. Animals were mere responders to stimuli. Concurrently, the field of *ethology*, without those sorts of judgments, elevated the study of behavior for its own sake. And it is in these studies that the complexity of the term becomes clear.

Konrad Lorenz, the ethologist and Nobel Prize winner, wrote that "description is the foundation of all science" (1988, 7), by which he meant description of *behavior*. To design an experiment, to even begin to generate a viable hypothesis, one needs to begin with observation of behavior and description of what one observes.

Behavioral description is especially common in—though not exclusive to—investigation of nonverbal subjects, such as animals and infant humans. Ethology, the study of animal behavior in natural settings, is more formally described as aiming to address the "what, when, how and why" of a behavior—while the "where" and "who" are premised, though not trivial (Lehner 1996). Broadly, the aim is to describe, detail, explain, and sometimes predict behavior. Niko Tinbergen (also a winner of the Nobel Prize with Lorenz) characterized four questions that an ethologist

2. *Mind* is associated with *brain*; while *behavior*, while often derivative of mental activity, arises from muscular movement. Mind requires the central nervous system; behavior may be due to just peripheral nervous system activation. Humans have minds; even a jellyfish shows behavior.

3. *Oxford English Dictionary*, 2nd ed., s.v. "behaviour."

might ask of behavior: its ontogeny (how it developed in an individual), its evolution (how it developed in the species), its causation (the mechanisms giving rise to the behavior), and its function, what he originally called its "survival value" (the behavior's consequences and adaptive value; Tinbergen 1963). Thus, beyond describing the behavior, its origins (proximate and ultimate) are being investigated through detailed observation. The "to what end" nature of the behavior can vary, as illustrated by the varying emphasis that could be brought to a question of behavior:

> *Why* do dogs wag their tails? (How is it useful, what purpose does it serve?)
> Why do *dogs* wag their tails? (How did this behavior arise in the species?)
> Why do dogs *wag* their tails? (What mechanism causes the behavior?)
> Why do dogs wag their *tails*? (Why tails and not some other body part?)
> (After David McFarland)[4]

It is an open question whether a complete reckoning of a behavior along the above-described dimensions is possible or necessary to achieve a sense of "understanding" of the behavior.

Nondefinitive Definitions

What counts as behavior to be studied is variously formulated—but in most cases, the formulation is, at its core, quite broad. "It will include all types of activities in which animals engage, such as locomotion, grooming, reproduction, caring for young, communication, etc." (Jensen 2009, 3); "The blinking of an eye, the sudden movement of a limb as it is withdrawn from a sharp object, the cacophonous display of a male blue bird of paradise . . . the care shown by a female chimpanzee . . . to a distressed infant, our own cultural etiquettes" (Barnard 2003, 3). It spans the internal and the external, "the bridge between the molecular and physiological aspects of biology and the ecological . . . the link between organisms and environment and between the nervous system and the ecosystem" (Snowdon 1991). In some sense, the gloss given as nearly a parenthetical in a methodology handbook—"the actions and reactions of whole organisms" (Martin and Bateson 1993, 3)—is the simplest summary. Behavior is "everything we do that can be directly observed" (King 2013, 2), "what an animal does"

---

4. Based on David McFarland's example cited in Lehner (1996, 9).

(Levitis, Lidicker, and Freund 2009, 106)[5]—but note that what "do" or "does" means is left unexplored.

These definitions assume the doer of the behavior—the "behaver"—to be an animal. While this may not be presumptuous for organisms on the human scale—no one questions whether a dog is an individual animal—eusocial species, such as bees, wasps, and ants, may verily be considered organisms at the group level: the hive, swarm, or army. In these cases, who is the behaver: the bee or the hive?

Implicitly, behavior is contrasted with unobservable internal processes, especially mental processes. Surprisingly, though, behavior is often seen as an inroad to "seeing" those mental processes insofar as a thought—such as a plan or an emotion—may be represented in the thinker's actions—such as voluntary activity toward a goal or physical expression of an emotion.

We might ask, is anything *not* behavior? Levitis, Lidicker, and Freund (2009) conducted a survey of 174 members of scientific societies that publish behavioral research (with expertise ranging from professional researchers on the one hand to those less well acquainted with behavioral science on the other) and found consensus in affirming some kinds of phenomena as "behaviors": "a spider builds a web," "flocks of geese fly in V formations," "a dog salivates in anticipation of feeding time." By contrast, "a rabbit grows thicker fur in the winter," "a cat produces insulin because of excess sugar in her blood," and "a person sweats in response to hot air" were all rejected.

While there seemed to be consensus around some doings, there was a terrific lack of consensus about the status of most plausible behaviors. "Ants that are physiologically capable of laying eggs do not do so because they are not queens" was controversial: there was no agreement about whether this constitutes "behavior." Similarly with a rat's "dislike for salty food" and a chameleon changing color. While the former, for instance, may be expressed *via* a behavior (choosing a sweet food over a salty one), the preference itself may be considered not a behavior. In the latter case, an appearance-based, involuntary physiological response, while it may prompt a behavior in another animal, is not necessarily a "behavior" by the bearer. If not, neither would aposematic markings or other evolved anatomical strategies generally considered to be behaviors.

5. Citing Davis 1966.

The contradiction inherent in the respondents' assessments is striking. For instance, while a rabbit's fur growing and a human's sweating, both visible if reflexive/automatic events, were not considered to be behaviors, a dog's reflexive salivation was. The ambivalence shown about a rat's dislike for salty food was not shown about a person's decision "not to do anything tomorrow if it rains," which was granted "behavior" status.

Psychologist Ogden Lindsley proposed the "dead-man test for behavior": if a dead person could do the act under consideration, "it wasn't behavior" (Lindsley 1991, 457). So fur growing, sweating, or camouflaging color changes would be animal behaviors; "acts" of "not doing anything tomorrow," lying still, failure to react, and so forth, would not.

While commonsensical, there is good reason to suspect that Lindsley's rule is an oversimplification, in particular as regards "absence of behavior." Paradoxical though it may seem, even the absence of a behavior is also veritably a behavioral response in some cases. For instance, a nonresponse to a stimulus, such as "playing dead," by a dog, a possum, or a hog-nosed snake—lying supine or on one's side, unmoving—is a suppression of other possible (mobile) behavioral acts. In the case of the dog, it is also, most probably, a learned response to specific training (Lindsay 2001). In other cases, an untrained absence of response—failure to respond to a play signal directed to one dog by another—is considered a communication and thus may be a behavior (Horowitz 2009a). Recently, some researchers have argued that "stillness" is an action that requires as much of muscles as movement (Noorani and Carpenter 2017). They have identified various kinds of stillness, including "active immobilization" (freezing), inhibition of responses to stimuli, and halting a movement in progress, all of which require inhibitory brain responses and thus could be considered behaviors.

As a result of their survey, Levitis et al. proposed that behavior could be defined as "the internally coordinated responses (actions or inactions) of whole living organisms (individual or groups) to internal and/or external stimuli," excluding developmental changes (Levitis, Lidicker, and Freund 2009, 108). Such responses would be "externally visible"—observable and measurable.

Acknowledging the conflicts that surround all attempts to define the term, another approach, to which I now turn, is to explicate the term through analysis of how behavioral science *in fact* observes and measures *behavior*.

Levels of Behavior

Two decades ago a group of researchers examined what chimpanzees and human children did after witnessing a demonstration of how to open a so-called artificial fruit: a locked box that held a fruit or candy inside (Whiten et al. 1996). The study was designed to gauge, through analysis of matching of behaviors specified by the experimenters, whether these subjects "imitated" the demonstration. Their result—that chimpanzees imitated to a limited degree, while children were much more imitative—was of interest to those who allege that imitation may be linked to an understanding of the another person's *intentions* (Tomasello 1996) insofar as those who imitate may be seeing the demonstrator's specific actions as necessary to her (unspoken) goal.

Before any such conclusion can be drawn, it is worth considering what counts as "imitation." Can we specify actions by the demonstrator that must be repeated in order for a subject to be considered to have imitated her demonstration? That is what Whiten et al. did in their coding, in which they enumerated the presence or absence of various behaviors by the subjects (1996).

When a different subject group—adult humans—was given the same task, however, the difficulty of such a specification became clear. By the Whiten et al. analysis, the adults often behaved more like the chimpanzees—less imitatively—than like the children (Horowitz 2003). Certainly, though, normally functioning human adults are not expected to have any trouble imitating a simple task; similarly, they would be expected to be the highest-performing group in their understanding of the rough intentions of the demonstrator. Highlighted is the fact that what part of the demonstrators' behavior should be imitated is seen differently by these different age groups and species. Indeed the very scope of the demonstrator's "behavioral act" is in question: a number of adult subjects not only opened the box to retrieve the candy inside but also reclosed the box, as the demonstrator presumably had to do after her demonstration.

These results should prompt us to consider that the subjects' actions can be seen at many different levels of specificity. In ethology, the "levels" on which a behavioral act is analyzed must be clearly identified—and are nontrivial (Lehner 1996). While "imitation" of the Whiten demonstration might be gauged, at the broadest level, by whether the subject opened the box or not, one might also look at more detailed levels of analysis

(Horowitz 2003): does the subject match the demonstrator's hand or finger shape, handgrip, repeat the same sequence of actions, and mimic the precise number of times each subaction is done?

Similarly, the level of behavior on which any behavioral study is conducted can vary from macro to micro (Lehner 1996): from species or population-level behaviors (as the V-formation of geese) to muscular or neuronal behaviors (those implicated in each goose in the coordination and flying in such a formation). An analysis of a greeting between dogs could take place on the level of "dyad," where two dogs come proximate to each other (or not); or on levels of "behavioral type," such as the presence or absence of "tail-wagging" and other high-level behaviors; or on the level of "behavioral act," such as "lateral and sometimes rotational movement of the tail, extended upward (dorsally)." Depending on the level that is specified in observation and analysis, a very different record of the animals' behavior could appear. For instance, a proximate approach could be friendly, examinatorial, or aggressive; a "wagging tail" could be either an affiliative greeting or a measure of excitement of a nongreeting sort, such as precopulation or between bouts of fighting or play, or even used in nonsocial situations, such as waiting for food or hunting (Kiley-Worthington 1976).

That there are many levels on which to consider "a behavior" reflects the varying descriptions even a single behavior may have. The level reflects the methods being used: a microscope "would be useless . . . for reading a novel" (Martin and Bateson 1993, 9). Since much observation is of animals of recognizable size behaving in ways visible to humans with our sensory equipment, we are naturally limited in the behaviors that we see. Also, the level chosen will determine the data that are gathered.

Naming

After a behavioral level has been specified in an observational study, "sampling" methods are used to reduce a lifetime of behaviors to a measurable subset (Altmann 1974): of only a single, recognizable animal; of instances of a single kind of behavior; of the actions happening at a fixed period. Such an approach implies that one can take a snapshot (of varying lengths, frequencies, and so forth) and have it "stand in" for the behavior of the same individual at another time or for other members of the species.

Research aiming to describe the form or context of "dyadic play," for instance, would aim to record behaviors of or around play bouts and would include an ethogram that lists the relevant behaviors for that context. An ethogram is a listing of all the relevant behavioral patterns of the subjects and a description of each. No ethogram is exhaustive; that is, given the levels of behavior and the varieties of behaviors in a species' repertoire on each level, the ethogram does not aim to delineate all. Instead, a study designates a behavior or set of behaviors of focus.

For this purpose, a continuous stream of behaviors is subdivided into bouts of behaviors, which themselves are describable using the ethogram (Lehner 1996). The behaviors may be considered "events" (with a measurable duration) or "states" (conditions): "standing up" is an event; "standing" is a state. They are both "behaviors" (Altmann 1974; Lehner 1996). In addition to describing a movement or posture, a named behavior may include a function: a primate's "pant threat" or "fear grimace," for instance. What were once undifferentiated phenomena become coherent through giving names to the types of behavioral events seen.

Names need not be without humor, originality, or subjectivity. But the definition of what "counts" as an example of that behavior should be clear. In a study of dyadic dog play, I included an attention-getting behavior, *in-your-face*, glossed as "position body or face inches from other's face" (Horowitz 2009a). While droll, the use of the term was specific. The caution here is not to let the particular characterization of the behavior slip into the interpretation of the behavior. For example, one ethogram, for a study of stickleback fish, includes the entry "sneaky swimming," defined as "smooth swimming usually along the bottom using caudal and pectoral fins with dorsal and ventral spines lowered" (Lehner 1996, 116). Given the research's aim to examine the fish's antipredatory behaviors, the adjective "sneaky" may be apt. But care needs to be taken to ensure that one not slip into saying that the fish, therefore, "are sneaky."

With functional descriptions, which give the proximate or ultimate use of the otherwise simply physical act, it is easiest to see the danger of naming. Inclusion of an alleged "function" of an act can be presumptuous: if a certain musculature of a primate's face is described not as such but as a "grimace" of "fear," some of the question of the emotion, experience, or intent of the animal is begged. Is a dog described as "urinating" or as "marking"—thus elevating his relieving himself to a social act?[6]

6. In "marking," an animal leaves urine apparently intended for others—including territory challengers, predators, and possible mates—to investigate (which "intent" is determined by the subsequent olfactory investigation by conspecifics).

Familiarity with an animal's behavior could be described as the ability, earned over time and repeated viewings, to carve behaviors into functional and empirical types. The expert dog-behavior observer recognizes the actions "mark" and "urinate" as having different physical or spatial features, different durations, appearing in differing contexts, and having, over time, different consequences. The expert primatologist identifies the grimace that appears in contexts of aggression by others as distinguishable from similar grimaces in other contexts. Thus, one could argue, there is nothing inappropriate about functional descriptions of behavior if one has some (unspecified) amount of experience observing—presumably guided by other ethograms. Knowing "what the animal is doing" thus relies on being not just a person with eyes and an ethogram, but a person with a certain type of expertise wrought of hours of watching guided by certain characterizations of the types of behaviors that could be seen. These characterizations must, therefore, be revised over time as experience is gained. By the same token, novice observers and expert observers may not be equally well equipped to use the same ethogram. Behavior is, in part, observer dependent.

Effect of Observer and Means of Observation

As used, the ethogram is the observational science's means of operationally defining behavior units: an identification of what counts as the behavior. Such identification is necessary for interobserver and also intraobserver reliability. For two observers to agree that they have both witnessed "dyadic play" behavior, the action or set of actions—including either necessary or sufficient—that count as play must be clearly defined. The necessity of gauges of "observer agreement" (see, e.g., Altmann 1974) highlights the fact that it is quite likely that two observers, looking at the same behavioral acts in context and given the same behavioral units in an ethogram for which to look, may still disagree about what was seen. Even in cases in which two observers agree sufficiently for analysis of the data to continue, agreement is often not 100 percent.

"The role played by the observer in biological research," Lewis Thomas wrote, is that "he or she simply observes, describes, interprets, maybe once in a while emits a hoarse shout, but that is that; the act of observing does not alter fundamental aspects of the things observed, or anyway isn't supposed to" (Thomas 1979, 88). This professed "simplicity" is undermined by an examination of the role that technologies, both uncomplicated and sophisticated, play in framing and revealing behavior. What an observer will see as "behavior" relies on a spatial and a time component. The spatial component—is the behaver close or far from the observer—circumscribes what can be observed. Even among conspecifics, interanimal distance also demarcates what they may consider the behavior to be. Indeed, the behaver himself, given his spatial identity with the behavior, may consider the behavior to be something different than any observer, human or non, does.

Any technology that changes the spatial resolution of a behavior to observers thus has a part in changing *the behavior* insofar as its description serves as a substitute for the act. The ethologist Adrian Kortlandt suggested that when researchers could finally obtain inexpensive field glasses after the first world war, the practice of ethology changed substantially: distant, small, or distant and small animals could be observed; their behavior was now accessible and thus subject to scientific examination (Allen 2004).

A time component is also relevant to considering behavior. For along the vector of time, life is a continuous stream of "behaviors," overlapping and continuous, within organisms and among organisms. If one wants to speak of "a behavior," it must be extracted from that stream. A photograph, enabled by a camera, "stops" time, giving the illusion (or providing the opportunity for insight) that a behavior could be captured within a frame. Video playback of a videotaped set of actions allows one to slow down or speed up time, enabling the observer to see what one otherwise might not. Animal "trap" cameras, which are triggered to operate by nearby motion, are often set to capture images of animals too shy or too wary to allow humans to approach. Such cameras engage both the spatial and time components of observations. Relatedly, recent satellite technology and GPS radio tracking enable a macroscopic view allowing observers to watch movement from such a distance as to widen perception (Yong 2016).

Impediments to Observing Behavior

The methods of observational science come with built-in impediments to objective analysis of behavior. Each of these impediments in turn reveals intrinsic difficulties with any discussion of or definition of *behavior*. One might be looking at behavior at the wrong level of analysis to see what is happening; incorrect sampling may mischaracterize behavior. Naming itself is an art that can enable or hamper viewing.

Humans' anthropocentric perspective may also hinder our ability to see the behavior of nonhumans. An animal's *Umwelt*, or worldview, defined by her sensory and cognitive capacities as well as the environmental niche she fills, differs from that of humans (Von Uexküll [1934] 1957). What appears to be a seen behavior may thereby be incorrectly interpreted, its context misunderstood, or its meaning or extent misperceived. A dog's appearing to stare blankly into space, for instance, may be characterized as "doing nothing"—or that dog, who can detect high-frequency sounds, may be hearing something and being "vigilant" or smelling the odor of another dog drifting toward him on the breeze.

Relatedly, anthropomorphisms can be impediments to seeing behavior clearly: what looks like a dog's "guilty look" to the human companion who is sure of her dog's understanding of right and wrong, and similarity of emotional experience, is actually revealed to be a learned submissive behavior on closer examination (Horowitz 2009b). But, too, avoidance of anthropomorphism in interpreting behavior can be as much of a bane: in the absence of an identifying "behavior" correlating with guilt, one might declare the absence of the emotional experience. Such a declaration would be hasty.

Indeed, the mere fact of human presence may change the behavior seen—a kind of uncertainty principle for ethology. This "observer effect" can be seen in studies where a behavior is noted only when an observer is either present or absent. For instance, researchers working with non-human primates found that observer presence correlated with a "decrease in appetitive behavior" and an increase in "rest" in macaque monkeys (Iredale, Nevill, and Lutz 2010). This is called the "Hawthorne effect" in sociology and psychology, where a person's awareness of being observed leads to a behavioral change in accord with the behaver's expectations about what the researcher is looking for.

The Science of Behavior

Curiously, even after decades of success of behaviorism, B. F. Skinner (1953) felt the need to defend behavior "as a scientific subject matter": he titled the first section of *Science and Human Behavior* "The Possibility of a Science of Human Behavior." This "possibility," seen sixty years hence, after psychological science has flourished, seems to be underplaying the case. Of course the study of behavior could be scientific. On the other hand, its seeming familiarity—after all, we experience, witness, and produce behavior all the time—may make it seem at times less scientific than something—neural or other microscopic activity—that requires at minimum a tool (microscope, MRI machine) to view and perhaps an intervening level of interpretation (as with MRI, in which brain activity—brain behavior—is *deduced* by virtue of the flow of blood in the brain when in the presence of specially situated magnets). "Science," a popular conception holds, involves laboratories, controlled settings, lab coats. Science is present when there are extractions, controls, conditions, test tubes, high-powered visualization technology.

When performed with awareness of the impediments to accurate observation, though, the study of behavior is not only an art, it is truly the most straightforward of sciences.

Suggestions for Further Reading

Horowitz, Alexandra. 2010. *Inside a Dog: What Dogs See, Smell, and Know*. New York: Scribner.

Lehner, Philip N. 1996. *Handbook of Ethological Methods*. 2nd ed. Cambridge: Cambridge University Press.

Levitis, Daniel A., William Z. Lidicker Jr., and Glenn Freund. 2009. "Behavioural Biologists Don't Agree on What Constitutes Behaviour." *Animal Behaviour* 78 (1): 103–10.

References

Allen, Colin. 2004. "Is Anyone a Cognitive Ethologist?" *Biology and Philosophy* 19: 589–607.

Altmann, Jeanne. 1974. "Observational Study of Behavior: Sampling Methods." *Behaviour* 49(3/4): 227–67.

Barnard, Christopher J. 2003. *Animal Behaviour: Mechanism, Development, Function, and Evolution*. Canada: Pearson Education.

Darwin, Charles. 1872. *The Expression of the Emotions in Man and Animals*. London: John Murray.

Davis, D. E. 1966. *Integral Animal Behavior*. New York: Macmillan.

Horowitz, Alexandra C. 2003. "Do Humans Ape? Or Do Apes Human? Imitation and Intention in Humans and Other Animals." *Journal of Comparative Psychology* 117(3): 325–36.

———. 2009a. "Attention to Attention in Domestic Dog (*Canis familiaris*) Dyadic Play." *Animal Cognition* 12 (1): 107–18.

———. 2009b. "Disambiguating the 'Guilty Look': Salient Prompts to a Familiar Dog Behavior." *Behavioural Processes* 81 (2009): 447–52.

Iredale, Steven K., Christian H. Nevill, and Corrine K. Lutz. 2010. "The Influence of Observer Presence on Baboon (*Papio* spp.) and Rhesus Macaque (*Macaca mulatta*) Behavior." *Applied Animal Behaviour Science* 122 (1): 53–57.

Jensen, Per. 2009. *The Ethology of Domestic Animals: An Introductory Text*. 2nd ed. Oxfordshire: CAB International.

Kiley-Worthington, M. 1976. "The Tail Movements of Ungulates, Canids and Felids with Particular Reference to Their Causation and Function as Displays." *Behaviour* 56 (1): 69–114.

King, Laura A. 2013. *Experience Psychology*. 2nd ed. New York: McGraw-Hill.

Lehner, Philip N. 1996. *Handbook of Ethological Methods*. 2nd ed. Cambridge: Cambridge University Press.

Levitis, Daniel A., William Z. Lidicker Jr., and Glenn Freund. 2009. "Behavioural Biologists Don't Agree on What Constitutes Behaviour." *Animal Behaviour* 78 (1): 103–10.

Lindsay, Steven R. 2001. *Handbook of Applied Dog Behavior and Training: Etiology and Assessment of Behavior Problems*. Ames: Iowa State University Press.

Lindsley, Ogden R. 1991. "From Technical Jargon to Plain English for Application." *Journal of Applied Behavior Analysis* 24 (3): 449–58.

Lorenz, Konrad. 1988. *Here I Am, Where Are You? The Behavior of the Greylag Goose*. New York: Harper Collins.

Martin, Paul, and Patrick Bateson. 1993. *Measuring Behaviour: An Introductory Guide*. Cambridge: Cambridge University Press.

Noorani, Imran, and R. H. S. Carpenter. 2017. "Not Moving: The Fundamental but Neglected Motor Function." *Philosophical Transactions of the Royal Society of London B: Biological Sciences* 372 (1718).

Romanes, George. 1882. *Animal Intelligence*. London: K. Paul, Trench.

Ramachandran, V. S., ed. 2012. *Encyclopedia of Human Behavior*. 2nd ed. London: Elsevier.

Skinner, B. F. *Science and Human Behavior*. New York: Free Press, 1953.

Snowdon, Charles T. 1991. "Significance of Animal Behavior Research." Presentation at the Animal Behavior Society Annual Meeting, Wilmington, NC, June 1–6.

Thomas, Lewis. 1979. *The Medusa and the Snail: More Notes of a Biology Watcher*. New York: Viking.

Tinbergen, Niko. 1963. "On Aims and Methods of Ethology." *Zeitschrift für Tierpsychologie* 20: 410–33.

Tomasello, Michael. 1996. "Do Apes Ape?" In *Social Learning in Animals: The Roots of Culture*, edited by B. G. Galef Jr. and C. M. Heyes, 319–46. San Diego, CA: Academic Press.

Uexküll, Jakob von. (1934) 1957. "A Stroll Through the Worlds of Animals and Men." In *Instinctive Behavior: The Development of a Modern Concept*, edited by C.H. Schiller, 5–80. New York: International Universities Press.

Whiten, Andrew, Deborah M. Custance, Juan-Carlos Gomez, Patricia Teixidor, and Kim A. Bard. 1996. "Imitative Learning of Artificial Fruit Processing in Children (*Homo sapiens*) and Chimpanzees (*Pan troglodytes*)." *Journal of Comparative Psychology* 110 (1): 3–14.

Yong, Ed. 2016. "The Space Station Is Becoming A Spy Satellite for Wildlife." *Atlantic*, July 6. http://www.theatlantic.com/science/archive/2016/07/the-international -space-station-becoming-a-spy-satellite-for-tiny-animals/490112/.

5 BIOPOLITICS

Dinesh Joseph Wadiwel

Through much of human history, fish needed to be hunted down in order to be captured. During the twentieth century, this hunting effort was increasingly industrialized so that it operated as a form of large-scale mechanized predation. However, over the last thirty years there has been a radical shift in the production of fish for food. Aquaculture—or fish farming—is rapidly taking the place of wild fish capture; aquaculture today accounts for approximately 40 percent of fish consumed (Food and Agriculture Organization of the United Nations 2014, 19).

The shift from hunting down fish to farming them has altered the technologies used to harness, contain, and manage the lives of these animals. Today, rather than capture devices such as the hook or the net, fish farming instead operates according to the logics of domestication. Here, animals are contained, fed, reproduced, and made to die through the application of technologies that aim to manage populations, make life flourish where necessary, and expose to death in a timely way in order to facilitate the economic and logistic demands of meat production. Where the older violence of wild fish hunting was *episodic* in nature (fish populations were subject to sporadic violence from human fishing boats), aquaculture is instead *continuous* in scope in such a way as almost every facet of the lives of fish in the farm has some direct relationality with human systems of domination. In this regard, fish farming offers the opportunity for producers to exert maximal surveillance and control to intervene in the lives of the populations they manage; as one aquaculture producer exclaims,: "the entire life cycle has been rigorously controlled. We know where it was born, where it died, and what it ate throughout its entire life" (Wharton, University of Pennsylvania 2006).

Michel Foucault suggests that the emergence of biopower was associated with a movement away from the rule by the sword that traditionally characterized sovereign power in the west (Foucault 1998). In this story, Foucault suggests that the traditional organization of power took the form of explosive violence, including bloodletting and terror against internal and external enemies in order to demonstrate the sovereign's authority and capture resources and territory (Foucault 1998; see also Foucault 1991). However, against this traditional association of power with an overt, possessive, and spectacular violence, Foucault argues that in the modern period, a shift occurs that alters the logics of power toward a focus on biological life and population. Here, Foucault (1998) suggests that power today is less associated with the older logic of "the ancient right to take life or let live" but is instead resonant with a "power to foster life or disallow it to the point of death" (138). In this reading, rather than viewing power as using episodic violence as the main mechanism of control, power instead is seen through instruments that continuously regulate the biological life of populations, just as aquaculture does.

While Foucault was not directly interested in the application of his concept of biopower to understanding political relations between humans and animals, there is certainly a strong case for developing such an analysis. In this chapter, I will provide an overview of the concept of "biopolitics" and its role in Animal Studies. I will initially offer an outline of the concept, exploring Foucault's understanding, Italian philosopher Giorgio Agamben's influential intervention, and the critiques and enhancements that have followed. I will then provide two tangents of analysis for the application of a biopolitical approach within thinking about animals. In the first, I conceptualize biopolitics as a method or frame with which to understand human and nonhuman relations, in particular, the rationalities and techniques utilized by humans in their management of animal life. In the second, I will explore biopolitics as a way to understand the "ontological" or structural divide between humans and animals and discuss the possibilities of this approach for interrogating and challenging anthropocentricism.

Conceptualizing Biopolitics

The origins of the word *biopolitics* predate the Foucauldian applications of the term. For example, Roberto Esposito associates the origins of the term with Rudolf Kjellén (Esposito 2008, 16–17). However, it was Foucault's formulation of biopolitics in both the first volume of *The History of Sexuality* (1998) and in his lectures from the mid-1970s onward (see esp. Foucault 2004, 2007, 2010a) that have provided much of the foundational material for contemporary conceptualization of biopolitics within political and social theory.

In identifying the emergence of a "biopower," Foucault narrates a movement away from a traditional "sanguinary" force associated with the brutal, episodic violence of sovereignty that aimed at the capture of territory and population (a power exercised through death or "thanatopolitics") toward a new rationality or focus of power that becomes concerned with life itself (biopower). Here, we might pause to note a potential distinction between "biopower" and "biopolitics." In Foucault's conceptualization offered in *The History of Sexuality*, volume 1, he suggests two "poles" of operation that make up "biopower." The first pole relates to modes of individualization, surveillance, and regimes of training and acculturation that focus on the body: Foucault (1998) describes this as an *"anatomo-politics of the human body"* (139). This first pole of power is a nod toward Foucault's previous work, *Discipline and Punish*, where he describes the emergence of the prison system through the prism of surveillance and discipline as modes of power (see Foucault 1991). The second pole relates to the development of regulatory mechanisms that focus on the "species of the body"—its composition, its characteristics, its tendencies—as an object of management. This second pole Foucault (1998) describes as a "biopolitics of the population" (139).[1] In line with this understanding of a shift in the focus of power, in my discussion below I will refer to biopolitics as a modality of power that utilizes an array of technologies and discourses to address biological life and population as an object.

Foucault outlines a trajectory of biopolitics that is closely aligned with the genealogy of sovereignty and law and elaborates a link between

1. Note that in this conceptualization, Foucault is essentially treating biopolitics as a subset of a broader tendency or trajectory in the evolution of power, which he gives the term *biopower*. In many respects this conceptualization offered in *History of Sexuality*, volume 1, is unstable and illustrates some of the challenges in pinning down "biopolitics" as a concept. Part of this instability is the lack of clarity between the concepts of "biopolitics" and "biopower" in Foucault's work (see Lemke 2011, 34). Another instability relates to the further developments that Foucault's concept of "biopolitics" undergoes after *The History of Sexuality*, volume 1, including, as I shall touch on below, the focused analysis of biopolitics, sovereignty and racism, the imagination of pastoral power, and then governmentality, all of which complicate how we might understand the distinction between biopower and biopolitics.

biopolitics and state racism (see Foucault 2004). Foucault argues that sovereignty, far from being the product of a benign social contract, is instead the product of deep antagonisms. Sovereignty, Foucault argues, is founded on war and conquest: "The vanquished are at the disposal of the victors. In other words the victors can kill them. If they kill them, the problem obviously goes away: the Sovereignty of the State disappears simply because the individuals who make up that State are dead. But what happens if the victors spare the lives of the vanquished? . . . The will to prefer life to death; that is what founds sovereignty" (Foucault 2004, 95). From this standpoint Foucault puts forward the view that the biopolitical divisions that are produced within societies carry the imprint of this original violence. Here Foucault describes biopolitics as precisely the production of populations that will be made to thrive against those that will be made to diminish. With this comes the peculiar rationality found in state racism that generates a logic of species survival as a driving rationale: "the more, 'I—as species not individual—can live, the stronger I will be, the more vigorous I will be. I can proliferate.' . . . The death of the other, the death of the bad race, of the inferior race (or the degenerate or the abnormal) is something that will make life in general healthier: healthier and purer" (Foucault 2004, 255). It is this rendering of biopower that produces a biopolitical violence that aims, seemingly at all costs, to secure a population against threat (invasion, contamination, the enemy within).

But there is also a different trajectory of biopolitics that can be traced through Foucault's concept of "governmentality" (see Foucault 2007). Here, Foucault is interested in tracking both the development in biopolitics of techniques for coming to terms with, understanding, and regulating biological populations and, simultaneously, the pastoral techniques of power that generate a rationality of government that seeks to influence the ways in which individuals govern themselves, or the "conduct of conduct." These broad questions associated with how one conducts oneself and the rationalities by which we are governed continue to arise once power is interested in the lives and internal characteristics of biological populations and the appropriate interventions available to (in a positive sense) foster these lives through beneficent techniques and conversely (in a negative sense) make them diminish through abandonment and violence.

An influential negative reading of biopolitics is offered by Giorgio Agamben (see esp. Agamben 1998, 1999, 2004, 2005). Drawing from Foucault's conception outlined above as well as from the work of Carl Schmitt (1988), Hannah Arendt (1976), and Walter Benjamin (1996), Agamben puts forward the view that political sovereignty is based on the capacity to place life within a zone of legal exception; this "bare life" is the "bearer of the link between violence and law" (Agamben 1998, 65). Here, biopolitics is quite directly related to the exercise of state violence in managing who belongs and who does not within the civil political sphere (in some respects this is a developed version of Foucault's discussion of state racism outlined above). Agamben's prominent example for understanding this relationship is the European concentration camp and the event of the Holocaust, which he argues represents a foundational moment within the twentieth century in the development of biopolitical sovereignty (Agamben 1998, 166–80; see also 1999). Again, state sovereignty here does not rest on a social contract: instead it is founded on the capacity to generate spaces where the law is suspended and life within this zone of exception is subject to extraordinary forms of violence: "the original political relation is the ban (the state of exception as a zone of indistinction between outside and inside, exclusion and inclusion)" (Agamben 1998, 181). Importantly, the "bare life" produced within this zone of exception—for example, the detainee in an internment camp—is held at a threshold between life and death: "The decisive activity of biopower in our time consists not of life or death, but rather of a mutable and virtually infinite survival. Biopower's supreme ambition is to produce, in a human body, the absolute separation of the living being and the speaking being, zoé and bios, the inhuman and the human—survival" (Agamben, 1999, 155–56). Here, Agamben offers a distinct twist in his understanding of biopolitics that is relevant for Animal Studies in his view that the "bare life" that is produced by the zone of exception is in essence the result of a separation of the animal from within the human: "the Jew, that is the non-man produced within the man, or the néomort and the overcomatose person, that is, the animal separated within the human body itself" (Agamben 2004, 37). In this view biopolitics is essentially an "anthropological machine" that constantly produces a violent division between the human and the nonhuman: "the decisive political conflict, which governs every other conflict, is that between the animality and the humanity of man. That is to say, in its origin Western politics is also biopolitics" (Agamben 2004, 80).

Although Foucault's concept of biopolitics and Agamben's influential interventions have proved immensely productive for different political analyses, it is worth noting that these accounts have been subject to controversies, disagreements, and supplementations. First, the concept of biopolitics is compromised by a degree of conceptual instability, particularly in relation to understanding the history of biopolitics, and whether biopower can be said to emerge as a *modern* relation. Jacques Derrida, for example, was to an extent dismissive of the concept of biopolitics, taking issue with Agamben's inconsistency in proclaiming biopolitics as both a modern trajectory of politics and simultaneously as an ancient one. Derrida effectively argues that all politics is to a degree biopolitical: "So I am not saying there is no 'new bio-power,' I am suggesting that 'bio-power' itself is not new. There are incredible novelties in bio-power, but bio-power or zoo-power are not new" (Derrida 2009, 330). Second, Agamben offers what has been described as a "negative" reading of biopolitics (see esp. Campbell 2008; Lemke 2011). The reading Agamben offers suggests biopolitics represents a contamination of classical politics and that the restrictive effects of this politics are, to an extent, inescapable (in a sense, according to this negative view, we are simply waiting for the next concentration camp). However more "positive" or "affirmative" visions of biopolitics are possible, notably, Michael Hardt and Antonio Negri's association of biopower with the political vitality of collective resistance against capitalism and domination (see, e.g., Hardt and Negri 2000; see also Lemke 2011, 65–76; Chrulew 2012) and Esposito's suggestion that we can aspire to a politics that seeks to safeguard the common and the communal rather than attempting to immunize or securitize populations through the use of violence (Esposito 2008, 2013; see also Tierney 2016). Third, there has been an ongoing critique of Foucault and Agamben's Eurocentric conceptualization of biopolitics, in particular, their neglect of the history of colonialism and the specific violence related to the enterprise of European conquest. An influential alternative account was developed by Achille Mbembe in his conceptualization of "necropolitics"—that is, the "subjugation of life to the power of death" (Mbembe 2003, 39; see also Mbembe 2001)—as a counter to Foucault and Agamben's understanding of "biopolitics." As discussed above, both Foucault and Agamben cite the origins of biopolitical violence within the confines of Europe; against this, Mbembe points out that this conceptualization misses the relationship of slavery and the colony in informing the historical development of the European camp: "What one witnesses in World War II is the extension to the 'civilized' peoples of Europe of the methods previously reserved for the 'savages'" (Mbembe 2003, 23). Finally, there have been a range of analyses that have attempted to describe systematically fields of power that were not theorized adequately in Foucault's original conception, for example, work on the relationship between gender and biopolitics (e.g., Smith 2010; Spade 2011; Repo 2015) and scholarship exploring the relationship between biopolitics and disability (e.g., Campbell 2009; Mitchell and Snyder 2015), including in relation to understanding the history of eugenics (Hughes 2012; Mitchel and Synder 2003).

As I shall discuss, some of these trajectories of critique about understanding biopolitics have shaped responses from Animal Studies scholars. But it is important, in this context, to stress that both Foucault and Agamben are silent on the implications of biopower for animals despite implicitly—or explicitly, as in the case of Agamben's (2004) discussion in *The Open*—putting forward a perspective on biopolitics that has ramifications for thinking about animals. In some respects, mainstream study of biopolitics across sociology, philosophy, and political theory has exacerbated this problem by failing to take seriously animals as political objects of biopolitical interventions.[2] This is despite the clear sense, in the words of Matthew Chrulew, that the "modern apparatuses of government administer not only human life, but *all* life: seeds and crops, animal individuals and populations, ecosystems, the earth itself" (Chrulew 2012, 54).

I describe below two tendencies for biopolitical thinking around human-animal relations that at least partially address the silences around animals that characterize much scholarship on biopower. First, I discuss work that deploys biopolitics as a methodological approach to understand material practices and rationalities of human utilization of, and violence against, nonhuman animals, including in the large-scale use of animals for food or in the logics of animal population management. Second, I describe scholarship that draws on biopolitical understandings to conceptualize the structural or ontological divide between humans and animals.

Animals and Biopolitics: Practices and Rationalities

Classic proanimal theory (Singer 2002; Regan 1983; Cavalieri 2001) has concerned itself with assessing ethical justifications for human use of animals. The philosophical project involves determining a consistent set of norms according to which beings—that is, human or nonhuman—with identifiable shared characteristics—for example, sentience, intelligence, or "subjects-of-a-life" —should be treated and providing demonstrations

2. This is perhaps a symptom of the generalized sidelining of Animal Studies within mainstream politics, philosophy, and sociology. Against this, there has been a growing a range of scholars that have attempted to correct how biopolitics is understood by addressing this silence (see, e.g., Wadiwel 2002; Wolfe 2012; Taylor 2013; Shukin 2009; Cole 2011; Thierman 2010; and Chrulew and Wadiwel 2016).

that current practices are in contradiction with these rational ethical norms (see chap. 10).

A biopolitical approach offers a different strategy: it is attentive to practices and relations of power that shape human and animal interactions and is particularly interested in how power is consumed by the regulation of biological life as a governing rationality. One of the benefits of a biopolitical approach is understanding how the rationalities that shape the production of animals resonate strongly with the relations of power that enfold humans, indeed, that the forms of biopolitical violence that we might track across interhuman affairs resonate with the techniques that shape human engagement with animals: "we are all, after all, potentially animals before the law" (Wolfe 2012, 105). In this sense, biopolitical approaches differ from classic proanimal theory at least insofar as scholars are less concerned with demonstrating that the human treatment of animals is in *contradiction* with prevailing and agreed ethical norms. Instead, they are interested in demonstrating that the treatment of animals is in *conformity* with prevailing rationalities of power, indeed, that the treatment of animals may often serve as a useful exemplar of the guiding logic by which power relations shape human and nonhuman actors.

Factory farms serve as a productive illustration of the rationality at the core of biopolitics and are useful for understanding the dynamics of human violence against other humans (see Wolfe 2012; Wadiwel 2004 and Pugliese 2013).[3] It is true that much work on factory farms, from a range of disciplines, has already occurred within proanimal theory and Animal Studies, including problematizations of the conditions of confinement, breeding, commodification, and slaughter that attend animal industries (see, e.g., Singer 2002; Adams 1990; Foer 2009; Gruen 2011; Noske 1997; Nibert 2013; and Pachirat 2011). However, scholars engaged with biopolitical thought have provided a different methodological frame for understanding and problematizing these relations of "agricultural power" (see Taylor 2013).

For some scholars, this is a question of understanding in detail the

3. On the face of it, a biopolitical approach is not initially engaged with questions about the ethics of animal utilization and appears *methodological* rather than moral in approach. Perhaps as such, work by Foucault scholars on the biopolitics of human-animal relations might be interpreted as lacking a normative commitment (see, e.g., the critique advanced by Steiner 2013). However, there is more than enough biopolitical scholarship in Animal Studies that takes a strong normative position in relation to human-animal relations for this interpretation to be questioned (see, e.g., Taylor 2013; Stanescu 2013; Dutkiewicz 2013; Cole 2011; Chrulew 2012; Wadiwel 2002, 2015; Wolfe 2012).

deployment of technologies and practices of disciplinary power (or anatamo-politics) that aim to make animals docile and productive (see, e.g., Coppin 2003; Novek 2005; and Thierman 2010). This work aims to demonstrate that power relations that have been integral to the arrangement of human institutions—such as the emergence of disciplinary mechanisms in prisons, schools, and health care facilities (see Foucault 1991)—are mirrored in the use of surveillance, individualization, and discipline related to production processes in animal agriculture. Dawn Coppin, for example, notes that disciplinary power requires the distribution of human subjects in space in order to attend to and make the individual docile (e.g., the prison); this same arrangement of power is essential for contemporary large-scale pig farming, which similarly relies on separations of animals into crates and pens to produce docility (see Coppin 2003, 604).

But beyond these individual disciplinary techniques, there are also a range of measures utilized to manage animals within large-scale agriculture that focus intently on the biological life of the animal and seek to manage groups of animals at a population level. Animal agriculture involves both intense controls over biological production and a scrupulous focus on managing cycles of reproduction: animals must be bred in great numbers, often through the large-scale use of reproductive technologies; maintained through their lives utilizing strict regimes of control over diet and movement; and then slaughtered at the right time and under the right conditions to produce a commodity that has value. Richard Twine explains this process specifically with reference to the biopolitics of science in the context of animal agriculture:

> If we think about the biopower of animal science broadly (not just genetics), we see an approach to the bodies of pigs, cattle, chicken and sheep that is about the precise and economically efficient control of inputs in order to try and control a particular qualitatively standardized meat, milk or fibre product. The sequencing of the genomes of these animals is an attempt to widen the toolkit and to heighten the degree of control in breeding. Just as important as genetics is research into feed, feed efficiency and animal health. (Twine 2010, 93–94).

These processes are closely tied to the logics of capitalism itself, and the drive to produce and reproduce life, something that was inherent to Foucault's original understanding of biopower: "capitalism . . . would not have been impossible without the controlled insertion of bodies into the machinery of production and the adjustment of the phenomena of population to economic processes" (Foucault 1998, 140–41). For animals, this means deep controls over aspects of the life cycle, such as breed societies that maintain and regulate animal populations (see Morris and Holloway 2009; Holloway and Morris 2016). Insofar as the use of animals and animal products for food functions as a mechanism for the biological reproduction of human populations (in a Marxist sense, the reproduction of animals as a means of subsistence for the reproduction of the human labor force), there is a conjoined politics of life involved in understanding the relationship between animal and human populations under contemporary conditions of capitalist exchange.

The production of animals for food is characterized by not only the intensive use of beneficent techniques to "foster" life but simultaneously techniques that produce death; notably, in the mass slaughter of animals for food. Here, life and death appear to oscillate together to produce the biopolitical relation. Animal advocates, for example, often express concern over the techniques some producers use to produce calf meat, which may require the use of close confinement of calves in "veal crates" and the restriction of diet to produce a pale colored flesh that is favored by some consumers. In *Animal Liberation*, Peter Singer (1975, 2002) took note of this to point out the way in which these calves were made to live in close proximity to mortality: "without any iron at all the calves would drop dead. With a normal intake their flesh will not fetch as much per pound. So a balance is struck which keeps the flesh pale and the calves—or most of them—on their feet long enough for them to reach their market weight" when they are three months old (133). This reveals an explicit biopolitical logic. In my earlier work, I note that the cruelty of veal production is tied closely to a rationality of power that deploys a range of techniques—control over feed, lighting, movement, interaction with other animals, length of life—that aims to hold these animals in a state of near death throughout their lives (Wadiwel 2002; see also Wadiwel 2004).

Indeed, the role of death in the production of animals for food looms as a vital and ever-present reality; animals are made to die in precise ways by production processes in order to produce profit, and as such death is a value-producing moment (Dutkiewicz 2013). This death also haunts a range of different economies in body parts, including the "rendering" of "waste products" such as gelatin (Shukin 2009) or the processes that are required for cow milk to allow it to be a saleable commodity (Nimmo 2010). As such, even slaughter—the business of making die—takes on a biopolitical aspect insofar as animal agriculture relies on making animals live in order for them to die, the two extremes coordinated closely with each other as part of the rationality of power. As such, the production of animals for food requires the "necroavailability" of animals (Reinert 2007, 2013). As James Stanescu notes, animals who are slaughtered for food have their lives fabricated with the explicit purpose of becoming corpses: "The slaughter of animals is never simply the killing of animals, but rather the production of corpses for consumption . . . the fabrication and production of lives to be part of the fabrication and production of corpses" (Stanescu 2013, 153).

Outside of the intense violence of animal agriculture and the slaughter-house, there are other examples of the ways in which biopolitical ratio-nalities and practices are central to systems utilized to manage animal lives and populations. There is a range of scholarship addressing animal management within zoos (see Chrulew 2010, 2011a, 2013, 2014a; see also Braverman 2013b); relationships with companion and service animals (Srinivasan 2013, 2015; Hediger 2013; Braverman 2013a), management of zoonotic diseases (Blue and Rock 2011), and modes of animal governance (see Cole 2011; Taylor 2013; Shukin 2011, 2013; Braverman 2013b; Wadiwel 2015). Biopolitical perspectives have also been applied to understanding interactions with "nature" and wild animals (see Braverman 2015; Lorimer and Driessen 2013; Chrulew 2011b; Srinivasan 2014; and Collard 2012). As such, biopolitical approaches to studying practices and relations between humans and animals would appear to be growing a diverse trajectory of Animal Studies scholarship.

The Division between the Human and the Animal

The above approaches have utilized a biopolitical analysis as a methodology to understand rationalities and practices of power. A different tendency within biopolitical scholarship in Animal Studies is to interrogate the division between humans and animals, the extent to which this division reflects or produces anthropocentrism, and the case for moving beyond this division (Agamben 2004: 80).

As discussed above, Agamben provides at least part of the conceptual resources for thinking about this division or "biological caesura" that divides humans from animals: in Agamben's formulation, biopolitics merely represents the continual articulation of the boundary between human and nonhuman life (see Agamben 2004; see also LaCapra 2009).

This observation is potentially fruitful for Animal Studies work on the problem of anthropocentrism (see chap. 3). As Matthew Calarco (2008) observes, Agamben situates the divide between humans and animals as a political problem, one that is definitive of anthropocentrism. This means that Agamben is less interested in highlighting the distinctive natural qualities that mark out humans as unique or special, instead interrogating how it is that humans are defined in a relational sense with respect to animals and the implications this has for material relations between humans and others: "not only does the distinction . . . [between human and nonhuman] . . . create the opening for the exploitation of nonhuman animals and others considered not fully human (this is the point that is forcefully made by animal ethicists), but it also creates the conditions for contemporary biopolitics, in which more and more of the 'biological' and 'animal' aspects of human life are brought under the purview of the State and the juridical order" (Calarco 2008, 94). Here, Calarco points to the continuity or "indistinction" (see Calarco 2015) between humans, nonhumans, and those situated in between.

It is worth emphasizing that from this perspective, biopolitics intertwines humans and nonhumans within power relations, hence, the interest by some scholars in the relationship between biopolitics, the human, and the genealogy of race (see, e.g., Pugliese 2013; Weheliye 2014). As discussed above, while Foucault identifies state racism as a product of a biopolitics, he failed to adequately account for patterns and logics of colonialism in the construction of contemporary biopolitics (see Mbembe 2003). In this context, a range of thinkers have focused on the history of racialized slavery to understand the relationship between the genealogy of biopolitics and the foundational role of antiblack violence within modernity (see Wilderson 2010 and Sexton 2010; see also Mbembe 2003, 22; James 1996, Weheliye 2014). There is here a political case, too, for reimagining the human outside of the colonial imaginary. Alexander G. Weheliye (2014), for example, has utilized black feminist perspectives to offer a critique of the construction of human within the Western enlightenment project and explore the project of ushering "different genres of the human" (2–3).

This perspective on biopolitics that emphasizes conflict—where biopolitics reflects a somewhat foundational political division—is suggestive of a deep hostility or ongoing "war" in human relationships with animals (Wadiwel 2015, 18–20). I have argued that this perspective views biopolitics through Foucault's suggestion that sovereignty reflects historical conquest by violence and that biopolitics, as the ongoing hierarchical separation between populations, is an echo of this original violence. This framework might usefully be applied to thinking about human sovereignty over animals as an ongoing form of biopolitical violence that shapes almost all relations with animals, including the use of animals for food, experimentation, sport, and companionship: "We eat, hunt, torture, incarcerate and kill animals because it is our sovereign right won from total victory; our sovereign pleasure" (Wadiwel 2015, 29).

We might note that these approaches to biopolitics that emphasize conflict or division might imply a "negative" reading, at least insofar as the shift of power toward life and population introduces new forms of systemic violence toward animals that represent a significant and perhaps inescapable set of foundational relations of constraint. But, as suggested above, there are at least some readings of biopolitics that are affirmative. Here it is imagined that biopolitics might be democratized in some way (Chrulew 2012, 63), or that animal resistance to human power might provide positive opportunities "for life to burst through power's systematic operation in ways that are more and more difficult to anticipate" (Wolfe 2012, 32), or that we might imagine possibilities for a "posthumanist communitarianism" (Gabardi 2017, 153–74). This sort of reading of biopolitics, as both representing a set of constraints that limit life but simultaneously offer political opportunities for rupturing anthropocentrism, points us toward a different approach to thinking about biopolitics, one that is attentive to how we might observe new assemblies and relations between humans and animals, an "ontopolitical task" to create "debate on how other possibilities for human-nonhuman animals can be brought into existence" (Asdal, Druglitrø, and Hinchliffe 2016, 19).

All of this might prompt us to ask how can we move forward from biopolitics, and can we imagine a future after the biopolitical? At least some engagements in Animal Studies and posthumanism have focused on this question through a radical project of trying to reimagine the human (see Seshadri 2012; see also Chrulew 2014b) or generate new forms of care and "radical affection" (Chen 2012, 237). A different tangent has come from animal focused scholars who have explicitly argued for an end to biopolitics as a mechanism for the hierarchical organization of life (Calarco 2008; Oliver 2009). Here there has been an acknowledgment that arguing for legal rights is potentially a dead end. Because Agamben's reading of biopolitics suggests a continually evolving antagonism, granting rights or legal protections to those who are considered less than human is not necessarily enough, since the biopolitical mechanism will continue to reproduce a radical rupture between those who are part of the political community and those who have been excluded. For Calarco, this means the political challenge that biopolitics presents is the problem of how to "halt" or "jam" the "anthropological machine," that is, how to develop "another mode of relation and community with nonhuman life" (Calarco 2008, 102; see also Calarco 2015).[4] In this view, arresting the function of the machine itself is seen as necessary for political change.

Admittedly, all such proposals to "end biopolitics" or create a new human are abstract and extraordinarily radical in scope. However, I would argue that a biopolitical perspective offers us some very concrete political challenges, some of which extend proanimal action in new directions. First, as I have indicated above, Animal Studies scholars using biopolitical perspectives have highlighted and troubled in quite different ways forms of violence within animal agriculture, animal-based research, pet keeping, and animal conservation. Instead of a focus on ethical responses to animal sentience, suffering, and cognition, attention is drawn to modes of discipline, reproduction, containment, and regulation of biological processes as forms of biopolitical violence and governance. One of the innovations here is drawing attention to the way in which the same violent biopolitical forces that shape the governance of animals are closely mirrored in human relations of power. There is here the largely unexplored potential for a shared politics between human and nonhuman interests. For example, recognizing that the same biopolitical logics that produce the prison as a site for the violent management of human populations also operate to produce the factory farm as a model for control of animal populations. This offers a fruitful space for exploring the possibility of a shared politics that opposes mass incarceration as a modality of power. Second, insofar as anthropocentricism generally is a political concern for animal advocates, then biopolitics offers us ways to deepen our analysis, highlighting the ill effects of the "anthropological machine" for animals. If biopolitics represents the violent division between humans and animals, then there is new potential here to understand the relationship between anthropocentricism and other forms of structural violence and oppression. If it is true that biopolitics informs contemporary forms of violence against human

4. In a similar vein, Kelly Oliver also proposes "stopping" the anthropological machine at the heart of biopolitics through drawing attention to the pedagogic function of animals within the human-animal binary: "we learn something about the category *human* by exploiting its relation to the category *animal*" (Oliver 2009, 244).

populations—including logics of military intervention, mass incarceration, practices of torture and ill treatment, mass killings and differential uses of force, discipline and coercion against individuals on the basis of race, ability, gender, or sexuality—then Animal Studies scholars provide a crucial perspective in an understanding of how these same logics affect nonhuman animals and the role of anthropocentrism in generating this violence. From this view point, the study of animals and biopolitics is far from marginal but is instead crucial to understanding the contemporary logic of all political violence.

Suggestions for Further Reading

Agamben, Giorgio. 1998. *Homo Sacer: Sovereign Power and Bare Life*. Translated by Daniel Heller-Roazen. Stanford, CA: Stanford University Press.

Chrulew, Matthew. 2012. "Animals in Biopolitical Theory: Between Agamben and Negri." *New Formations* 76: 53–67.

Foucault, Michel. 1998. *The Will to Knowledge*. Vol. 1 of *The History of Sexuality*. London: Penguin Books.

Taylor, Chloë. 2013. "Foucault and Critical Animal Studies: Genealogies of Agricultural Power." *Philosophy Compass* 8 (6): 539–51.

Wadiwel, Dinesh Joseph. 2015. *The War against Animals*. Leiden: Brill.

Wolfe, Cary. 2012. *Before the Law: Human and Other Animals in a Biopolitical Frame*. Chicago: University of Chicago Press.

References

Adams, Carol J. 1990. *The Sexual Politics of Meat*. New York: Continuum.

Agamben, Giorgio. 1998. *Homo Sacer: Sovereign Power and Bare Life*. Translated by Daniel Heller-Roazen. Stanford, CA: Stanford University Press.

———. 1999. *Remnants of Auschwitz: The Witness and the Archive*. Translated by Daniel Heller-Roazen. New York: Zone Books.

———. 2004. *The Open: Man and Animal*. Translated by Kevin Attell. Stanford, CA: Stanford University Press.

———. 2005. *State of Exception*. Chicago: University of Chicago Press.

Arendt, Hannah. 1976. *The Origins of Totalitarianism*. San Diego, CA: Harcourt Brace.

Asdal, Kristin, Tone Druglitrø, and Steve Hinchliffe. 2016. "Introduction: The More Than Human Condition; Sentient Creatures and Versions of Biopolitics." In *Animals and Biopolitics: The More-Than-Human Condition*, edited by Kristin Asdal, Tone Druglitrø and Steve Hinchliffe, 11–24. London: Routledge.

Benjamin, Walter. 1996. "Critique of Violence." In *Selected Writings* vol. 1, *1913–1926*, edited by M. Bullock and M. W. Jennings, 236–52. Cambridge, MA: Belknap.

Blue, Gwendolyn, and Melanie J. Rock. 2011."Trans-biopolitics: Complexity in Interspecies Relations. *Health: An Interdisciplinary Journal* 15 (4): 353–68.

Braverman, Irus. 2013a. "Passing the Sniff Test: Police Dogs as Biotechnology." *Buffalo Law Review* 61: 81–168.

———. 2013b. *Zooland: The Institution of Captivity*. Stanford, CA: Stanford University Press.

———. 2014. "Governing the Wild: Databases, Algorithms, and Population Models as Biopolitics." *Surveillance and Society* 12 (1): 15–37.

———. 2015. *Wildlife: The Institution of Nature*. Stanford, CA: Stanford University Press.

Calarco, Matthew. 2008. *Zoographies: The Question of the Animal from Heidegger to Derrida*. New York: Columbia University Press.

———. 2015. *Thinking Through Animals: Identity, Difference, Indistinction*. Stanford, CA: Stanford University Press.

Campbell, Fiona Kumari. 2009. *Contours of Ableism: The Production of Disability and Ableness*. New York: Palgrave Macmillan.

Campbell, Timothy. 2008. "Translator's Introduction: Bios, Immunity and Life; The Thought of Roberto Esposito." In *Bios: Biopolitics and Philosophy*, by Roberto Esposito, translated by Timothy Campbell, vi–xlii. Minneapolis: University of Minnesota Press.

Cavalieri, Paola. 2001. *The Animal Question: Why Non-human Animals Deserve Human Rights*. Oxford: Oxford University Press.

Chen, Mel Y. 2012. *Animacies: Biopolitics, Racial Mattering, and Queer Affect*. Durham, NC: Duke University Press.

Chrulew, Matthew. 2010. "From Zoo to Zoöpolis: Effectively Enacting Eden." In *Metamorphoses of the Zoo: Animal Encounter after Noah*, edited by Ralph R. Acampora, 193–219. Lanham, MD: Lexington Books.

———. 2011a. "Managing Love and Death at the Zoo: The Biopolitics of Endangered Species Preservation." *Australian Humanities Review* 50. http://www.australian humanitiesreview.org/archive/Issue-May-2011/chrulew.html.

———. 2011b. "Reversing Extinction: Restoration and Resurrection in the Pleistocene Rewilding Projects." *Humanimalia* 2 (2): 4–27.

———. 2012. "Animals in Biopolitical Theory: Between Agamben and Negri." *New Formations* 76. 53–67.

———. 2013. "Preventing and Giving Death at the Zoo: Heini Hediger's 'Death Due to Behaviour.'" In *Animal Death*, edited by Fiona Probyn-Rapsey and Jay Johnston, 221–38. Sydney: Sydney University Press.

———. 2014a."'An Art of Both Caring and Locking Up': Biopolitical Thresholds in the Zoological Garden." *Sub-Stance* 43 (2): 124–47.

———. 2014b. "The Power of Silence." *Humanimalia* 5 (2): 155–61.

Chrulew, Matthew, and Dinesh Wadiwel, eds. 2016. *Foucault and Animals*. Boston: Brill.

Cole, Matthew. 2011. "From 'Animal Machines' to 'Happy Meat'? Foucault's Ideas of Disciplinary and Pastoral Power Applied to 'Animal-Centred' Welfare Discourse." *Animals* 1 (1): 83–101.

Collard, Rosemary-Claire. 2012. Cougar-Human Entanglements and the Biopolitical

Un/making of Safe Space. *Environment and Planning D: Society and Space* 30 (1): 23–42.

Coppin, Dawn. 2003. "Foucauldian Hog Futures: The Birth of Mega-Hog Farms." *Sociological Quarterly* 44 (4): 597–616.

Derrida, Jacques. 2009. *The Beast and the Sovereign*, vol. 1. Chicago: University of Chicago Press.

Dutkiewicz, Jan. 2013. "'Postmodernism,' Politics, and Pigs." *Phanex* 8 (2): 296–307.

Esposito, Roberto. 2008. *Bios: Biopolitics and Philosophy*. Translated by Timothy Campbell. Minneapolis: University of Minnesota.

———. 2013. "Community, Immunity, Biopolitics." *E-misférica* 10 (1). http://hemisphericinstitute.org/hemi/en/e-misferica-101/esposito.

Foer, Jonathon Safran. 2009. *Eating Animals*. New York: Little, Brown.

Food and Agriculture Organization of the United Nations. 2014. *The State of World Fisheries and Aquaculture, 2014*. Rome: Food and Agriculture Organization of the United Nations.

Foucault, Michel. 1990. *The Care of the Self*, vol. 3 of *The History of Sexuality*. London: Penguin.

———. 1991. *Discipline and Punish: The Birth of the Prison*. London: Penguin.

———.1998. *The Will to Knowledge*, vol. 1 of *The History of Sexuality*. London: Penguin, 1998.

———. 2004. *Society Must Be Defended: Lectures at the Collège de France, 1975–76*. London: Penguin.

———. 2007. *Security, Territory, Population: Lectures at the Collège de France, 1977–78*. London: Palgrave Macmillan, 2007.

———. 2010a. *The Birth of Biopolitics: Lectures at the Collège de France, 1978–1979*. New York: Palgrave Macmillan, 2010.

———. 2010b. *The Government of Self and Others: Lectures at the Collège de France 1982–1983*. New York: Palgrave Macmillan, 2010.

———. 2011. *The Courage of Truth: Lectures at the Collège de France 1983–1984*. New York: Palgrave MacMillan.

———. 2014. *On the Government of the Living: Lectures at the Collège de France, 1979–1980*. New York: Picador.

Gabardi, Wayne. 2017. *The Next Social Contract: Animals, the Anthropocene, and Biopolitics*. Philadelphia: Temple University Press.

Gruen, Lori. 2011. *Ethics and Animals: An Introduction*. Cambridge: Cambridge University Press.

Hardt, Michael, and Antonio Negri. 2000. *Empire*. Cambridge: Harvard University Press.

Hediger, Ryan. 2013. "Dogs of War: The Biopolitics of Loving and Leaving the U.S. Canine Forces in Vietnam." *Animal Studies Journal* 2 (1): 55–73.

Holloway, Lewis. 2015. "Biopower and an Ecology of Genes: Seeing Livestock as Meat via Genetics." In *Political Ecologies of Meat* edited by J. Emel and H. Neo, 178–94. London: Earthscan.

Holloway, Lewis, Christopher Bear, and Katy Wilkinson. 2014. "Re-capturing Bovine Life: Robot-Cow Relationships, Freedom and Control in Dairy Farming." *Journal of Rural Studies* 33: 131–40.

Holloway, Lewis, and Carol Morris. 2012. "Contesting Genetic Knowledge-Practices in Livestock Breeding: Biopower, Biosocial Collectivities and Heterogeneous Resistances." *Environment and Planning D: Society and Space* 30: 60–77.

———. 2016. "Biopower, Heterogeneous Biosocial Collectivities and Domestic Livestock Breeding." In *Foucault and Animals*, edited by Matthew Chrulew and Dinesh Wadiwel. Boston: Brill.

Holloway, Lewis, Carol Morris, Ben Gilna, and David Gibbs. 2009. "Biopower, Genetics and Livestock Breeding: (Re)constituting Animal Populations and Heterogeneous Biosocial Collectivities." *Transactions, Institute of British Geographers* (34): 394–407.

Hughes, Bill. 2012. "Civilizing Modernity and the Ontological Invalidation of Disabled People." *Disability and Social Theory: New Developments and Directions*, edited by D. Goodley, B. Hughes, and L. Davis, 17–32. Basingstoke: Palgrave Macmillan.

James, Joy. 1996. *Resisting State Violence: Radicalism, Gender, and Race in U.S. Culture*. Minneapolis: University of Minnesota Press.

LaCapra, Dominick. 2009. *History and Its Limits: Human, Animal, Violence*. Ithaca, NY: Cornell University Press.

Lemke, Thomas. 2011. *Biopolitics: An Advanced Introduction*. New York: New York University Press

Lorimer, James, and Clemens Driessen. 2013. "Bovine Biopolitics and the Promise of Monsters in the Rewilding of Heck Cattle." *Geoforum* 48: 249–59

Mbembe, Achille. 2001. *On the Postcolony*. Berkeley: University of California Press.

———. 2003. "Necropolitics." *Public Culture* 15 (1): 11–40.

Michell, David, and Sharon Snyder. 2003. "The Eugenic Atlantic: Race, Disability, and the Making of an International Eugenic Science, 1800–1945." *Disability and Society* 18 (7): 843–64.

———. 2015. *The Biopolitics of Disability: Neoliberalism, Ablenationalism, and Peripheral Embodiment*. Ann Arbor: University of Michigan Press.

Morris, Carol, and Holloway Lewis. 2009. "Genetic Technologies and the Transformation of the Geographies of UK Livestock Agriculture: A Research Agenda." *Progress in Human Geography* 33: 313–33.

Nibert, David. 2013. *Animal Oppression and Human Violence: Domesecration, Capitalism, and Global Conflict*. New York: Columbia University Press.

Nimmo, Richie. 2010, *Milk, Modernity and the Making of the Human: Purifying the Social*. London: Routledge.

Noske, Barbara. 1997. *Beyond Boundaries: Humans and Animals*. Montreal: Black Rose Books.

Novek, Joel, 2005. "Pigs and People: Sociological Perspectives on the Discipline of Nonhuman Animals in Intensive Confinement." *Society and Animals* (13): 221–44.

Oliver, Kelly. 2009. *Animal Lessons: How They Teach Us to Be Human*. New York: Columbia University Press.

Pachirat, Timothy. 2011. *Every Twelve Seconds: Industrialized Slaughter and the Politics of Sight*. New Haven, CT: Yale University Press.

Pugliese, Joseph. 2013. *State Violence and the Execution of Law: Biopolitical Caesurae of Torture, Black Sites, Drones*. Abingdon: Routledge, 2013

Regan, Tom. 1983. *The Case for Animal Rights*. Berkeley: University of California Press.

Reinert, Hugo. 2007. "The Pertinence of Sacrifice: Some Notes on Larry the Luckiest Lamb." *Borderlands e-Journal* 6 (3). http://borderlands.net.au/vol6no3_2007/reinert_larry.htm.

———. 2013. "The Disposable Surplus: Notes on Waste, Reindeer, and Biopolitics." *Laboratorium: Russian Review of Social Research*. 4. http://www.soclabo.org/index.php/laboratorium/article/view/52/955.

Repo, Jemima. 2015. *The Biopolitics of Gender*. Oxford: Oxford University Press.

Schmitt, Carl. 1988. *Political Theology: Four Chapters on the Concept of Sovereignty*. Cambridge, MA: MIT Press.

Seshadri, Kalpana Rahita. 2012. *HumAnimal: Race, Law, Language*. Minneapolis: University of Minnesota Press.

Sexton, Jared. 2010. "People-of-Color-Blindness: Notes on the Afterlife of Slavery." *Social Text* 28 (2): 31–56.

Shukin, Nicole. 2009. *Animal Capital*. Minneapolis: University of Minnesota Press.

———. 2011."Tense Animals: On Other Species of Pastoral Power." *CR: The New Centennial Review* 11 (2): 143–66.

———. 2013. "Security Bonds: On Feeling Power and the Fiction of an Animal Governmentality." *ESC: English Studies in Canada* 39 (1): 177–98.

Singer, Peter. 2002. *Animal Liberation*. New York: Harper Collins.

Smith, Anne Marie. 2010. "Neo-eugenics: A Feminist Critique of Agamben." *Occasion: Interdisciplinary Studies in the Humanities*. http://arcade.stanford.edu/occasion/neo-eugenics-feminist-critique-agamben

Spade, Dean. 2011. *Normal Life: Administrative Violence, Critical Trans Politics and the Limits of Law*. Brooklyn: South End Press.

Srinivasan, Krithika. 2013. "The Biopolitics of Animal Being and Welfare: Dog Control and Care in the UK and India." *Transactions of the Institute of British Geographers* 38: 106–19.

———. 2014. "Caring for the Collective: Biopower and Agential Subjectification in Wildlife Conservation." *Environment and Planning D: Society and Space* 32 (3): 501–17.

———. 2015. "The Welfare Episteme: Street Dog Biopolitics in the Anthropocene." In *Animals in the Anthropocene: Critical Perspectives on Non-Human Futures*, edited by N. Boyd, M. Chrulew, C. Degeling, A. Mrva-Montoya, F. Probyn-Rapsey, N. Savvides, and D. Wadiwel, 201–20. Sydney: Sydney University Press.

Stanescu, James. 2013. "Beyond Biopolitics: Animal Studies, Factory Farms, and the Advent of Deading Life." *Phaenex* 8 (2): 135–60.

Steiner, Gary. 2013. *Animals and the Limits of Postmodernism*. New York: Columbia University Press.

Taylor, Chloë. 2013. "Foucault and Critical Animal Studies: Genealogies of Agricultural Power." *Philosophy Compass* 8 (6): 539–51.

Thierman, Stephen. 2010. "Apparatuses of Animality: Foucault Goes to a Slaughter-house." *Foucault Studies* 9: 89–110

Tierney, Thomas F. 2016. "Roberto Esposito's Affirmative Biopolitics' and the Gift." *Theory, Culture, and Society* 33 (2): 53–76.

Twine, Richard. 2010. *Animals as Biotechnology: Ethics, Sustainability, and Critical Animal Studies*. London: Earthscan.

Wadiwel, Dinesh Joseph. 2002. "Cows and Sovereignty: Biopolitics and Bare Life." *Borderlands e-Journal* 1 (2). http://borderlands.net.au/vol1no2_2002/wadiwel_cows.html.

———. 2004. "Animal by Any Other Name? Patterson and Agamben Discuss Animal and Human Life." *Borderlands e-Journal* 3 (1). http://www.borderlands.net.au/vol3no1_2004/wadiwel_animal.htm.

———. 2015. *The War against Animals*. Leiden: Brill.Weheliye, Alexander G. 2014. *Habeas Viscus: Racializing Assemblages, Biopolitics, and Black Feminist Theories of the Human*. Durham, NC: Duke University Press.

Wharton, University of Pennsylvania. 2006. "Aquaculture Is Needed to Satisfy Global Demand for Fish." Knowledge@Wharton, October 18. http://knowledge.wharton.upenn.edu/article/aquaculture-is-needed-to-satisfy-global-demand-for-fish/.

Wilderson, Frank B. 2010. *Red, White and Black: Cinema and the Structure of U.S. Antagonisms*. Durham, NC: Duke University Press.

Wolfe, Cary. 2012 *Before the Law: Human and Other Animals in a Biopolitical Frame*. Chicago: University of Chicago Press.

6 CAPTIVITY

Lori Marino

If you look up *captivity* in a dictionary you will typically find that the definition refers to space. For example, Merriam-Webster defines *captivity* as "the state of being kept in a place (such as a prison or a cage) and not being able to leave or be free." But captivity represents more than the physical circumstance of being confined in space or in movement; captivity represents a *state of being*. And while it includes the fact that one's preferences for action are limited, it goes beyond that. Captivity is a persistent *psychological* state of extreme dependence, tedium, and anxiety. This is true for humans and other animals in captivity.

There are always two elements of captivity. First, there is a directional relationship between the captor and the captive. The captive is in a situation in which he or she is being held against his/her will and is dependent on the captor. Second, the captor benefits from the relationship at the expense of the captive. When these critical psychological dimensions are taken into account, a richer definition of *captivity* emerges as the following: "captivity occurs when there is a self-directed creature capable of independent intentional actions, and her movements, choices, and actions are subject to the control of another who benefits from this control. Captivity is a condition of powerlessness over one's options" (Rivera 2014, 249).

Captivity occurs in a range of contexts to a variety of different kinds of beings. Imprisonment of human beings as a result of punishment for a crime, abduction, and institutionalized slavery of various forms often come to mind. Some theorists have contemplated the idea that our species itself is held captive by a process of self-domestication (Brüne 2007). Arguably, one of the most extreme versions of captivity occurs when humans confine wild animals. Captivity is not only an infringement on the autonomy of wild sentient beings but it is, largely, an imposition of circumstances that are at odds with their very being. Captivity for wild animals is not just the "place" or "space" they are held; captivity fundamentally includes being forced into a lifestyle against one's very nature. This is further complicated by the fact that it is imposed on these animals by another species who has not the ability or the desire to understand what this limitation means for those held captive.

On Domestication of Pets, Farmed Animals, and "Laboratory" Animals

There has been much discussion about whether it makes sense to think of domestic animals, such as those who are considered pets, and farmed or "lab" animals as captive. Clearly, "captivity" for pets, for example, dogs and cats, is, in principle, a very different experience than for animals in factory farms or in laboratories. But the issue is controversial. Many objections to domestication have to do with the process itself and the ethics of creating other animals for our purpose. Reasonable arguments have been made that the very process of domestication, even of dogs and cats, for companionship is a form of captivity (Horowitz 2014). Some argue that the domestication of dogs and cats is unethical because we have made them dependent and subservient (Francione and Charlton 2016). But others see parallels with human interdependence (Donaldson and Kymlicka 2011). Certainly all children, for instance, are dependent on their parents, and all humans are interdependent with other humans to a large extent. Others view domestication of dogs and cats as not problem-free but claim that the very process of domestication is not a harm in and of itself (Cochrane 2014).

One can look at the issue of captivity for domestic animals in terms of the more practical components of what it means for the lives of the animals themselves. The process of domestication, I would suggest, assuming there isn't harmful genetic manipulation and overbreeding, may not be as morally objectionable as the ways in which we actually treat domesticated animals. Farmed animals and "lab" animals are domesticated precisely to be used as commodities and objects—all deliberate harms. And their very existence means their lives will be torturous and unnaturally short. There is nothing about being a domesticated chicken or rat that makes domestication any better for those animals to be in proximity to humans. However, dogs and cats are not generally brought into the world with the intention of harming and killing them. Surely these outcomes occur, but it is not an inherent part of having these animals as companions. Domesticated animals in factory farms and laboratories are there *in order to be* used and killed in ways that *necessarily* cause harm. That is not the case with pets. Now let us turn to wild animals in other forms of captivity, namely, entertainment, where the claim is often made that, not unlike pet animals, the captive wild animals can actually thrive.

Common Elements of Captivity for Wild Animals

Captivity for wild animals is experienced in a range of settings. These include zoos, marine parks and aquariums, circuses and traveling shows, and even private homes. All of these manifestations of captivity share common elements. All wild animals in captivity are subjected to (a) restrictions and loss of control, (b) forced interspecies interaction and intrusion either through performances or by being put on display, and (c) monotony, all while held in artificial settings that have little resemblance to habitats that support their evolutionary heritage and adaptations. These general characteristics have profound impacts on the animals' well-being (see chap. 28).

The effects of captivity on wild animals depend on the extent of the mismatch between the captive situation and species-specific adaptations. As Nussbaum (2011) articulates, "Each creature has a characteristic set of capabilities, or capacities for functioning, distinctive of that species, and that those rudimentary capacities need support from the material and social environment if the animal is to flourish in it characteristic way" (237). White (2015) suggests that flourishing is "full, healthy growth and development of the traits, skills and dispositions that allow a being to have a satisfying and successful life as a member of that species" (263). To thrive or flourish, then, is to be free to express *the characteristic nature of their species.* Thriving is not equivalent to the absence of suffering. It is the capacity or ability to live the life one is adapted to living. More often than not, captivity for wild animals, no matter how good the veterinary care or how well intentioned the captors, will, by definition, be incompatible with thriving.

While no animals thrive in zoos the way they would in nature, those who fare the worst in captivity share a certain set of characteristics—they are large in size, highly self-aware, cognitively complex, wide ranging, and socially complex (including the need for interpersonal space). This includes cetaceans (dolphins and whales), elephants, primates, bears, big cats, and others. All of these species have needs that cannot be loosely approximated, let alone met, in zoos and aquariums, *and* they are intelligent/aware enough to experience the "wrongness" of their situation. This level of awareness adds a layer of psychological complexity to the experience of captivity for these animals. Self-awareness, the continuing sense of self as an individual with a life, makes individuals more, not less, vulnerable to the challenges of captivity. In short, self-aware beings know that their life is not the way they want it to be, and this adds to their suffering (Wise 2005).

As an example, let's look at three groups of animals who share all of the above characteristics and, not surprisingly, are among the least well adapted to life in captivity—cetaceans, elephants, and great apes. Below I describe some of the better known impacts of captivity on members of these groups.

Physiological Deconditioning

Cetaceans, elephants, and great apes live very active lives in nature. In captivity they manifest numerous musculoskeletal and physiological health problems because of lack of exercise. For instance, cetaceans are, typically, wide ranging and adapted to traveling long distances and diving to vast depths. Wild orcas (killer whales) often swim over one hundred miles a day (Jett and Ventre 2011) and can dive to depths of 150 meters or more on a regular basis (Baird, Hanson, and Dill 2005). In captivity orcas (and many other cetaceans) tend to move slowly or stay suspended at the surface of their concrete tanks and do not have enough depth in the tanks to actively dive and remain submersed. The physical deconditioning associated with this inactive lifestyle in orcas is manifest in the well-known dorsal fin collapse phenomenon (the drooping over of the dorsal fin because of gravity). Dorsal fin collapse is observed in nearly 100 percent of male orcas in captivity and only 1 percent of free-ranging male orcas (Jett and Ventre 2011). The bent dorsal fin is not a health problem per se but indicates serious physical deconditioning in captives, and the effects of deconditioning, while still poorly understood, more than likely lead to a number of health problems associated with suffering and mortality.

Elephants in captivity suffer from common maladies of living in restricted spaces and, often, being forced to perform tricks that are stressful to their massive bodies. Studies of home-range sizes of Asian elephants, for instance, have all shown spatial needs of 100–200 square kilometers at a minimum, orders of magnitude more than what they are allotted in even the most spacious zoo exhibits (Lindsay 2008). Moreover, they are often confined and forced to stand on inappropriate substrate, for example, concrete surfaces. As a result, in addition to suffering severe musculoskeletal disorders such as arthritis, the vast majority of captive elephants are made lame by foot disorders (Fowler 2001; Lewis et al. 2010). Foot disease and arthritis are the major causes of euthanasia in captive elephants (Fowler, 2001).

Ever since the middle of the twentieth century, great apes in zoos have been known to have heart problems (especially idiopathic cardiomyopathy). Captive great apes of all kinds (chimpanzees, gorillas, orangutans, bonobos) suffer from cardiovascular disease—and it is one of the major causes of mortality (Strong et al. 2014). Because of the prevalence of heart disease in captive great apes, numerous research projects have sprung up around the world to address this medical issue for those kept in zoos and laboratories. While it is thought that the sedentary lifestyle (and perhaps partly the diet) in captivity contributes to cardiac deconditioning, there is still much to be learned about the pathophysiology (Gill 2012).

Behavioral Abnormalities

All elephants, cetaceans, and primates exhibit abnormal behavior in captivity to a greater or lesser degree. The aberrant behavior, often extreme and deadly, includes behavioral stereotypies, that is, repetitive purposeless behaviors, self-mutilation, and excessive aggressiveness toward conspecifics and humans (see Marino and Frohoff 2011 for a review of this literature in cetaceans). There are a number of factors that contribute to aberrant behavior in captivity. One of them is the lack of space. Confinement affects not only physical freedom but social relationships and the ability to deal with conflicts. As an example, while serious aggression is rare in nature, captive orcas have a long history of injuring and killing each other because there is nowhere for them to disperse in a concrete tank when tensions get high. Moreover, no orca has ever seriously injured a human in nature, but there are multiple cases of orcas attacking and killing humans in captivity (Marino and Frohoff 2011).

Elephants in captivity commonly exhibit stereotypies (swaying, rocking, and head bobbing) and hyperaggression. In a survey of thirty-five zoos, it was found that at least 28 percent of elephants were injured by conspecifics (Doyle 2014, 50). This percentage substantially exceeds the level of aggression among wild elephants in nature.

Chimpanzees in zoos often exhibit abnormal behaviors indicative of compromised mental health (Birkett and Newton-Fisher 2011). Chimpanzees reared from birth in restrictive captive conditions or traumatized in other ways show persistent abnormal behaviors that resemble the human pathologies of posttraumatic stress disorder and depression (Ferdowsian et al. 2011). Chimpanzees in particular suffer from the unique circumstances of being kept as "pets" in private homes. Largely due to the media falsely portraying chimpanzees as cute and cuddly "child surrogates," there is an epidemic of "privately owned" chimpanzees in the United States (Ross 2014). The effect on these apes is devastating (Ross 2014). Most of these "at home" chimpanzees were ripped from their mother's side at infancy and hand-reared by humans, producing a well-documented litany of negative psychological consequences (Kalcher-Sommersguter et al. 2015; Ross 2014). All chimpanzees bought as infants grow up to be "unmanageable" as adults and end up sequestered in dank cages, medicated, and/or killed by authorities after a violent incident (Shaffrey 2009).

Another component of captivity that has severe negative effects on well-being is the enforced presence, and often interaction with, humans. For instance, dolphins in "swim with the dolphin," "dolphin assisted therapy," and "dolphin petting pool" exhibit all of the behavioral abnormalities outlined above, including hyperaggression (Stewart and Marino 2009). Kyngdon, Minot, and Stafford (2003) found increased retreat behavior in dolphins forced to participate in a "swim" program. And the same harmful effects occur when dolphins, whales, and elephants are forced into performances of unnatural behaviors with trainers in zoos, aquariums, and circuses. The long list of cetacean and elephant trainers injured or killed during performances over many years attests to the fact that these animals often resent being forced to perform for food and approval.

It has been argued that visitors to zoos and aquariums may serve as sources of amusement and even "environmental enrichment" for some animals. But it is clear that most wild animals do not want bodily contact with or even proximity to humans. While some studies offer little evidence for a direct negative impact of visitors on captive animals (e.g., chimpanzees; Cook and Hosey 1995) there are several ominous clues that suggest being "put on display" causes distress and abnormal behavior.

There are numerous media-hyped instances of captive cetaceans—for example, beluga whales, orcas, bottlenose dolphins—responding to marine park visitors with behaviors misinterpreted as "play" and "amusement" when it is clear these reactions are aggressive. There are also documented cases of chimpanzees showing hostile reactions to zoo visitors. One that caught media attention was the case of Santino, a male chimpanzee in a Swedish zoo who regularly stashed stones and then threw them at visitors when they came around his display (Osvath 2009). There are numerous documented cases of visitor presence affecting behavior in a wide range of captive animals, and in many cases, also suppressing their natural behaviors (Morgan and Tromberg 2007).

For beings like elephants, great apes, and cetaceans, for instance, there is an additional component to the stress of being "put on display." Malamud (2017) describes one of the most demeaning aspects of captivity—knowing one is being ogled:

> The zoo fundamentally inscribes the looked-at animals inside their cages—or their "cageless enclosures," that is, cages that don't look like cages (to us)—as subaltern. We, the people, the spectators, are free; they are trapped. We are in our natural habitats (San Diego, or Hamburg, or London) and they are not. We stay in the zoo for as long as it amuses us to be there and they stay in there forever. We can move on to the next cage, or the gift shop, or the cafeteria, while they cannot. We can leave, we can escape, we can go to places we would rather be, we can achieve privacy from prying eyes; they cannot. We are powerful, and they are fundamentally, quintessentially disempowered. (Malamud 2017, 400)

Many of us have had the experience of going to a zoo and staring into the eyes of a chimpanzee on display and his returning our gaze with a confrontational stare that makes it seem he feels the inequity of the situation. When we recognize that, most of us can only deflect our eyes in shame and whisper an apology.

Immune System Dysfunction

In the past few years the strong relationship between chronic stress (i.e., distress) and immunosuppression has been established in humans and other animals. For captive animals, being denied freedom of movement, living in monotonous conditions, and being forced into unwanted social situations, lead to chronic distress. Unlike in a natural environment, where stress is experienced as peaks and troughs, the chronic nature of captive stress stays at an unnaturally high level, wearing down the immune system and leading to susceptibility to physical disease and a high mortality rate. Evidence for stress responses during a number of components of captivity in cetaceans supports this view (Jett and Ventre 2011; St. Aubin and Geraci 1988; Noda et al. 2007; Spoon and Romano 2011). The United States Marine Mammal Inventory Report published by the National Marine Fisheries Service lists numerous stress-related disorders—such as ulcerative gastritis, perforating ulcer, cardiogenic shock, and psychogenic shock—as "cause of death" along with immunodeficiency-based infections (National Marine Fisheries Service, Office of Protected Resources 2010). It has also been suggested that the high prevalence of tuberculosis in captive elephants is due to this chronic distress syndrome (Mikota 2009) and, as mentioned above, captive great apes are highly susceptible to cardiac problems that may be due to a number of elements of captivity, including chronic distress.

Those who hold animals captive regularly claim that the animals are protected from the rigors of the natural environment, captivity removes the "necessity" of having to look for food and shelter, and to avoid predators—in essence—providing captives with an "easy" life. But this claim profoundly misrepresents the lives of other animals. For cetaceans, elephants, apes, and many other animals, the challenge of searching for food and avoiding harm from predators and other environmental factors are necessary for psychological welfare. Wild animals evolved to cope with the rigors of the wild and, therefore, thrive by meeting the challenges they evolved to contend with. They have no such adaptive history with captivity. One has to wonder why those who hold animals in captivity are assuming that a life free of the ordinary activities one engages in is a good life. There are very few humans who will feel fulfilled sitting around doing nothing. Further, if a challenge-free life for wild animals is so healthful and agreeable, why the higher mortality rates, lower survival rates, prevalence of stress-related and infectious diseases, and abnormal behavior in captivity?

Failure to Thrive

Not only is sickness and high mortality characteristic of adults in captivity, there is also a high infant mortality rate in elephants, cetaceans, and great apes in captivity. While it is true that in some species infant mortality in nature is also high, it remains critical to ask the question of why this is the case in zoos and aquariums where there are no predators, supposedly no environmental pollutants, and first-class veterinary care. The answer may be that infants born into such artificial circumstances cannot thrive because the situation is just too alien.

There is a well-known syndrome associated with deprived environments called "failure to thrive." It is particularly prevalent in human infants who are institutionalized or abandoned, but also in very old adults left alone in geriatric wards. Without intervention, failure to thrive results in a "wasting away," depression, and ultimately death. It is thought to be caused by a combination of factors such as improper nutrition, lack of nurturance, lack of attention and normal social interaction, and so forth. And it is found in humans and nonhumans under the same circumstances (Black and Dubowitz 1991). These circumstances exist in captivity for elephants, cetaceans, primates, and many other animals. Their infants are born into highly artificial surroundings in which their social and psychological, as well as physiological, needs cannot be met. In captivity, females are often artificially inseminated at unusually young ages and are not afforded the proper social network in which to learn how to be a parent. Because of the artificial social groupings created in captivity, there is also very little support for new mothers from conspecifics, unlike in natural habitats. And, finally, in addition to all of these deprivations, there is often the intrusion of humans into the situation, with trainers and veterinarians inserting themselves into the tanks and displays in proximity to infants, even handling them. Not only are these situations unnatural, they are also alarming for mothers and their infants, often leading to abandonment on the part of these traumatized mothers and death for the infant.

What Does Keeping Other Animals Captive Say about Us?

The concept of captivity as it relates to wild animals, can only be understood by appreciating who these animals are as individual members of a species with an evolutionary history that creates an adaptive framework for thriving. Once forced outside of that framework, these individuals cannot thrive, and many don't even survive. Nonetheless, our species continues to place wild animals in these untenable situations for our benefit.

Common justifications for keeping wild animals confined and on display typically focus on the ostensible educational value of seeing wild animals up close. The hypothesis is that seeing wild animals "in person" leads to an emotional connection and commitment to their preservation in the natural setting. In other words, zoos and aquariums claim that there is positive cognitive and emotional learning and a subsequent attitude change when visitors see animals on display. This conjecture is the basis of the "ambassador" concept in zoos. Although at face value, this claim might seem reasonable, there is no compelling evidence to support it (Marino et al., 2010). And some would argue that even if there were convincing support for this hypothesis, it still does not justify the suffering many wild animal individuals endure for the so-called benefit of species. There are many other ways to educate people about wild animals and motivate conservation without holding animals captive.

In addition to the lack of substantive support for the notion that visiting zoos and aquariums leads to long-term positive changes in knowledge, attitudes, and behaviors, it has been argued, not unreasonably, that there may be a negative impact of wild animal displays on conservation. Malamud (2017) suggests that seeing wild animals in such a convenient way sends the message that we can commoditize and exploit them. In fact, there is some evidence showing that when wild animals are observed with humans in a human setting, concerns about conservation decrease (Ross et al. 2008). Therefore, wild animal displays can actually "backfire" when they are an attempt to generate conservation-minded attitudes.

Consideration of the concept of captivity has two sides. One is the effects of captivity on other animals as explored above. This requires us to consider who *they* are. But the other side of the issue requires us to examine what captivity, that is, being a captor, says about who *we* are. The unflattering picture is one of a species in denial about its own nature as an animal, a species unwilling to share the world and craving the ability to control and exploit nature.

There are deep-seated reasons for our need to separate ourselves from nature and feel superior to the other animals; our need for zoos, aquariums, and circuses goes beyond mere entertainment (see Marino and Mountain, 2014). Our species believes that by subjugating the other animals, we become something other than what we are, animals ourselves. And it is to our advantage to pretend, if even unconsciously, that we are above all of the vicissitudes that befall the rest of nature, including danger, illness, and mortality. It is clear that confining wild animals in captivity has much to do with our inability to see ourselves for who we are or to maintain our cherished, albeit flawed, view that we are superior. So we oppress, restrain, and sanitize nature in order to keep up the illusion of human exceptionalism. Unfortunately, we are a species with the technical ability to remake the world to conform to our neuroses about who we are. It is tragic that our fantasies are, for other animals, the stuff of which nightmares are made.

Suggestions for Further Reading

Gruen, Lori, ed. 2014. *The Ethics of Captivity*. New York: Oxford University Press.

Malamud, Randy. 2017. "The Problem with Zoos." In *The Oxford Handbook of Animal Studies*, edited by Linda Kalof, 397–410. New York: Oxford University Press.

Marino, L., and M. Mountain. 2014. "Denial of Death and the Relationship between Humans and Other Animals." *Anthrozoos* 28 (1): 5–21.

Nussbaum, Martha C. 2011. "The Capabilities Approach and Animal Entitlements." In *The Oxford Handbook of Animal Ethics*, edited by Tom L. Beauchamp and Raymond G. Frey, 237. Oxford: Oxford University Press.

References

Baird, Robin, M. Bradley Hanson, and Lawrence Dill. 2005. "Factors Influencing the Diving Behaviour of Fish-Eating Killer Whales: Sex Differences and Diel and Interannual Variation in Diving Rates." *Canadian Journal of Zoology* 83 (2): 257–67.

Birkett, Lucy P., and Nicholas E. Newton-Fisher. 2011. "How Abnormal Is the Behaviour of Captive, Zoo-Living Chimpanzees?" *PLoS ONE* 6 (6): e20101.

Black, Maureen, and Howard Dubowitz. 1991. "Failure-to-Thrive: Lessons from Animal Models and Developing Countries." *Journal of Developmental and Behavioral Pediatrics* 12 (4): 259–67.

Brüne, Martin. 2007. "On Human Self-Domestication, Psychiatry, and Eugenics." *Philosophy, Ethics, and Humanities in Medicine* 2 (1): 21.

Cochrane, Alasdair. 2014. "Born in Chains? The Ethics of Animal Domestication." In *The Ethics of Captivity*, edited by Lori Gruen, 156–73. New York: Oxford University Press.

Cook, Shelley, and Geoffrey R. Hosey. 1995. "Interaction Sequences between Chimpanzees and Human Visitors at the Zoo." *Zoo Biology* 14 (5): 431–40.

Donaldson, Sue, and Will Kymlicka. 2011. *Zoopolis: A Political Theory of Animal Rights*. Oxford: Oxford University Press.

Doyle, Catherine, 2014. "Captive Elephants." In *The Ethics of Captivity*, edited by Lori Gruen. New York: Oxford University Press.

Ferdowsian, Hope R., Debra L. Durham, Charles Kimwele, Godelieve Kranendonk, Emily Otali, Timothy Akugizibwe, J. B. Mulcahy, Lily Ajarova, and Cassie Meré Johnson. 2011. "Signs of Mood and Anxiety Disorders in Chimpanzees." *PLoS ONE* 6 (6): e19855.

Fowler, Murray E. 2001. "An Overview of Foot Conditions in Asian and African Elephants." In *The Elephant's Foot: Prevention and Care of Foot Conditions in Captive Asian and African Elephants*, edited by B. Csuti, E. L. Sargent, and U. S. Becher, 3–7. Ames: Iowa State University.

Francione, Gary L., and Anna E. Charlton. 2016. "The Case against Pets." *AEON Magazine*. https://aeon.co/essays/why-keeping-a-pet-is-fundamentally-unethical.

Gill, Victoria 2012. "Why Do Zoo Apes Get Heart Disease?" BBC Nature. http://www.bbc
.co.uk/nature/17542031.

Horowitz, Alexandra. 2014. *Canis familiaris*: Companion and Captive. In *The Ethics of Captivity*, edited by Lori Gruen. New York: Oxford University Press.

Jett, John S., and Jeffrey M. Ventre. 2011. "Keto and Tilikum Express the Stress of Orca Captivity." The Orca Project, January 20. https://theorcaproject.wordpress.com /2011/01/20/keto-tilikum-express-stress-of-orca-captivity/.

Kalcher-Sommersguter, Elfriede, Signe Preuschoft, Cornelia Franz-Schaider, Char-lotte K. Hemelrijk, Karl Crailsheim, and Jorg J. M. Massen. 2015. "Early Maternal Loss Affects Social Integration of Chimpanzees throughout Their Lifetime." *Scientific Reports* 5: 16439.

Kyngdon, D. J., E. O. Minot, and K. J. Stafford. 2003. "Behavioural Responses of Captive Common Dolphins *Delphinus delphis* to a 'Swim-with-Dolphin' Programme." *Applied Animal Behavior Science* 81: 163–70.

Lewis, Karen D., David J. Shepherdson, Terrah M. Owens, and Mike Keele. 2010. "A Survey of Elephant Husbandry and Foot Health in North American Zoos." *Zoo Biology* 29 (2): 221–36.

Malamud, Randy. 2017. "The Problem with Zoos." In *The Oxford Handbook of Animal Studies*, edited by Linda Kalof, 397–410. Oxford University Press.

Marino, Lori, and Toni Frohoff. 2011. "Towards a New Paradigm of Non-captive Research on Cetacean Cognition." *PLoS ONE* 6 (9): e24121.

Marino, L., S. Lilienfeld, R. Malamud, N. Nobis, and R. Broglio. 2010. "Do Zoos and Aquariums Promote Attitude Change in Visitors? A Critical Evaluation of the Ameri-can Zoo and Aquarium Study." *Society and Animals* 18: 126–38

Marino, L., and M. Mountain. 2014. "Denial of Death and the Relationship between Humans and Other Animals. *Anthrozoos* 28 (1): 5–21.

Mikota, Susan K. 2009. "Stress, Disease and Tuberculosis in Elephants." In *An Ele-phant in the Room: The Science and Well-Being of Elephants in Captivity*, edited by Debra L. Forthman, Lisa F. Kane, D. Hancocks, and P. F. Waldau. North Grafton, MA: Tufts Center for Animals and Public Policy.

Morgan, Kathleen N., and Chris T. Tromborg. 2007. "Sources of Stress in Captivity." *Applied Animal Behaviour Science* 102 (3): 262–302.

National Marine Fisheries Service, Office of Protected Resources. 2010. United States Marine Mammal Inventory Report.

Noda, Katsura, Hideo Akiyoshi, Mica Aoki, Terumasa Shimada, Fumihito Ohashi. 2007. "Relationship between Transportation Stress and Polymorphonuclear Cell Functions of Bottlenose Dolphins, *Tursiops truncatus*." *Journal of Veterinary Medical Science* 69 (4): 379–83.

Nussbaum, Martha C. 2011. "The Capabilities Approach and Animal Entitlements." In *The Oxford Handbook of Animal Ethics*, edited by Tom L. Beauchamp and Ray-mond G. Frey, 237. Oxford: Oxford University Press.

Osvath, Mathias. 2009. "Spontaneous Planning for Future Stone Throwing by a Male Chimpanzee." *Current Biology* 19 (5): R190–91.

Rivera, Lisa. 2014. "Coercion and Captivity." In *The Ethics of Captivity*, edited by Lori Gruen. New York: Oxford University Press.

Ross, Stephen R. 2014. "Captive Chimpanzees." In *The Ethics of Captivity*, edited by Lori Gruen. New York: Oxford University Press.

Ross, S. R., K. E. Lukas, E. V. Lonsdorf, T. S. Stoinski, B. Hare, R. Shumaker, and J. Goodall. 2008. "Inappropriate Use and Portrayal of Chimpanzees. *Science* 319 (5869): 1487.

Shaffrey, Ted. 2009. "Chimp, Owner Had Unusual Bond." *USA Today*, February 19. https://usatoday30.usatoday.com/news/nation/2009-02-19-chimpanzee-previous-attack_N.htm.

Spoon, Tracey R., and Tracy A. Romano. 2011. "Neuroimmunological Response of Beluga Whales (*Delphinapterus leucas*) to Translocation and a Novel Social Environment." *Brain, Behavior and Immunity* 26 (1): 122–31.

St. Aubin, David J., and Joseph R. Geraci. 1988. "Capture and Handling Stress Suppresses Circulating Levels of Thyroxine (T4) and Triiodothyronine (T3) in Beluga Whales *Delphinapterus leucas*." *Physiological Zoology* 61 (2): 170–75.

Stewart, Kristin L., and Lori Marino. 2009. *Dolphin-Human Interaction Programs: Policies, Problems and Alternatives*. Policy paper, Animals and Society Institute. Ann Arbor, MI: Animals and Society Institute.

Strong, Victoria, Malcolm Cobb, Kate White, and Sharon Redrobe. 2014. "Great Ape Mortality Study." *Veterinary Record* 174 (4): 102.

White, Thomas I. 2015. "Whales, Dolphins and Ethics: A Primer." In *Dolphin Communication and Cognition: Past, Present, Future*, edited by Denise L. Herzing and Christine M. Johnson, 257–70. Cambridge, MA: MIT Press.

Wise, Steven. 2005. "Entitling Non-human Animals to Fundamental Legal Rights on the Basis of Practical Autonomy." In *Animals, Ethics and Trade: The Challenge of Animal Sentience*, edited by Jacky Turner and Joyce D'Silva, 87–100. London: Earthscan.

7 DIFFERENCE

Kari Weil

What is the difference between humans and animals, and what difference does that difference make? These questions are urgent ones not only for science but for politics and ethics, because lines of difference have been used to justify varying legal and moral standing for humans and animals as for different species and types of animals—whether demarcating domestic from wild animals or farm animals (who are more often factory animals) from the pets we take into our homes. The first question, moreover, is wrongly stated. It should ask, what is the difference between human and nonhuman animals, and the second question should ask, what difference, if any, does and should that make? We are animals, of course, although we humans have spent centuries trying to deny it. *Animal* is a term and a reproach that we humans have given to others in order to separate ourselves from them and to treat them differently. It is increasingly clear, however, that there is no rational foundation for human exceptionalism. The very capacities that had previously been thought to distinguish humans from nonhuman animals—language, consciousness, thought—are known to be possessed by at least some species of nonhumans.

Paying attention to differences among species (and among individuals) is necessary for understanding what can help them flourish and what they need for their livelihood and for their distinct ways of being in the world. But difference, history shows us, almost immediately implies hierarchy—whether in terms of a difference between humans and nonhumans or in terms of the differences within each of those categories—differences between species, as between genders, breeds, colors, anatomies, and so on. This is evident in the Great Chain of Being established by Plato and Aristotle and continued within religious traditions that regarded all life and matter as existing within a natural and necessary order (see also Lovejoy 2009). God occupies the top realm, followed by the angels, humanity, animals, plants, and ending with minerals at the bottom. Each link of the chain, moreover, had its own subdivisions of sexes, families, and species such that males were regarded as superior to females, kings superior to peasants, and lions superior to insects.

Aristotle made this hierarchy, along with its intersecting political consequences, explicit:

> It is clear that the rule of the soul over the body, and of the mind and the rational element over the passionate, is natural and expedient; whereas the equality of the two or the rule of the inferior is always hurtful.
>
> The same holds good of animals in relation to men. . . . The male is by nature superior, and the female inferior; and the one rules, and the other is ruled. . . .
>
> When there is such a difference between soul and body, or between men and animals . . . the lower sort are by nature slaves, and it is better for them as for all inferiors that they should be under the rule of a master. (Aristotle 1989, 4)

The Chain of Being constructs a series of binary oppositions (human-animal, mind-body, male-female) that determine who is on top and who or what is beneath. Thus, women and slaves are regarded as more like animals, or "closer to nature" as Sherry Ortner described it, insofar as women are understood to be more bodily, more driven by instincts than by reason, hence less rational and less free than men. Because of this, it was "natural" that men should master women, just as men should also master animals and slaves. This difference between man and animal as constituted by an essential difference between mind and body or between spirit and matter was most forcefully articulated by Descartes, who described the animal as a machine who can react to a stimulus out of an unthinking bodily reflex but could never offer a thinking response. Moreover, the comparison to machines, Descartes argued, "does not show that brutes have less reason than men but that they have none at all" (Descartes 1989, 14). The political and ethical ramifications of Descartes designation of animal difference are concise and clear. "Thus my opinion is not so much cruel to animals as indulgent to men . . . since it absolves them from the suspicions of a crime when they eat or kill animals" (19)[1]

Is there a way of thinking difference outside the binaries, outside the hierarchies and unjust behaviors they justify in order to think and appreciate instead what Kelly Oliver (2009) refers to as a "multitude of differences" (292)? This is a question that feminists posed with regard to matters of sexual difference, and their writing offers a helpful background to the problem of animal difference and *différance*.

Difference and Saming

Feminist attention to questions and matters of sexual difference can be instructive for questions of animal difference. In the latter part of the last century, French feminist Julia Kristeva described three "generations" of women who tackled this problem. The first generation of women, often represented by Simone de Beauvoir, downplayed sexual difference in order to demand sexual equality and a dismantling of patriarchal hierarchy.[2] A second generation rejected the first stage's implicit upholding of the male as a norm to which the female must be assimilated in order to be acknowledged as a full subject. Affirming difference outside of opposition or hierarchy, it wanted to acknowledge how different bodies and bodily functions give rise to different but no less valuable ways of being and of knowing. To acknowledge woman's different body and bodily functions, moreover, was also to acknowledge the need for different kinds of support and protection. Identifying this generation with French feminist Luce Irigaray, Naomi Schor described the contest between these two stages as that between othering and saming:

> Just as Beauvoir lays bare the mechanisms of othering, Irigaray exposes those of what we might call, by analogy, "saming." If othering involves attributing to the objectified other a difference that serves to legitimate her oppression, saming denies the objectified other the right to her difference, submitting the other to the laws of phallic specularity. If othering assumes that the other is

1. For an alternate view that argues that the Western tradition promoted respect for the differences represented by other animals and compassion to them, see Preece (2006).

2. It was Julia Kristeva who described the stages I outline as "generations," each one implying less a chronology than a "'signifying space,' a corporeal and desiring space" such that each may remain affectively alive and interwoven with the next and not simply opposed to it (Kristeva 1986, 187).

knowable, saming precludes any knowledge of the other in her otherness. (Schor 1989, 45)[3]

These two mechanisms have their counterparts with regard to non-human animals, and each carries strategic gains and risks. The mechanism and strategy of saming, or what Matthew Callarco recently describes as the claim of "identity," is fundamental to the Nonhuman Rights Project that seeks to change the status of at least some animals from property to persons on the basis that they share certain basic human capacities such as self-consciousness, intelligence, and language. Such calls for animal rights work to combat the particular othering of nonhuman animals that Frans de Waal calls "anthropodenial," or the refusal to see that some share basic human capacities. Anthropodenial is the counter charge to the scientific sin of anthropomorphism, or the act of "saming" by projecting human characteristics onto other animals who may not share them (de Waal 2006, 59–68). Saming and othering are thus comparable to the mechanisms of anthropomorphism and anthropodenial. Anthropodenial errs by refusing to acknowledge the many ways in which nonhuman and human animals are alike. Anthropomorphism errs by disallowing nonhuman animals their differences or by seeing these differences as politically and ethically valuable. Both anthropomorphism and anthropodenial may stem from a pervasive anthropocentrism that upholds the human as the normative subject of rights and worth much as the masculine was regarded as the normative model of subjectivity.[4]

The *Animot*: Differences between and Differences Within

Within critical theory, the charge of othering was rightly waged not only against sexism but also against racism, ableism, and a series of prejudicial views that challenged normative definitions of the human. This indictment was made in conjunction with the call for a right to or acknowledgment of difference. Animal Studies thus follows women's studies, critical race studies, ethnic studies, disability studies, and other academic fields that have drawn attention to a plurality of differences and have demanded the dismantling of the Chain of Being and its hierarchy. But there is at

3. See also the discussion of othering and saming in Gruen and Weil (2012, 480–81).

4. It is because of the argument for rights on the basis of their likeness to humans that some are critical of the great ape or Nonhuman Rights Project.

least one caveat. Even if we acknowledge that nonhuman animals have linguistic capacities, we are forced to speak and write for them within the academy, having no choice but to employ the very language whose politics and prejudices we also wish to deconstruct. What then of animal identity? "Animal, what a word!" writes Derrida (2008); "Men would be first and foremost those living creatures who have given themselves the word that enables them to speak of the animal with a single voice and to designate it as the single being that remains without a response, without a word with which to respond" (32), contesting both the singular of the word, *animal*, that covers over the many differences between species, and the way this falsely singular identity has been used in opposition to the human capacity to respond. Derrida will write *animot* instead, a French word that resists translation as it also subverts the relation between sound and meaning in French. *Animot* sounds like the plural of animals (*animaux*) even as that plural is hidden within the singular word or *mot*. As he puts the word *animal* under scrutiny, so Derrida wants to suggest that the differences that exist between species and between individual animals are far greater than the difference presumed to exist between what we call human and what we call animal.

Derrida's own concern is less for the animals and more toward showing how humans are not all we think ourselves to be. The multiplication and elaboration of the many differences within and between species—differences that cannot simply be opposed—was the project of the German biologist and founder of the field of ethology, Jakob von Uexküll. Uexküll coined the term *Umwelt* to call attention to the fact that different species inhabit different phenomenal worlds. In so doing, he presented a direct challenge to Heidegger's description of "the animal" as "poor in world" by virtue of having no language with which to know and name what is felt or seen or touched. For Uexküll, there is not just one but there are many worlds, and he emphasizes that all animals exist within their own particular world or environment "to the same degree of perfection" in order to guard against hierarchical evaluations between them. More significantly, Uexküll refutes the inability of animals to respond by emphasizing that the relation between each animal and world is a meaningful one. Animals perceive and interpret stimuli as signs. In his most famous example, the tick may wait what might seem like years to us—but perhaps only minutes to it—before finding the right signals of scent, temperature, and touch that tell it to drop now and latch on. Uexküll's descriptions and drawings of these *Umwelten* are wondrous. Reveling with awe at a multiplicity of differences, he details the varying and seemingly strange capacities of different species, such as the spider who spins her web in such a way that the anticipated and imagined fly will not see it (Uexküll 2010).

Uexküll's ethology turns all animals into subjects in their worlds, if not "subjects of a life," the expression that Tom Regan uses to also describe the status of all human and nonhuman animals as individuals (Regan 1983, 113). Indeed, Uexküll, like Regan, helps us to see the value of all animals, if not of their lives to them, but he offers little guidance for comparing such unique lives. Should there be no difference, then, between killing a tick and slaughtering a cow? Even if their lives are equally valuable to them, are they to the larger world and should they be to me? Perhaps the flip side to the accusations of a humanist "sameness," as a ground for inclusion and exclusion for purposes of rights and protections, is the posthumanist difficulty of determining other grounds for ethical and political choices. Matthew Callarco is one among many who are critical of the "difference view" for having little to offer in terms of strategy or policy, suggesting furthermore that it is a "parasitic mode of political thinking" (Calarco 2015, 44).[5] But parasites, I would suggest, can help to change the way we view the shape and function of the organisms and systems they depend on. If nothing else, parasites help us to rethink the bounded identities on which identity and so differences may still depend.

It is in this regard that the third generation of feminists imagined by Julia Kristeva will have specific importance for the question of animal difference. This generation rejects the very idea of "identity," which must be rejected as unreal metaphysics. Similarly, a new generation of animal scholars makes the plea to reject the human—and the humanism it subtends—as tired metaphysics. This generation does not ground its thinking and politics on identities defined through different capacities or abilities (consciousness, thought, feeling, etc.). Instead it is developing new ethics and political practices according to those incapacities and vulnerabilities—a matter of "not being able" rather than of ability—that we may share without claiming identity (Derrida 2008, 28).[6] Responsibility and responsiveness to injustice take the focus away from matters of

5. It will become apparent in what follows that my three stages are similarly conceived to Calarco's except that he distinguishes difference from *indistinction*, a term that he uses to refer to the Derridean understanding of *différance*.

6. See also Anat Pick's essay on vulnerability in this volume. Judith Butler makes similar arguments for a condition of "precariousness" as essential to thinking ethically outside a humanist or anthropocentric framework (Butler 2009).

difference. Kristeva means to shake her readers into a sense of responsibility, if not complicity, and she charges that in spite of efforts to the contrary, we have interiorized the process of identity construction, making each of us the composite of apparently opposed identities, including "victim/executioner." We need "a new ethics," she claims (Kristeva 1986, 209, 211).

Even as we shift focus from the questions of identity and difference to those of responsibility and to the obligation to be responsive to an other's call—human and/or animal—we nevertheless return to a (different) question of difference. For it was, of course, the difference between the capacity to respond (with intention) and to react (automatically) that, for Descartes, distinguished humans from animals, and it is to that difference as *différance* that I now want to turn.

Différance: From Tracking Response to Bearing Responsibility

Jacques Derrida coined the verbal noun "différance" (with an *a*) to call attention to the way that identity is constructed through processes of differentiation or through deferral: the French verb, *différer*, meaning "to differ" and "to defer" (Derrida 1982, 3–27). First used in 1963, it became a central concept in deconstruction and focused on how and why thinking precedes through the construction of binary and hierarchical oppositions that, if examined, cannot be held (Derrida 1982, 3–27). To deconstruct identity is also to show how it always bears the trace of what it is not or will at some point become. In a late interview, Derrida returns to the importance of the term within his own oeuvre and reflects that it "became linked for me early on with the immense problematic of animality" (Derrida and Roudinesco 2004, 21). It is thus with hindsight (or through deferral) that Derrida recasts his entire oeuvre as what David Wood calls a "zoophilosophy" (Wood 2004, 129). By 2006, when he published *L'Animal que donc je suis (The Animal That Therefore I Am)*, the notion of deferral is regarded as inherent to the definition of being human as being and coming after the animal; "je suis" meaning both I am and I follow. What it means to be after the animal, whether in evolutionary terms or in the sense of trying to grasp what it means to be animal (and so human), is at the core of Derrida's zoophilosophy.

Whether we become human out of animality in evolutionary terms, through transcendence or overcoming of animal/nature as Hegel describes, or in Freudian terms by repressing our instincts, it is as human, Derrida explains, that we claim to know what the state of being animal is. It is as human that we give the name "animal" to that from which we want to claim our difference. And that human, like the animal it distinguishes itself from, is drawn from a space of fundamental "indistinction." This is the space that Giogio Agamben describes in *The Open: Man and Animal* with the phrase, "the open wound that is my life" (borrowed from Georges Bataille) in order to describe the existential trauma of life as always caught between man and animal, caught in "the central emptiness, the hiatus that—within man—separates man and animal" (Agamben 2004, 7). Caught in this state of betweenness, the human must separate, name, and exclude the animal in order to claim "his" (and sometimes her) humanity. This is the work of what Agamben calls the anthropological machine—a machine of differentiation that functions symbolically and materially to produce the "nonman at the heart of man" as that which must be excluded. The premodern version of this machine, he suggests, worked by humanizing previously excluded forms of life—werewolves, barbarians, slaves, even women—even as they remain less than human. The modern version works inversely by animalizing others in order to exclude nonhuman animals from the properly human as well as certain races, ethnicities, and sexes, thus leaving them vulnerable and forsaken. The anthropological machine is thus a political machine that shows how difference is never a neutral matter even as it obscures its own agency for naming things. Indeed, as a naming machine, it exposes differentiation as something of a compulsion. Language and naming are marks of a reactive process more than any response, reminding us that in his animality, as Derrida writes, "Man is less a beast of prey than is prey to language" (Derrida 2008, 121). In other words, we humans are not always the thinking beings we presume to be, and our so-called reasoned responses might really be unthought reactions, reactions fashioned by passions or habits if not by the machine of language itself.

Can we "jam the machine," or is there a way of thinking the production of difference otherwise, outside the hierarchical and violent reactions described by Derrida and Agamben? This, it seems to me, has been one of the aims of Elizabeth Grosz's recent work that attempts to view Derridean difference within a different "genealogy" that stretches from Darwinian evolution to Deleuzian notions of "becoming"(Deleuze and Guattari 1988).

What I am interested in is an understanding of difference as the generative force of the world. . . . It is the inhuman work of difference—rather than its embodiment in human "identity," "subjectivity," or "consciousness," rather than its reflection in and through identity—that interests me now, the ways in which difference stretches, transforms, and opens up any identity to its provisional vicissitudes, its shimmering self-variations that enable it to become other than what it is. (Grosz 2011, 91)

Sexual difference is one form of this generative force, as Grosz explains in her readings of Luce Irigaray, because it is "the condition for the independent emergence of all other living differences" within nature (Grosz 2011, 105). It is neither a sufficient nor perhaps the most important condition for this emergence, either for Irigaray or for Darwin, Grosz explains, nor does it have a fixed identity of its own, "no one location, no one organ or condition" (Grosz 2011, 110). Aligning difference and sexual difference, Grosz turns attention to the dynamic processes through which new forms of life are constantly emerging through the "intra-actions" of organisms with each other, with matter, and with the environment.[7] Hers is thus an evolutionary perspective but one that "renders the human a temporary species" within an "infinite elaboration . . . of new forms of body and also new forms of culture, new modes of social organization, new arts, new species" (Grosz 2011, 119). Matter and life become undone and transformed as they also direct us toward a future in which location of "the human," "the animal," or any singular identity can only be regarded as retrograde, a conservative resistance to life itself.

Grosz's anticipatory model of *différance*, celebrating in advance the "shimmering" of emergent life forms not yet known, might remind readers of Donna Haraway's 1985 "Manifesto for Cyborgs." Part imagined future, part present reality, the postgender cyborg that Haraway envisions there and hopes to inhabit has already breached boundaries of difference between human and animal and between human/animal and machine, especially as both are acted on and acting as "coded texts for writing and reading the world" (Haraway 1989, 176). Cyborg imagery, she suggests in that piece, is a way out of the destructive dualisms of difference, attempting if not to replace them, then at least to have them share the stage with a present, material reality of construction through "pleasurably tight coupling[s])" (Haraway 1989 176). Such couplings with nonhuman animals as love objects and relationalities that affectively and genetically change who we are become the focus of her later *Companion Species Manifesto* and *When*

---

7. *Intra-action* and *intra-activity* are terms coined by Karen Barad to propose agency as enactment rather than something that someone has (Barad 2007).

Species Meet. "Human beings have always been in partnership," she writes. "To be human is to be a congeries of relationalities, even if you are talking about *Homo erectus*. So it's relationalities all the way down" (Gane 2006, 147). Built on relationalities rather than differences, Haraway's couplings are to be understood otherwise than the becomings of Deleuze and Guattari who, she says adamantly, "don't give a flying damn about animals" (Gane 2006, 143). Indeed, it is in her attention to situated relatings, to our actual and ongoing engagement with other animals, that Haraway changes the very way we must think about response.

> Response, of course, grows with the capacity to respond, that is, responsibility. Such a capacity can be shaped only in and for multidirectional relationships, in which always more than one responsive entity is in the process of becoming. That means that human beings are not uniquely obligated to and gifted with responsibility; animals as workers in labs, animals in all their worlds, are response-able in the same sense as people are; that is responsibility is a relationship crafted in intra-action through which entities, subjects and objects come into being. (Haraway 2007, 71)

Haraway thus thinks notions of emergent or deferred identities—the fact that "partners do not precede their relating" (Haraway 2007, 17)—in terms of an ethical demand. The capacity to respond is dependent on and thus should be responsive to those relatings by which it is constituted in the first place. Differences between identities are and will continue to be differences within any given identity.

The semiotic concept of the trace thus returns as the material, affective, and potentially ethical force that ties me to who precedes me, who has made me who I am, and to whom I am thus indebted. This sense of response as responsive and responsible to another calls attention to the lie of moral self-determination and the dependent nature of a response that is also a reaction. Responsibility, we might then say, reveals all moral action as necessarily parasitic because we are all, in Haraway's terms, "becoming with" other animals in the multispecies web of life. Parasites, too, Michel Serres reminds us, can become major players in life. But the questions of who is feeding off whom, to what extent and for what purposes, are important to ask given the unequal nature of different relations. This is a different way of thinking about what Grosz calls "becoming undone" and the fact that we may each respond differently to different others and only understand the importance of that response after the fact, when it is too late. This is why it feels necessary but also impossible to construct a universal law of response, one that indicates which animals to respond to, which we can kill with impunity—mosquitos, for instance, but not cows—and whether there is a difference between such killing and "letting die" in factory farms or polluted air or waters. Cary Wolfe answers the question to who and how many must we respond, by paying heed to the implacability of deferral: "We *must* choose, and by definition we cannot choose everyone and everything at once. But this is precisely what ensures that, *in* the future, we *will have been wrong*. Our 'determinate' act of justice now will have been shown to be too determinate, revealed to have left someone or something out" (Wolfe 2013, 103).

Attending to difference, then, is both necessary and impossible, the same words that Derrida uses to describe translation (Derrida 2009, 336). It is necessary because we are always bound to an other who precedes us and awakens our sense of responsibility to their difference. It is impossible because of what in the other and ourselves remains opaque to translation and in excess of what we willingly or unwillingly perceive. Translation across difference is what is fundamental to any ethical obligation, according to Judith Butler: "Otherwise we are ethically bound only to those who already speak as we do, in the language we already know" (Butler 2012, 17). Difference, then, is the very site of the ethical, the space across which we must attempt to hear and to translate and to respond without assimilating to the same. Such translation may be impossible, but even the effort it takes can make all the difference in the world.

Suggestions for Further Reading

Calarco, Matthew. 2015. *Thinking Through Animals: Identity, Difference, Indistinction*. Stanford, CA: Stanford University Press.

Derrida, Jacques. 2008. *The Animal That Therefore I Am*. Translated by David Wills. New York: Fordham University Press.

Grosz, Elizabeth. 2011. *Becoming Undone: Darwinian Reflections on Lie, Politics, and Art*. Durham, NC: Duke University Press, 2011.

Gruen, Lori. and Kari Weil. 2012. Introduction to "Animal Others." *Hypatia* 27 (3).

Weil, Kari. 2012. *Thinking Animals: Why Animal Studies Now*. New York: Columbia University Press.

References

Agamben, Giorgio. 2004. *The Open: Man and Animal*. Translated by Kevin Attell. Stanford, CA: Stanford University Press.

Antonello, Pierpaolo, and Roberto Farneti. 2009. "Antigone's Claim: A Conversation with Judith Butler." *Theory and Event* 12 (1).

Aristotle. 1989. "Animals and Slavery." In *Animal Rights and Human Obligations*, edited by Tom Regan and Peter Singer, 109–10. Englewood Cliffs, NJ: Prentice Hall.

Barad, Karen. 2007. *Meeting the Universe Halfway: Quantum Physics and the Entanglement of Matter and Meaning*. Durham, NC: Duke University Press.

Butler, Judith. 2012. *Parting Ways: Jewishness and the Critique of Zionism*. New York: Columbia University Press.

Calarco, Matthew. 2015. *Thinking Through Animals: Identity, Difference, Indistinction*. Stanford, CA: Stanford University Press.

Deleuze, Gilles, and Félix Guattari. 1988. *A Thousand Plateaus: Capitalism and Schizophrenia*. Translated by Brian Massumi. Minneapolis: University of Minnesota Press.

Derrida, Jacques. 1982. "Différance." In *Margins of Philosophy*, translated by Alan Bass, 1–28. Chicago: University of Chicago Press, 1982.

———. 2008. *The Animal That Therefore I Am*. Edited by Mary-Louise Mallet and translated by David Wills. New York: Fordham University Press.

———. 2009. *The Beast and the Sovereign*. Vol. 1. Translated by Geoffrey Bennington. Chicago: University of Chicago Press.

Derrida, Jacques, and Elizabeth Roudinesco. 2004. *For What Tomorrow: A Dialogue*. Translated by Jeff Fort. Stanford, CA: Stanford University Press.

Descartes, René. 1989. "Animals Are Machines." In *Animal Rights and Human Obligations*, edited by Tom Regan and Peter Singer, 60–66. Englewood Cliffs, NJ: Prentice Hall.

De Waal, Frans. 2006. *Primates and Philosophers*. Princeton, NJ: Princeton University Press.

Gane, Nicholas. 2006. "When We Have Never Been Human What is to Be Done? Interview with Donna Haraway." *Theory, Culture and Society* 23 (7/8): 135–58.

Grosz, Elizabeth. 2011. *Becoming Undone: Darwinian Reflections on Lie, Politics, and Art*. Durham, NC: Duke University Press.

Gruen, Lori, and Kari Weil. 2012. "Animal Others: Editors' Introduction." *Hypatia* 27 (3): 477–487.

Haraway, Donna. 1989. "A Manifesto for Cyborgs: Science, Technology, and Socialist Feminism in the 1980s." In *Coming to Terms: Feminism, Theory, Politics*, edited by Elizabeth Weed, 173–204. New York: Routledge.

———. 2007. *When Species Meet*. Minneapolis: University of Minnesota Press.

Kristeva, Julia. 1986. "Women's Time," *The Kristeva Reader*, edited by Toril Moi. New York: Columbia University Press.

Lovejoy, Arthur O. 2009. *The Great Chain of Being: A Study of the History of an Idea*. Cambridge, MA: Harvard University Press.

Oliver, Kelly. 2009. "Sexual Difference, Animal Difference: Derrida and Difference Worthy of Its Name." *Hypatia* 24 (2): 290–312.

Preece, Rod. 2006. *Brute Souls, Happy Beasts, and Evolution: The Historical Status of Animals*. Vancouver: University of British Columbia Press.

Regan, Tom. 1983. *The Case for Animal Rights*. Berkeley: University of California Press.

Schor, Naomi. 1989. "This Essentialism Which Is Not One: Coming to Grips with Irigaray." *Differences* 1 (2): 38–58.

Uexküll, Jacob von. 2010. *A Foray into the Worlds of Animals and Humans*. Translated by Joseph D. O'Neil. Minneapolis: University of Minnesota Press.

Wolfe, Cary. 2013. *Before the Law: Humans and Other Animals in a Biopolitical Frame*. Chicago: University of Chicago Press.

Wood, David. 2004. "Thinking with Cats." In *Animal Philosophy*, edited by Peter Atterton and Matthew Calarco, 129–44. New York: Continuum.

8 EMOTION

Barbara J. King

Ever since Charles Darwin described jealous orangutans and loving dogs in *The Expression of Emotions in Man and Animals* (1872), vigorous debate has ensued about which reports of animal emotion are warranted and reliable and which might merely be anthropomorphic projections of what we humans feel and only *wish* to see in the animals who surround us.

Darwin observed evidence of emotion in numerous species, including invertebrates. Today, the science of animal emotion includes an ethical dimension: if it's not only our closest living primate relatives and domesticated animal companions who feel love and joy and fear and grief but a wide variety of wild animals and farmed animals, then animal welfare issues take on fresh and pressing dimensions. In order to think comprehensively about animal welfare, we should take into consideration animals' emotional as well as physical well-being (see chap. 29).

Yet assessment of animal emotion poses significant challenges. Five vignettes will help ground my discussion of these challenges in the real lives of flesh-and-blood animals.

Five Vignettes

Chimpanzees in Zambia

At the Chimfunshi Wildlife Orphanage Trust in Zambia in 2010, nine-year-old chimpanzee Thomas died of a massive infection. Edwin van Leeuwen et al. (2016) report that his body became a magnet for his group members. Orphaned at age five when his mother died, Thomas had become a highly social adolescent who had developed a close bond with the adult male Pan.

On the day of his death, his body lay on the ground in an open patch outside the densest part of the forest near to the sanctuary's perimeter fence. The absence of his body from that location earlier and his body's stiffness when sanctuary workers removed the body half an hour after they first spotted it indicates that Thomas had died in denser forest and was then dragged to the more open and visible location. (This act by the chimpanzees raises fascinating questions all on its own: did the apes realize that the body would be seen more readily by humans in that perimeter location? Did they want Thomas's body to be found by the staff?)

During the twenty-minute period before staff removed Thomas's body, over half the chimpanzees in this group of forty-three were filmed interacting at and around the body in some way. Observers were powerfully struck by the "quiet attendance" at the body. The chimpanzees' sitting so close to each other was notable, given the tendency for these apes to explode into noisy excitement at unusual events. Pan visited and inspected the body more than the other two adult males and carried out a sort of gatekeeping of what went on around it. As one example, when twelve chimpanzees were gathered within three meters of the body, Pan "grabbed a branch and lunged with high speed over the body." The gathered apes scattered. The most prolonged attention to Thomas came from a female, Noel, who had been Thomas's friend. Noel used a piece of grass to clean Thomas's teeth.

Were Pan, Noel, and the other chimpanzees sitting in quiet attendance expressing emotion at the sight of Thomas's body? Were the apes' responses rooted in curiosity and exploration? Were any or all them expressing grief?

Deploying a definition of *grief* that requires a "before" and "after" comparison of the survivors' behavior, we see that, according to van Leeuwen et al. (2016), the chimpanzees' behavior was indeed altered from their species-typical patterns. Only with a report about subsequent behaviors could it be determined whether Pan, Noel, or any of the other apes exhibited signs of sustained distress in the form of social withdrawal or altered eating and sleeping patterns. In numerous cases these patterns are present in animal survivors, including wild chimpanzees, leading to a strong claim for the presence of grief (King 2013).

Van Leeuwen et al. (2016) do not use the terms *emotion* or *grief* in their report of the chimpanzees' responses to Thomas's death. Yet they do hint at a role for emotion in the behaviors they observed, suggesting that the interest in Thomas's body was "not random but related to prior social relationships." They also note the possibility that Noel's teeth cleaning was a type of "empathetic care" for Thomas.

Gelada baboons in Ethiopia

Over a three-and-a-half-year period, Peter Fashing et al. (2011) observed fourteen females in an Ethiopian highlands gelada baboon population carrying the corpses of their dead infants for periods ranging from one hour to forty-eight days. Yet these monkey mothers exhibited nothing that Fashing et al. (2011) could interpret as emotional upset or grief; they acted in wholly different ways than did the chimpanzees who responded to Thomas's death. Indeed, the mother baboon who corpse carried for forty-eight days even mated with a male while holding her infant's body in one hand.

Might that mother, or any other of the mothers who had lost their infants, have been experiencing an internal state of distress or grief? Other reports from the wild do suggest emotional responses to death in monkeys (e.g., Campbell et al. 2016; Yang, Anderson, and Li 2016). It's possible, even likely, that wild animals subject to predator pressure have been fine-tuned by natural selection to mask, in at least some situations, signs of debilitating emotions just as they have been selected to hide signs of visible weakness, injury, or pain. However, the lack of visible expression of distress by the gelada mothers certainly supports the conclusion that they are not experiencing emotional distress due to the loss of their young.

Goats in California and Kent, England

A goat named Mr. G at Animal Place sanctuary in Grass Valley, California, listlessly reclined in his enclosure. Well cared for in his new home, he remained indifferent even to his favorite foods and chose to move minimally. He was physically fine upon checkup; sanctuary workers concluded that Mr. G was depressed.

This goat had been rescued from poor conditions on a woman's property along with other animals, including a burro named Jellybean, who had been taken to another sanctuary. After six days and mounting worry for Mr. G's state, Animal Place arranged to bring Jellybean to their sanctuary. Within minutes of introducing Jellybean to Mr. G's enclosure, Mr. G. transformed. The shift in his emotional state is unmistakable on video: he greets the burro with enthusiasm, and the two animals touch muzzles. Within twenty minutes, Mr. G begins to eat, standing right next to Jellybean.

Compared to the situation with the chimpanzees who gathered around Thomas's body, the situation with this single goat can be understood more straightforwardly by reading a set of embodied signals: Mr. G expressed sadness when he was taken from his familiar (though inadequate) location and from his friend. Strictly speaking, there's no way to tell whether Mr. G's sadness meant he was missing his friend any more than we can tell the source of an emotion in a prelinquistic human infant. What's clear is that the reunion with Jellybean altered Mr. G's emotional state for the better.

Ethologists Elodie Briefer and Alan McElligott (2013) study goat moods, defining these as long-term diffuse emotional states that arise from accumulating, shorter-term emotional states rather than from a specific event. Working with eighteen sanctuary goats in Kent, England, they first trained goats to distinguish between positive maze corridors (food baited with apples and carrots in a blue bucket) and negative ones (food absent with only an empty blue bucket). Then they tested goats in an "ambiguous" corridor they had not seen before, also food absent with an empty blue bucket. Speed was used as a proxy for positive anticipation of what would be found in each corridor.

Among females, the goats who came from a past poor welfare condition were more optimistic than the females from the control group, meaning that they were faster in the corridors (though this difference did not reach statistical significance). For males, the control group individuals were faster than the others, but again, not to significance. Even though one group of goats had been abused, their expectations were not pessimistic compared to that of the others. This combination of case study—when Mr. G's sadness lightened—and systematic study—when sanctuary goats expressed positive anticipation despite past trauma—tells us that goats' emotions and moods alter according to their surroundings.

A Goose in Idaho

In writing about the behavior of certain birds, the biologist Bernd Heinrich uses the word *love* without qualification or apology. He's careful to acknowledge species variation: many birds, he notes, carry out regular extra-pair mating as part of what otherwise looks like a monogamous mating. Heinrich (2010) cites storks as an example of birds who end up in monogamous pairs because the male and female each develop an attachment to the nest, "where they meet and mate with whomever is there" (19).

With Canada geese it's different because a bond emerges between two individuals. One goose, named Tinker Belle, raised by a woman in Idaho from infancy, mated with a male, then split her time between him and her human friend. One day the mate simply disappeared. Tinker Belle searched for the male for three days, flying around and calling "in a frantic voice," but then for a week refused to eat, became very weak, and sat with her bill under her wing. She did recover. "Perhaps," Heinrich (2010) says, "such feelings are not possible without first having a need, a prerequisite for love" (27).

In his chapter on blue jays in *One Wild Bird at a Time*, Heinrich suggests that in some cases these birds vocalize out of a wish to connect with each other. Blue jay screams may mean "Here I am. How are you?" and then birds in hearing distance "can then feel reassured that they are not alone in the area and can either ignore it or reply "Here I am too" (Heinrich 2016,104–5). Variations in call features and in context may add emotional nuance to that basic "I'm here" message, modifying it to mean "I'm excited" or "This is scary." On this view, explanations for bird calls rooted in specific adaptive motivations related to mating, predator avoidance, and foraging won't always be enough, because like many mammals, birds, too, have feelings about their experiences.

Octopuses in Seattle

At the Seattle Aquarium, one year's annual Valentine's Day Octopus Blind Date involved the introduction of Rain, a female, and Squirt, a male, Pacific octopuses who previously had experienced each other through a barrier—breached only by their sucker-to-sucker contact—between their two tanks. At the event's start, Rain is colored dark orange in her tank, Squirt a grayish color in his. Sy Montgomery (2015) recounts what happens when the two are introduced. Both octopuses turn red. Rain reaches out with three arms, and Squirt moves into them; she presents her underside to him. "They embrace mouth-to-mouth, thousands of glistening, exquisitely sensitive suckers tasting, pulling, sucking on each other. Both of them flush with excitement" (99).

Chromatophores lining the eight arms of octopuses cause these color changes and also, according to Montgomery, offer a color-coded map to the animals' level of arousal. Red signals excitement, white is calm and contentment. Octopuses also turn white when they are old and the chromoatophore muscles lose their tone.

Octopuses are the sole invertebrate to be included in the Cambridge Declaration of Consciousness. The attempt to move from octopus sentience and cognition to octopus emotion is a challenge to cephalopod science.

These brief examples taken together—representing three mammals, one bird, and one mollusk, or four vertebrates and one invertebrate depending on how you cut it—show that animals may express emotion and also that significant challenges exist to the study of animal emotion. How can emotion in a wide variety of animals be explored scientifically without succumbing to unreasonable fear of anthropomorphism, a specter that continues to loom over discussions of emotion in nonhuman animals?

What Is Emotion?

Let's begin with an understatement: *emotion* is a contested term. David Anderson and Ralph Adolphs (2014) lay out the matter this way: "The paradox of emotions is that, on the one hand, they seem self-evident and obvious when examined introspectively; on the other hand, they have been extremely difficult to define in objective scientific terms" (187–88).

In *Affective Neuroscience* (1998), Jaak Panksepp says that experiencing an emotion is intense and overwhelming, whereas a mood is "more tidal" or "weak but persistent" (47). This distinction is relatively straightforward and fits nicely with the research on goat mood described above. Does an emotion, though, differ also from a feeling?

Antonio Damasio and Gil Carvalho (2013) consider emotions to be bodily states "triggered by the perception or recall of exteroreceptive stimuli" (144), apparently meaning stimuli from outside the body itself. Examples include joy, shame, sadness, disgust, and anger. By contrast, feelings "are mental experiences that accompany a change in body state" (ibid.), thus, they are emotions that are consciously experienced. When a prey animal sights a predator, then, that prey animal may experience only a bodily state marked by adrenaline release, increased heart rate, and muscular changes that produce a facial expression of fear, or that animal might also *consciously* experience what his or her body is undergoing as a feeling.

Panksepp (1998) preserves a similar distinction in noting that in the brain, "emotional circuits must be able to generate affective feelings" (49). He refers to a series of what he calls "genetically ingrained brain emotional operating systems" that he marks with capitalized terms such as RAGE, SORROW, FEAR, PANIC, LUST, CARE, and PLAY. Panksepp allows that "probably" (51) emotional systems change as an individual develops through the life courses and acquires experiences, but his work is representative of an approach to animal emotion that is deeply rooted in analyzing homologous brain circuits across mammalian species.

RAGE provides a good example. This emotion, according to Panksepp, evolved as a way to motivate individuals to compete against others when attempting to acquire resources in the environment. It is mediated by certain regions of the amygdala and the hypothalamus plus part of the midbrain, and he notes that "subcortical anatomies and major neurochemistries for the feeling of anger are remarkably similar in all mammals" (Panksepp 1998, 190). Certainly, rage in a chimpanzee wouldn't exactly mirror rage in a rat, but species' variation emerges in relatively subtle ways in what triggers the anger, and this in turn will depend on which resources are deemed most worth fighting for.

By contrast, scholars in developmental-systems psychology and in anthropology emphasize, instead, that emotions in humans and at least some other animals emerge and develop in social relationships. Certainly the relationships themselves are rooted in brain circuits, but according to this perspective, the most profound pathway is one of co-construction of emotion during exchanges between social partners. In work on seven-month-old infants, Elena Geangu et al. (2016) show that very young children discriminate between facial expressions of fear and happiness using culturally mediated strategies. Tracking infants' eye movements showed that "Western Caucasian" infants raised in the United Kingdom relied significantly more on the mouth compared with "East Asian" infants raised in Japan, who focused on the eyes—tracking the adult strategy in each culture. Going further, Alan Fogel (1993) makes the case that emotion itself develops from a dynamic coregulated sharing where meaning emerges from a continuous process of mutual adjustment and creativity.

Catherine Lutz (1988), who also refers to emotion as "an emergent product of social life" (5), exemplifies the strong tendency by ethnographers of the human experience not only to question models of emotion rooted in the individual but also to highlight "the argument for viewing psychological phenomena such as emotions as a form of discourse rather than as things to be discovered beneath the skin or under the hat" (7). As with all discourse, then, there must be a translation process when considering emotion across contexts and cultures: how certain emotions are constructed and expressed in one sociocultural tradition is quite unlikely to be identical to how it is constructed and expressed in another. Despite this, emotions such as anger, fear, and happiness, Lutz says, are routinely objectified as if one can just go and look for them within an individual.

Lutz's work shows that the way we talk about emotion may become a barrier to actually understanding emotional experience. When we label emotions one by one in a reifying way, it's natural—but, Lutz says, wholly inadequate—to put the focus on emotions as the key unit of analysis instead of the "human, cultural, and historical inventions for viewing self and relations with others" (Lutz 1988, 9) that emotions genuinely are.

There's remarkable convergence between the view from anthropology that Lutz represents and the work of neuroscientists such as Lisa Feldman Barrett. In an article for the magazine *Nautilus* (Barrett 2017a) based on her newest book (2017b), Barrett invites us to examine and contest our commonsense conclusions about emotions, including the expectation that we can each read others' emotions accurately:

> Hundreds of studies show that instances of the same emotion involve differ-
> ent heart rates, breathing, blood pressure, sweat, and other factors, rather
> than a single, consistent response. Even in the brain, we see that instances
> of a single emotion, such as fear, are handled by different brain patterns at
> different times, both in the same individual and in different people. This
> diversity isn't random. It's tied to the situation you're in. (Barrett 2017a)

Real-world events, the science shows, *don't* trigger invariant emotions and we *can't* read each other's emotions accurately. At one level, the revisioning of emotions in these ways I have been discussing through the work of Lutz and Barrett introduces a sober note into a discussion of animal emotion. For one thing, if it's so fraught a process to recognize emotions in our own species and to translate emotions across cultures, how might we reliably assess emotions across species?

At a foundational level, though, these approaches underscore the experiences of animals with intense attention to and care for social partners— co-constructed life experiences are an important factor to foreground in developmental studies, including developmental studies of emotion. Here, Christine Harris and Caroline Prouvost's (2014) case study on the development of jealousy in dogs is instructive. It's an interesting claim right from the start that dogs can feel jealousy: Darwin had no hesitation in making such a claim, but in modern ethology it's basic emotions like fear, happiness, and sadness (see Bekoff 2007) that are the typical focus of study in nonhuman animals; when jealousy has been considered, as Harris and Prouvost point out, it's been in the context of sexual bonds in order to evaluate fitness consequences when a third party intervenes.

Jealousy is considered to be a protective response by an individual to a perceived threat to the primacy of a key social relationship. Working outside the traditional sexual-jealousy framework, Harris and Prouvost aimed to construct a social-triangle experimental situation that built on anecdotal reports of jealousy felt by dogs when their human companions ("owners") paid attention to another animal. They videotaped thirty-six dogs as they were tested in their homes with the owners unaware of the specific jealousy hypotheses under consideration. The dogs encountered three conditions: *stuffed dog*, or the jealousy condition, when the owners ignored them but fussed affectionately over a stuffed dog that briefly barked and wagged its tail; *jack-o'-lantern*, where the owner fussed in the same way but over an object, a jack-o'-lantern pail; and *book*, the control condition where owners read aloud a children's book that featured musical melodies and pop-up pages.

The dogs were more aggressive and attention seeking when their owners interacted with the stuffed dog than in the other two conditions. For example, pushing and touching of the owners/objects occurred significantly more often by the dogs in the jealousy (stuffed dog) condition than the other two conditions. Interestingly, 86 percent of the dogs sniffed the stuffed dog's anal region during the experiment, which Harris and Prouvost take to mean that they saw the stuffed dog as real, though it seems to me more likely that dogs would conclude from a sniff test that the stuffed dogs *weren't* real. In the remaining dogs who did not carry out the exploratory sniffing, jealous behaviors were rare. Perhaps doglike objects do elicit jealousy and nondoglike objects don't. In any case, the authors conclude that while all dogs have the neurobiological capacity for jealousy, the test situation "failed to induce the emotional state in some dogs" (Harris and Prouvost 2014).

This experiment brings to the fore the notion that brain circuits *prepare* an animal to express an emotion, but in many cases the social context is key, not just in a trivial sense that there's a social trigger but also in the more profound sense that the emotion of jealousy (or grief or love) cannot be said to exist outside the unfolding social event that reflects the emotional nature of the social relationship. What would have happened in the experiment if half the owners had been instructed to stop acting affectionately toward the stuffed dog and begin acting affectionately toward the real (test) dog? In some way, the dogs' behavior would have shifted. The dogs' emotional states can be understood not as entities in and of themselves lodged concretely within an individual but as contingent upon the unfolding events that the individual himself or herself co-constructs.

Certainly the lack of consensus among researchers when talking or writing about emotions may result in moments of disorientation. For years I have discussed animal grief as an emotion that may be felt by individuals who survive the death of a relative or other social partner. Following my presentation on that topic to a group of family-systems therapists in 2016, the other keynote speaker, psychologist Anthony Papa, opened his discussant remarks by noting that grief is not an emotion! This was startling to me, though welcome because it provided an opportunity for dialogue. Later, Papa (e-mail message to author, July 7, 2016) explained his remark this way: "Research with recently bereaved [people] finds that grieving consists of sadness, anger, anxiety, etc.—a whole bundle of different emotions. You could think of grief as the metaexperience of this bundle of emotions all triggered by different aspects of a single loss but tied thematically by that loss."

Papa's perspective reminds us that when it comes to studying the emotions of nonhuman animals, scientists still work at a very basic stage, dealing with emotions parsed one by one. What is perhaps most important for science and ethics is that a working definition be adopted in order to explore the ways in which living flesh-and-blood animals do (and do not) consciously experience a range of emotions and tell us so through their movements, behaviors, vocalizations, and expressions. A good candidate for such a definition comes from Carl Safina (e-mail message to author, August 16, 2015; see Safina 2015) who says that "emotion is how we feel about our perceptions," a phrase that in its "we" embraces nonhuman as well as human animals and invites rigorous description of animals' visible responses' to their perceptions while not dwelling on all the numerous distinctions that spice an academic treatment of this topic.

How Do We Recognize Emotion in Other Animals?

Anne Engh et al. (2006) measured levels of glucocorticoid, a steroid which is responsive to experiences of stress, in the feces of twenty-two free-ranging female gelada baboons in Botswana who had witnessed a predator attack and kill close kin. These levels spiked higher than those measured in control females who had witnessed a predator attack but lost no close relatives. Engh et al. refer to "bereaved" baboons, and indeed it is possible to see from their behavior more than a stress response alone. After the first four weeks, the baboons worked to ameliorate their loss by extending their social grooming networks. They increased the number of grooming partners and also the rate at which they participated in grooming.

Complementary sources of information—in this case hormonal assay plus extended visual scrutiny of behaviors—allow evaluation of both an internal body state and the visible manifestation of that state. Panksepp (1998, 9) goes so far as to say that all visible manifestations of emotions are "only vague approximations of the underlying neural dynamics." For this reason, Panksepp insists (4) that the study of animal emotion must proceed by "focusing on the shared emotional and motivational processes of the mammalian brain" (4).

Yet field biologists and anthropologists may press the point that close behavioral observation of wild and domestic animals in natural conditions (or as close to natural as possible) is required for fully understanding animal emotion. If we accept Fogel's view that emotions develop in relationships, then the real-time, moment by moment, contingent, and unpredictable unfolding of behavioral events that make up those relationships may yield critical information. Data from noninvasive techniques of hormonal analysis or comparative brain imaging may be significant as well.

Might researchers merely be mapping onto animals' behaviors a projection of the emotions humans might experience in similar situations? On this view, rooted in worries about anthropomorphism, the chimpanzees at Chimfunshi wildlife orphanage in Zambia discussed above were certainly aroused at the sight of Thomas's body. They might well have been curious about Thomas's unprecedented stillness as he lay on the ground, but to invoke emotion in their quiet vigil is misguided. The dogs in the jealousy experiment similarly might have been more engaged and curious when their owners interacted with an object that looked somewhat like themselves and that could move its tail and make sounds, but it's too big a leap to go from that to jealousy.

Pushback against anthropomorphic criticism may hinge on a question parallel with the one that de Waal asks in the title of his latest book (2016), *Are We Smart Enough to Know How Smart Animals Are*? That is, are we smart enough to recognize when animals are expressing their emotions? "Anthropomorphism!" charges may be wielded without attending to a distinction offered by Gordon Burghardt (1991) over a quarter century ago: *critical anthropomorphism* is a careful and useful tool in decoding animal behavior. When we combine knowledge of an animal's natural history with our own insights, as sentient animals ourselves, about the behavior we see, we'll come out ahead of the game with good hypotheses to test: anthropomorphism *is* scientific when done right.

In a related but nonetheless slightly different vein, animal emotion researchers (e.g., Bekoff 2007; Balcombe 2016b) insist that if we spend time with animals and watch them closely and persistently, our own eyes will tell us that grief and jealousy, joy and anger, are real because the animals themselves tell us so. Anthropomorphism from this perspective is not just misguided but fatally anthropocentric; it takes as a starting point that emotions are *human* emotions when in fact our species has no rights of ownership over the kinds of emotions we are discussing here.

Linking Animal Emotion and Animal Welfare

Lori Gruen (2015) points out in *Entangled Empathy* that ethical analyses of animal welfare too often are rooted in discussion of abstract principles. If abstract moral reasoning is the basis for ethical engagement with questions of our treatment of animals, we risk alienating ourselves, Gruen (2015) says, from the very animals lives we are already in relationship with: "Many, perhaps most, current discussions of what we owe animals fail to attend to the particularity of individual animal lives and the very different sort of relationships we are in with them" (67). When we interact with, come to understand something about, and empathize with the lives of chimpanzees, that's a good thing, and yet it gets us no closer to understanding something about and empathizing with the lives of chickens and chipmunks.

Context matters. To practice entangled empathy in our relationships with other animals means, for Gruen, "to be responsive and responsible in these relationships by attending to another's needs, interests, desires, vulnerabilities, hopes, and sensitivities" (3). Heightened appreciation to the expression of emotion by animals as they live their day to day lives may play a crucial role in this process.

Certainly empathy for animals and scientific research on animal emotion may strengthen each other. Close attention to emotional expression at not just the species but also the population level (because animals have culture and their behaviors may vary culturally; Whitehead and Rendell 2014) and individual level (because animals have personalities; King 2017) is an approach to animal behavior that embraces rather than fears critical anthropomorphism as an ethological tool.

To understand more about how fish feel physical pain and possibly emotions as a result of that pain, for instance, neurobiological data may be combined with close observations of fish behavior. Writing in the inaugural issue of the journal *Animal Sentience*, Brian Key (2016) argues that fish lack the necessary neocortical circuitry to feel pain (see chap. 17). Asking what do "noxious stimuli" feel like to fish, Key answers his own question: "The evidence best supports the idea that they don't feel like anything to a fish." In peer-commentary response, scientists like Culum Brown (2016), Jonathan Balcombe (2016a), Gordon Burghardt (2016), and Jennifer Mather (2016) reach the polar opposite conclusion. Fish brain areas (primarily the telencelphalon) almost certainly function to process pain; smart, flexible, sometimes playful behavior of fish lend support to the growing certainty that fish do feel pain. Balcombe (2016b,101) concludes that some fish may feel "a range of emotions" that include fear, stress, playfulness, joy, and curiosity. This work on fish reinforces a central but often underappreciated point: the study of animal emotion produces the most reliable conclusions when it proceeds from the bottom up, that is, by close and noninvasive observation of animals devoid of a priori assumptions.

Much more is known about expression of emotions in elephants than in fish; grief rituals of elephants are well documented. We even know that individuals in elephant family groups may experience acute posttraumatic stress disorder (PTSD) as a result of the violence that comes along with war and poaching. In the breakthrough paper "Elephant Breakdown," Gay Bradshaw et al. (2005) reported that "wild elephants are displaying symptoms associated with human PTSD: abnormal startle response, depression, unpredictable asocial behavior and hyper-aggression." Using this scholarly elephant work as a touchstone, other scientists have begun to look more closely at the expression of emotion in animals who must cope with trauma from habitat destruction, changes associated with global warming, or confinement by humans in biomedical laboratories, theme parks, and zoos.

Despite the challenges we encounter in understanding emotion in other animals, individuals of many animal species, living in many wild and captive environments, *do* tell us through their behavior something about their emotional experiences. Surely it is our responsibility to watch and listen closely in order to see and hear when animals miss their friends because of death or separation or long for social companionship of their own choosing or become frightened or sad when put into situations in which they suffer—and respond as ethically as we can.

Suggestions for Further Reading

Barrett, Lisa Feldman. 2017b. *How Emotions Are Made: The Secret Life of the Brain*. New York: Houghton Mifflin Harcourt.

Burghardt, Gordon. 1991. "Cognitive Ethology and Critical Anthropomorphism: A Snake with Two Heads and Hognose Snakes That Play Dead." In *Cognitive Ethology: The Minds of Other Animals*, edited by C. A. Ristau, 53–90. Hillsdale, NJ: Lawrence Erlbaum.

Gruen, Lori. 2015. *Entangled Empathy: An Alternative Ethic for Our Relationships with Animals*. New York: Lantern Books.

King, Barbara J. 2017. *Personalities on the Plate: The Lives and Minds of Animals We Eat*. Chicago: University of Chicago Press.

King, Barbara J. 2013. *How Animals Grieve*. Chicago: University of Chicago Press.

Panksepp, Jaak. 1998. *Affective Neuroscience: The Foundations of Human and Animal Emotions*. Oxford: Oxford University Press.

Solomon, Robert C., ed. 2003.*What Is an Emotion: Classic and Contemporary Readings*. Oxford: Oxford University Press.

References

Anderson, David J., and Ralph Adolphs. 2014. "A Framework for Studying Emotions across Phylogeny. *Cell* 157 (1): 187–200.

Balcombe, Jonathan. 2016a. "Cognitive Evidence of Fish Sentience." *Animal Sentience*. http://animalstudiesrepository.org/cgi/viewcontent.cgi?article=1059&context=animsent/

———. 2016b. *What a Fish Knows: The Inner Lives of Our Underwater Cousins*. New York: Scientific American / Farrar, Straus and Giroux.

Barrett, Lisa Feldman. 2017a. "Emotional Intelligence Rewrite." *Nautilus*, August 3. http://nautil.us/issue/51/limits/emotional-intelligence-needs-a-rewrite.

Barrett, Lisa Feldman. 2017b. *How Emotions Are Made: The Secret Life of the Brain*. New York: Houghton Mifflin Harcourt.

Bekoff, Marc. 2007. *The Emotional Lives of Animals: A Leading Scientist Explores Animal Joy, Sorrow, and Empathy—and Why They Matter*. Novato, CA: New World Library.

Bradshaw, G. A., Allan L. Schore, Janine N. Brown, Joyce H. Poole, and Cynthia J. Moss. 2005. "Elephant Breakdown." *Nature* 443: 807.

Briefer, Elodie F., and Alan G. McElligott. 2013. "Rescued Goats at a Sanctuary Display Positive Mood after Former Neglect." *Applied Animal Behaviour Science* 146 (1): 45–55.

Brown, Culum. 2016. "Comparative Evolutionary Approach to Pain Perception in Fishes." *Animal Sentience*. http://animalstudiesrepository.org/cgi/viewcontent.cgi?article=1029&context=animsent.

Burghardt, Gordon. 1991. "Cognitive Ethology and Critical Anthropomorphism: A Snake with Two Heads and Hognose Snakes That Play Dead." In *Cognitive Ethology: The Minds of Other Animals*, edited by C. A. Ristau, 53–90. Hillsdale, NJ: Lawrence Erlbaum.

———. 2016. "Mediating Claims through Critical Anthropomorphism." *Animal*

Sentience. http://animalstudiesrepository.org/cgi/viewcontent.cgi?article=1063 &context=animsent.

Campbell, Liz A. D., Patrick J. Tkaczynski, Mohamed Mouna, Mohamed Qarro, James Waterman, and Bonaventura Majolo. 2016. "Behavioral Responses to Injury and Death in Wild Barbary Macaques (*Macaca syvanus*)." *Primates* 57 (3): 309–15.

Damasio, Antonio, and Gil B. Carvalho. 2013. "The Nature of Feelings: Evolutionary and Neurobiological Origins." *Nature Reviews Neuroscience* 14: 143–52.

Darwin, Charles. 1872. *The Expression of the Emotions in Man and Animals.* London: John Murray.

de Waal, Frans. 2016. *Are We Smart Enough to Know How Smart Animals Are?* New York: W. W. Norton.

Engh, Anne L., Jacinta C. Beehner, Thore J. Bergman, Patricia L. Whitten, Rebekah R. Hoffmeier, Robert M. Seyfarth, and Dorothy L. Cheney. 2006. "Behavioural and Hormonal Responses to Predation in Female Chacma Baboons (*Papio hhamadryas ursinus*). *Proceedings of the Royal Society B* 273: 707–12.

Fashing, Peter J., Nga Nguyen, Tyler S. Barry, C. Barret Goodale, Ryan J. Burke, Sorrel C. Z. Jones, Jeffrey T. Kerby, Laura M. Lee, Niina O. Nurmi, and Vivek V. Venkataraman. 2011. "Death among Geladas (*Theropithecus gelada*): A Broader Perspective on Mummified Infants and Primate Thanatology." *American Journal of Primatology* 73: 405–9.

Fogel, Alan. 1993. *Developing through Relationships: Origins of Communication, Self, and Culture.* Chicago: University of Chicago Press.

Geangu, Elena, Hiroko Ichikawa, Junpeng Lao, So Kanazawa, Masami K. Yamaguchi, Roberto Caldara, and Chiara Turati. 2016. "Culture Shapes 7-Month-Olds' Perceptual Strategies in Discriminating Facial Expressions of Emotion." *Current Biology* 26: R633–34.

Gruen, Lori. 2015. *Entangled Empathy: An Alternative Ethic for our Relationships with Animals.* New York: Lantern Books.

Harris, Christine R., and Caroline Prouvost. 2014. "Jealousy in Dogs." *PLoS ONE* 9 (7): e94597.

Heinrich, Bernd. 2010. *The Nesting Season: Cuckoos, Cuckolds, and the Invention of Monogamy.* Cambridge, MA: Belknap.

———. 2016. *One Wild Bird at a Time: Portraits of Individual Lives.* Boston: Houghton Mifflin Harcourt.

Key, Brian. 2016. "Why Fish Do Not Feel Pain." *Animal Sentience.* http://animalstudies repository.org/cgi/viewcontent.cgi?article=1011&context=animsent.

King, Barbara J. 2013. *How Animals Grieve.* Chicago: University of Chicago Press.

———.2017. *Personalities on the Plate: The Lives and Minds of Animals We Eat.* Chicago: University of Chicago Press.

Lutz, Catherine A. 1988. *Unnatural Emotions: Everyday Sentiments on a Micronesian Atoll and Their Challenge to Western Theory.* Chicago: University of Chicago Press.

Mather, Jennifer. 2016. "An Invertebrate Perspective on Pain." *Animal Sentience.* http://animalstudiesrepository.org/cgi/viewcontent.cgi?article=1046&context= animsent.

Montgomery, Sy. 2015. *The Soul of an Octopus.* New York: Simon and Schuster.

Panksepp, Jaak. 1982. "Toward a General Psychobiological Theory of Emotions." *Behavioral and Brain Sciences* 5: 407–67.

———. 1998. *Affective Neuroscience: The Foundations of Human and Animal Emotions.* Oxford: Oxford University Press.

———. 2015. *Beyond Words: What Animals Think and Feel.* New York: Henry Holt.

Safina, Carl, 2015. *Beyond Words: What Animals Think and Feel.* New York: Henry Holt.

Van Leeuwen, Edwin J. C., Innocent Chitalu Mulenga, Mark D. Bodamer, and Katherine A. Cronin. 2016. "Chimpanzees' Responses to the Dead Body of a 9-Year-Old Group Member." *American Journal of Primatology* 78 (9): 914–22.

Whitehead, Hal, and Luke Rendell. 2014. *The Cultural Lives of Whales and Dolphins.* Chicago: University of Chicago Press.

Yang, Bin., James R. Anderson, and Bao-Guo Li. 2016. "Tending a Dying Adult in a Wild Multi-level Primate Society." *Current Biology* 26: R387–407.

Videos

Animal Place. 2014. "Mr. G and Jellybean." YouTube, May 20. https://www.youtube.com/watch?v=bv2OGph5Kec.

Walker, Matt. 2016. "Chimps Filmed Grieving for Dead Friend." BBC Earth, May 18. http://www.bbc.com/earth/story/20160517-chimps-grieve-for-dead-friend.

9 EMPATHY

Lori Gruen

Stories about animals helping other animals, even humans, appear regularly in the news and on social media. There is a tale of elephants saving a blind woman who was lost; humpback whales reportedly moving seals out of the water to protect them from orcas; male chimpanzees adopting and caring for young orphan chimps not their own; dolphins helping human swimmers who are being approached by sharks; meerkats risking their own safety to stay with a sick or dying member of the group. There are even laboratory studies of what gets called *animal empathy* that are widely reported, for example, voles who have been seen consoling each other after one has been experimentally shocked and rats who will forego their favorite treats to rescue other rats who are trapped.[1] Part of the attraction of these stories may be that they question whether nature is really red in tooth and claw. Finding kindness, compassion, and concern in the animal kingdom challenges our notions of what other animals are like and the kinds of relationships they can have with others, sometimes even across species.

Those of us who live with other animals may not be particularly surprised to hear of prosocial behaviors across the animal kingdom. Our dogs and even cats can be quite heroic. Stories about dogs pulling children from burning buildings and protecting humans from all sorts of dangers

1. Simon Worrall, "Yes, Animals Think and Feel," *National Geographic*, July 15, 2015, http://news.nationalgeographic.com/2015/07/150714-animal-dog-thinking-feelings-brain-science/; Adam Boult, "Animals more capable of empathy than previously thought, study finds," *Telegraph*, January 23, 2016, http://www.telegraph.co.uk/news/science/science-news/12117501/Animals-more-capable-of-empathy-than-previously-thought-study-finds.html; Brian Resnick, "Do Animals Feel Empathy?" *Vox*, August 5, 2016, http://www.vox.com/science-and-health/2016/2/8/10925098/animals-have-empathy.

abound. And there is a remarkable story of a cat named Tara who became an Internet sensation when her bold act of saving a child from a dog attack was caught on video. Apparently over 16.8 million viewers tuned in to watch.[2] Even though stories of cats coming to the aid of others are less common, there is something that all of these tales elicit—that humans aren't alone in feeling for and acting on behalf of others. But does it make sense to call this empathy? An answer to this question depends on what is meant by *empathy*, and there are a lot of different ways of understanding both the term and the phenomenon.

Empathy unto Death

In their important work on empathy, Stephanie Preston and Frans de Waal introduced what is now a widely cited instance of what they called *cognitive empathy*:

> Kuni, a female bonobo at the Twycross Zoo in England, once captured a starling. She took the bird outside and set it onto its feet, the right way up, where it stayed shaking. When the bird didn't move, Kuni threw it a little, but it just fluttered. Kuni then picked up the starling, climbed to the highest point on the highest tree, and carefully unfolded the bird's wings, one wing in each hand, before throwing it into the air. (Preston and de Waal 2002, 19)

In his 2009 book, *The Age of Empathy*, de Waal revisits this story, adding to the previous description of Kuni's actions—"she spread its wings as if it were a little airplane, and sent it out into the air, thus showing helping action geared to the needs of a bird." He also recounts a parallel story from the Arnhem Zoo,

> where chimpanzees live on an island surrounded by a moat. The moat is full of life, such as fish, frogs, turtles, and ducks. One day, a couple of juvenile chimps had picked up a little duckling and were swinging it around, being far too rough with it. When they tried to grab one of the other ducklings, which were wisely hurrying back to the water, an adult male ran over in an intimidating manner and scattered the young apes. Before leaving the scene, he walked over to the last ducking still on land. With a quick hand movement, like that of a child playing with marbles, he flicked it into the moat. (de Waal 2009, 91–2)

2. "Tara (cat)," Wikipedia, https://en.wikipedia.org/wiki/Tara_(cat).

De Waal describes both cases as "targeted helping" that depends on an empathetic understanding of the interests and needs of another. And despite the language he uses depicting sentient birds as toy airplanes and marbles, he is suggesting that the other apes here see the starling and duckling as in need of their help. But is Kuni really empathetically responding to the starling's needs? Is the male chimp genuinely empathizing with the interests of the young duck? Throwing a bird that cannot fly off the highest branch of the highest tree is not a particularly caring act of empathy, and flicking a baby bird into a moat does not seem to be an action that is best described as empathetic. Perhaps in both cases, what the apes are thinking is that something is where it doesn't belong. Starlings belong in the air, ducks belong on the water, so they were in the wrong place, and this mistake needed to be corrected, as when a book falls off the shelf. Correctly interpreting behavior is always complicated (see chap. 4), and perhaps neither of these suggestions is what was really going on. We can, and often do, make mistakes when we are trying to be empathetic, and perhaps that is what happened here: the apes were genuinely attempting to empathize but, unfortunately, their actions weren't particularly successful and led to harm.

Sometimes alleged acts of empathy purposely lead to harm, even death. Consider Temple Grandin's work. She uses her self-professed autistic form of empathy to design more efficient devices for slaughtering animals. Grandin differentiates visual empathy, sensory empathy, and emotional empathy and says she can empathize in the first two ways, but not through emotional empathy. That is consistent with prominent psychological views of autism that suggest that the capacities for emotional engagement with others are underdeveloped in most individuals with autism.[3] But sensory perceptions can often be highly developed. Visual and sensory empathy are what Grandin says connects her to animals. She writes, "Using my visual thinking skills, it is easy for me to imagine myself in an animal's body and see things from their perspective" (Grandin 1995, 176). And this type of "empathy" is what allows her to develop methods to calmly and quickly lead animals to their deaths. Since, she says, "We're responsible for slaughterhouse animals; they wouldn't exist if it weren't for us" (Worsham and Olson 2012, 24). "It is better for them to not be afraid as they die. Fear is so bad for animals I think it's worse than pain" (Grandin 1995, 189).

3. Simon Baron-Cohen (1995) discusses the various ways that autistic people are unable to empathize and socially connect with others. These conclusions have been challenged, however. See Davidson and Smith (2009) and Cohen-Rottenberg (2009), for example.

Empathy, according to psychologist Paul Bloom (2016), can actually be the "gas" for causing fear, pain, and even war. He writes,

> When scholars think about atrocities, such as the lynching of blacks in the American South or the Holocaust in Europe, they typically think of hatred and racial ideology and dehumanization, and they are right to do so. But empathy also plays a role. Not empathy for those who are lynched or put in the gas chambers, of course, but empathy that is sparked by stories told about innocent victims of these hated groups, about white women raped by black men or German children preyed upon by Jewish pedophiles. . . . Empathy tilts the scale too much in favor of violent action. It directs us to think about the benefits of war—avenging those who have suffered, rescuing those who are at further risk. (Bloom 2016, 192–93)

This view of empathy is akin to one that is often thought of when people talk about the "good torturer"—that person who understands the feelings of his victim so well that he knows exactly how to cause the most pain to extract what he wants.

If empathy is consistent with harming others—or worse, aids in the development of efficient killing machines—or undergirds violence, war, and torture, it really is hard to see why it would be of interest for Animal Studies. But as most scholars of empathy note, empathy represents a variety of ways of being in relation to others, most usually in a helping capacity, in relations of compassion and care.

Varieties of Empathy

In the psychological literature, *empathy* is alternatively used to mean a state of "knowing what another person or being is feeling" (an epistemic state in the empathizer) or "feeling what another person or being is feeling" (an affective state) or "responding compassionately to another's distress" (perception/action state; Levenson and Ruef 1992, 234).

Neuroscientist Jean Decety and psychologist Megan Meyer (2008) propose a different list of ways that empathy has been understood:

- The ability to put oneself into the mental shoes of another person to understand her emotions and feelings (a form of simulation or inner imitation; Goldman 1993)
- A complex form of psychological inference in which observation,

memory, knowledge, and reasoning are combined to yield insights into
the thoughts and feelings of others (Ickes 1997)

- An affective response more appropriate to someone else's situation than
  to one's own (Hoffman 1975)
- An other-oriented emotional response congruent with the other's
  perceived welfare (Batson et al. 1997)
- An affective response that stems from the apprehension or comprehension of another's emotional state or condition and that is similar to
  what the other person is feeling or would be expected to feel in the given
  situation (Eisenberg 2000)

Empathy has also been described in developmental terms, starting with
automatic emotional transfer, sometimes called *contagion*, that is thought
to be rather spontaneous. It involves a kind of mimicry of the individu-
al(s) in one's immediate environment, a direct perception of the emotions
of others that automatically triggers the same emotion. Then there is a
slightly more sophisticated empathy, sometimes called *personal empathy*,
characterized by "me too-isms," where the one empathizing imagines what
they would feel like in the place of the other but doesn't distinguish what
the other might actually be experiencing from what the empathizer is feeling. This type of empathy involves a type of projection. Once one has the
capacity to distinguish self and other, when they can knowingly take the
perspective of another being as another being and when they can make
rudimentary causal inferences, then a type of cognitive empathy can occur.
The primary difference between earlier forms of emotional or affective
empathy and cognitive empathy is that in the latter the empathizer is not
merely mimicking or projecting onto the emotions of the person they are
empathizing with but is engaged in a reflective act of imagination that puts
her into the other's situation and/or frame of mind.[4]

There is also a rich history of philosophical discussions of empathy
used to explore art and aesthetics, embodiment and phenomenology, and
in what is called the sentimentalist tradition of ethics, associated most

---

4. There is a literature on "mirroring" and "mirror neurons" that sometimes gets linked to
discussions of empathy, but there are a number of problems with this linking. On the milder
side, there is the problem that plagues most discussions of empathy, being clear about what is
meant by empathy and empathetic understanding (see Debes 2017). A more serious problem
is that the data on mirror neurons has been over interpreted and extended far beyond their
appropriate application, making it hard to see what, if anything, they have to do with shared
emotions or empathy (see Hickok 2014).

notably with David Hume and Adam Smith, though they used the term *sympathy* to mean what we usually now mean by *empathy*. Within this historical literature, as with the more contemporary psychological literature, *empathy* means different things and may be deployed for different reasons.[5] But there are two important ideas that link these philosophical discussions of empathy to each other as well as to the psychological investigations. First, empathy involves an imaginative resonance across differences; this is akin to the affective part of empathy. The second is the intention to shift perspectives or simulate/embody the perspective of another; this is the cognitive part of empathy.

Both imaginative resonance and perspective taking require fairly elaborate mental capacities, sometimes referred to as "theory of mind." One is thought to have a theory of mind when they are able to recognize the mental states or experiences of someone else and further recognize that they are distinct from the other; that their own experiences may be different from the experiences, perceptions, and thoughts of someone else. Most humans over the age of three have theory of mind, but can other animals recognize the distinct perspectives of others?

The first people to pose this question were David Premack and Guy Woodruff in their 1978 article "Does the Chimpanzee Have a Theory of Mind?" According to them, to have a theory of mind, a being "imputes mental states to himself and to others (either to conspecifics or to other species as well" (515). Premack and Woodruff (1978) suggested that the types of mental states that would be attributed to others might include "purpose, intention, beliefs, thinking, knowledge, pretending, liking, and doubt" (518). In order to answer their question, Premack and Woodruff asked chimpanzee Sarah to solve a series of problems for a human who was videotaped in a variety of circumstances, for example, trying to open a can of soup, trying to reach a banana, trying to turn on a light that was missing a bulb, and so forth. Sarah was shown each video until the last five seconds, at which point the video was put on hold. Sarah was then shown two photographs, one of which represented a solution to the problem. For example, perhaps she would be shown a screw driver and a can opener after watching the video in which the subject was trying to open the can of soup. Then the experimenter would leave the room, and Sarah would select one of the two photographs by placing her selection in a designated location. The right answer to this task was, obviously, the can opener. Sarah picked

---

5. For a wide-ranging resource, see Maibom (2017).

the right answer over 85 percent of the time. It seems she understood that the human was trying to achieve a goal; that goal wasn't her goal, but she was able to determine what the human needed in order to achieve the goal. (She also understood how this test worked. I can easily imagine a chimpanzee watching part of a video and then losing interest in the whole thing.) This suggested that Sarah was able to attribute "at least two states of mind to the human actor, namely, intention or purpose on the one hand, and knowledge or belief on the other" (Premack and Woodruff 1978, 518).

That Sarah arguably has a theory of mind suggests that she can empathize, but that she can doesn't yet show that she will. In addition to being able to imagine the state of the other and take that perspective, she would also have to have concern for the other and be motivated to do something to help. Premack and Woodruff tested this by using two different actors, one a caretaker that Sarah liked and one whom she didn't particularly care for. She would do better on tests that helped the person she liked.[6]

What this suggests is that empathy involves a combination of mental states—an awareness of the distinction between self and other, an understanding of the interests and needs of the other, and a feeling that motivates one to act to help meet those needs.

Entangled Empathy

I prefer to think of empathy as a process that includes perception, reflection, and concern. Although the process may not be linear, we can think of the various parts of the process as going something like this. The well-being of another grabs the empathizer's attention. The empathizer reflectively imagines himself in the position of the other, making sure to more or less partition his own perspective from that of the other. Then he makes a judgment about how the conditions that the other finds herself in contribute to her state of mind or well-being. The empathizer will then carefully assess the situation and figure out what information is pertinent to effectively help the person in question.

This sort of empathy doesn't, indeed it can't, separate emotion and cognition. It also blends ethical and epistemological concerns. And it leads

6. I've known Sarah for over a decade (although not when she was performing these particular tests) and am quite fond of her. She has a very distinct personality, and it is pretty clear whom she likes and whom she doesn't. It doesn't surprise me at all that she would be willing to solve a problem for a person (or a chimpanzee) she likes and not help one she doesn't.

to action because what draws our attention in the first place is another's experiential well-being. Once our perception starts the process, we will want to pay critical attention to the broader conditions that affect the well-being or flourishing of those with whom we are empathizing. This requires us to attend to things we might not have otherwise. Empathy of this sort requires gaining perspective and usually motivates the empathizer to act.

This motivational potential is one of the ways empathy is important for ethics. One of the shortcomings of most accounts of ethics is that they construe ethics too much in terms of beliefs and knowledge, leaving open the question, why care? (see chap. 10). Even if I know what I ought to do, why should I do it? It's one thing to know what another wants, needs, or experiences; it is another thing to feel concern and be moved to help.

I call this process entangled empathy, in which we are attentive to both similarities and differences between ourselves and our own situation and that of the fellow creature with whom we are empathizing (Gruen 2015a). It is an experiential process in which we recognize we are in relationships with others and are called on to be responsive and responsible in these relationships by attending to another's needs, interests, desires, vulnerabilities, hopes, and sensitivities. We alternate between our perspective and the perspective of the one we are empathizing with, and in this we are able to preserve the sense that we are in relationship to and not merged into the same perspective. To do it well we have to try to understand the individual's particular experiences and situation and her individual personality. Very often this is not easy to do without expertise and careful observation, particularly in the case of nonlinguistic or linguistically inaccessible others.

The process of figuring out what the perspective of a very different other might be is centrally important for Animal Studies. Since an empathizer's evaluations, deliberations, intentions, and choices are shaped, at least in part, by the human social context in which he or she lives, various human social institutions and norms will inform expectations (see chap. 3). Empathetic engagement will often involve examining the conditions under which both the empathizer and the one with whom she is empathizing's desires are shaped and consider both the internal and external factors that helped shape deliberative capacities. In the human case, entangled empathizers will try to answer a host of questions. What psychological predispositions do I have and does she have and how do they affect our different levels of confidence in choosing? Were certain deliberative paths closed to me or to her by familial, educational, social, political, racial, gendered, economic, or religious barriers or prohibitions? How successful do

I think I will be in my empathizing with her and will my success or failure depend on the social position I occupy? Do people like me culturally or historically have less chance of empathically succeeding and of "making a difference"? In the case of other animals, empathizers might ask themselves, what were the early rearing conditions this animal experienced and how did that shape her current experiences? What sort of species-typical behaviors does a creature of this kind usually engage in, and does she have opportunities to engage in those behaviors? What sorts of social relationships are important, whether they be with conspecifics or animals from other species, including humans?

Entangled empathy would not endorse the slaughter methods that Grandin's empathy helped her develop, nor would it condone the sort of violence that Paul Bloom thinks empathy leads to. Given the complexity of entangled empathy, it isn't likely that other animals, like Kuni or the male chimpanzee in the zoo or even Sarah, engage in this sort of empathy, even if they are able to understand the perspectives and interests of others and can be motivated to help, at least some of the time. But most of us humans can work to come to understand, as far as possible, the complex and often quite different situations of others, including other animals, and to answer the variety of questions that help us to accurately empathize.

Empathy's Others

Whether it is possible to really know what another is experiencing in order to empathize with them is quite challenging. Often we think we understand when we don't. Much of the time we are unaware of the way that our implicit biases distort our thinking. Research has found that the formation of our identities as parts of particular groups, for example, racial or species groups, affects our capacities to empathize across group lines. There is both neuroscientific research and social psychological research that shows that in-group biases are deeply ingrained in our psyches and lead to judgments that support commitments to flawed stereotypes and prejudices. This is sometimes referred to as "the racial empathy gap."

In a series of studies, people from different races were presented with images of subjects having their skin pricked with a needle.[7] White people

---

7. For a summary of the studies and the problems that the racial empathy gap poses for policing, see https://aeon.co/essays/unconscious-racism-is-pervasive-starts-early-and-can-be -deadly. This has interesting resonances with work on what animals can feel pain (see chap. 17).

consistently expressed greater empathy for other white people, Asian people for other Asian people, whereas most, although not all black people were more egalitarian in their empathy. To complicate matters, in one 2012 study, both white and black people rated black people as feeling less pain than white people and thus less in need of empathy (Ojiaku 2016).

The biases can be corrected by bringing what is implicit into consciousness and working regularly to close the empathy gap. But here, too, there are challenges. In educational research, there is a worry about "false empathy," a process "in which a White person believes he or she is identifying with a person of color, but in fact is doing so only in a slight, superficial way" (Duncan 2002, 89).[8] Empathy in these cases is suspect at best and plays out often as a form of feel-good voyeurism akin to what Saidiya Hartman queries about engagement with slave narratives in *Scenes of Subjection* (1997):

> Are we witnesses who confirm the truth of what happened in the face of the world-destroying capacities of pain, the distortions of torture, the sheer unrepresentability of terror, and the repression of the dominant accounts? Or are we voyeurs fascinated with and repelled by exhibitions of terror and sufferance? . . . At issue here is the precariousness of empathy and the uncertain line between witness and spectator. (Hartman 1997, 3–4)

And the problem goes deeper—to put oneself, as a white person, into the position of the black person or the slave, in order to understand the experience and suffering, amounts to a further erasure of the black body. In this sense, Hartman writes, "it becomes clear that empathy is double-edged, for in making the other's suffering one's own, this suffering is occluded by the other's obliteration" (19).

Frank Wilderson (2013), in "'Raw Life' and the Ruse of Empathy," interrogates "an optimism that assumes relationality within and between all sentient beings" (182). His analysis is that there are some beings who are beyond relationality. "The explanatory powers of empathy and analysis are scandalized when confronted with the Black position, a paradigmatic location synonymous with slavery" (184). Following on the definition of slavery provided by Orlando Patterson (1988) as a permanent, violent domination of natally alienated and generally dishonored persons, Wilderson sees blackness as a form of social death, a state of being deprived of

8. Duncan is citing Delgado (1996): 12.

relationality. So "even perceived moments of empathic identification with the Slave are ruses" (189) as one cannot empathize with objects or beings that are not in the relation.

If white people in a culture of antiblack racism cannot understand the burdens of racism without erasing black people, what hope is there that humans will be able to empathize with a chimpanzee, a dairy cow, or a lab rat? Are there some who will forever remain empathy's others?

Entangled empathizers will try to work though complicated processes of understanding others, human and non, in situations of differential social, political, and species-based power. These are complex, sometimes dangerous processes in which we may only get a "glimpse" of the other, and we are likely to make mistakes.[9] But given that we are entangled in deep, co-constituting relationships, this work to understand is not just something desirable, I believe it is central to our very agency. Our agency is relational in a robust sense; it is co-constituted by our social and material entanglements. Social entanglements often extend beyond the human and far beyond our geographical location. Material entanglements include our socioeconomic opportunities and barriers to opportunity shaped by race and class. They also include our entanglement with the food we have access to, the safety of our physical environments, the animals and humans whose labor and bodies are exploited in what we consume, our greenhouse gas emitting activities that are creating climate refugees, and so forth. All of these actions, in part, constitute who we are. Our agency at any particular time is an expression of entanglements in multiple relations across space, species, and substance.

On this entangled relational account of agency, there is no place beyond relations; antiblackness or speciesism, for example, are political and ethical

---

9. Claire Jean Kim (2015), in her work *Dangerous Crossings*, recommends an "ethics of avowal." In contrast to disavowal, the act of rejection or dissociation that often leads to perpetuating patterns of social injury, she suggests that we recognize the ways that our struggles are linked and that we attend to the concerns of other subordinated groups, particularly when their concerns are embattled. In an op-ed following the killing of Cecil the Lion and Samuel DuBose, I suggested that we should empathize with the pain and indignities of others who are disempowered and avow, rather than belittle, their search for justice. Historically, disregard for the lives and bodies of black people has been justified through a process of dehumanization that specifically compares them to animals. If it were no longer acceptable to treat animals as animals and violate and kill them, the animalization process that serves to justify structures of white male power would be weakened. Weakening that structure is one way to avow the lives of those who were wantonly killed and perhaps allow more just social relations to develop from our grief and anger (Gruen 2015b).

relations that construct black people and animals as fungible, disposable, and perhaps paradoxically, outside of relationality. The relations we are in are not always, perhaps not even often, the sorts of things we chose. Some relations I am forced into, some I seek to develop, some are unjust, some are harmful, some may even work to deprive me of my subjectivity. And since we are constituted in various ways by these relations, "others" are never beyond empathy.

Suggestions for Further Reading

Aaltola, Elisa. 2018. *Varieties of Empathy*. London: Rowman & Littlefield International.

Andrews, Kristin, and Lori Gruen. 2014. "Empathy in Other Apes." In *Empathy and Morality*, edited by Heidi Lene Maibom, 193–209. Oxford: Oxford University Press.

Donovan, Josephine. 2007. "Attention to Suffering: Sympathy as a Basis for Ethical Treatment of Animals." In *The Feminist Care Tradition in Animal Ethics*, edited by Josephine Donovan and Carol Adams, 58–86. New York: Columbia University Press.

Gruen, Lori. 2015. *Entangled Empathy: An Alternative Ethic for our Relationships with Animals*. New York: Lantern Books.

References

Baron-Cohen, Simon. 1995. *Mindblindness: An Essay on Autism and Theory of Mind*. Cambridge, MA: MIT Press.

Batson, C. D., K. Sager, E. Garst, M. Kang, K. Rubchinsky, and K. Dawson. 1997. "Is Empathy-Induced Helping Because of Self-Other Merging? *Journal of Personality and Social Psychology* 73: 495–509.

Bloom, Paul. 2016. *Against Empathy*. New York: Ecco Books.

Cohen-Rottenberg, Rachel. 2009. "A Critique of the Extreme-Male-Brain Theory of Autism." Autism and Empathy: Dispelling Myths and Breaking Stereotypes. https://autismandempathyblog.wordpress.com/a-critique-of-the-extreme-male-brain-theory-of-autism/.

Davidson, Joyce, and Mick Smith. 2009. "Autistic Autobiographies and More-than-Human Emotional Geographies." *Environment and Planning D: Society and Space* 27: 898–916.

Debes, Remy. 2017. "Empathy and Mirror Neurons." In *Routledge Handbook of Philosophy of Empathy*, edited by Heidi Maibom, 54–63. New York: Routledge.

Decety, Jean, and M. Meyer. 2008. "From Emotion Resonance to Empathic Understanding: A Social Developmental Neuroscience Account." *Developmental Psychopathology* 20 (4): 1053–80.

Delgado, R. 1996. *The Coming Race War? And Other Apocalyptic Tales of America after Affirmative Action and Welfare*. New York: New York University Press.

de Waal, Frans. 2009. *The Age of Empathy: Nature's Lessons for a Kinder Society*. New York: Harmony Books.

Duncan, Garret. 2002. "Critical Race Theory and Method: Rendering Race in Urban Ethnographic Research." *Qualitative Inquiry* 8 (1): 85–104.

Eisenberg, N. 2000. "Emotion, Regulation, and Moral Development." *Annual Review of Psychology* 51: 665–97.

Goldman, A. 1993. "Ethics and Cognitive Science." *Ethics* 103: 337–60.

Grandin, Temple. 1995. *Thinking in Pictures and Other Reports from My Life with Autism*. New York: Vintage.

Gruen, Lori. 2015a. *Entangled Empathy: An Alternative Ethic for our Relationships with Animals*. New York: Lantern Books.

———. 2015b. Samuel Dubose, Cecil the lion and the ethics of avowal. *Al Jazeera*, July 31. http://america.aljazeera.com/opinions/2015/7/samuel-dubose-cecil-the -lion-and-the-ethics-of-avowal.html.

Hartman, Saidiya. 1997. *Scenes of Subjection: Terror, Slavery, and Self-Making in Nineteenth-Century America*. New York: Oxford University Press.

Hickok, Gregory. 2014. *The Myth of Mirror Neurons*. New York: W. W. Norton.

Hoffman, M. L. 1975. "Developmental Synthesis of Affect and Cognition and Its Implications of Altruistic Motivation. *Developmental Psychology* 23: 97–104.

Ickes, W. 1997. *Empathic Accuracy*. New York: Guilford Press.

Kim, Claire Jean. 2015 *Dangerous Crossings: Race, Species, and Nature in a Multicultural Age*. Cambridge: Cambridge University Press.

Levenson, R. W., and A. M. Ruef. 1992. *Empathy: A Physiological Substrate. Journal of Personality and Social Psychology* 63 (2):234–46.

Maibom, Heidi, ed. 2017. *Routledge Handbook of Philosophy of Empathy*. New York: Routledge.

Ojiaku, Princess. 2016. "Is Everybody a Racist?" https://aeon.co/essays/unconscious-racism-is-pervasive-starts-early-and-can-be-deadly.

Patterson, Orlando. 1988. *Slavery and Social Death*. Cambridge, MA: Harvard University Press.

Premack, David, and Guy Woodruff. 1978. "Does the Chimpanzee Have a Theory of Mind?" *Behavioral and Brain Sciences* 1 (4): 515–26.

Preston, S. D., and F. de Waal. 2002. "Empathy: Its Ultimate and Proximate Bases." *Behavioral and Brain Sciences* 25(1): 1–20.

Wilderson, Frank, III. 2013. "'Raw Life' and the Ruse of Empathy." In *Performance, Politics and Activism*, edited by Peter Lichtenfels and John Rouse, 181–206. Hampshire, UK: Palgrave Macmillan.

Worsham, Lynn, and Gary Olson. 2012. "Temple Grandin, Translator: Sounding Autism, Seeing Animals, Making a Difference." *JAC: A Journal of Rhetoric, Culture, and Politics* 32 (1/2): 11–56.

10 ETHICS

Alice Crary

Anyone who sets out to grapple with questions about animals and ethics is likely to feel pressure to justify themselves. When children are starving, journalists are beheaded, and young women are kidnapped and forced into marriage, it can seem frivolous to take an interest in how animals are treated. There are long-standing ethical traditions that may seem to support this dismissive attitude—traditions that represent animals as morally indifferent objects, that is, as things that do not in themselves call for respect and whose flesh is therefore available to be cultivated, caged, cut, measured, prodded, maimed, dismembered, displaced, displayed, harvested, and devoured in any manner as long as it suits recognized human purposes (Regan and Singer 1989). Today ethical approaches that in this way deny animals moral standing are increasingly on the defensive. Yet they still have outspoken defenders (e.g., Oderberg 2000), and the image they bequeath to us of animals as mere disposables that don't in themselves call for solicitude remains operative on a massive scale in settings such as confined feeding operations, industrial slaughterhouses, aquafarms, hunting grounds on land and in the oceans, zoos, laboratories, and sites of the large-scale conversion of tropical and other forests.

Struck by the need to challenge not only the handling of animals in these settings but also the bodies of thought that seem to support it, people who are concerned with the mistreatment of animals frequently start by trying to show simply that animals do in fact impose on us direct claims for particular forms of treatment and so do have moral standing. The same thinkers then sometimes go on to discuss how, in the nonideal messiness of actual cases, animals should be treated. Part of what makes many real-world cases messy has to do with the fact that human-animal interactions often occur in contexts in which human beings are themselves subject to serious and sometimes intersecting forms of bias (e.g., racist and classist bias). These contexts may well confront us with difficult questions about the possibility of unsentimental and clear-sighted concern for animals that does not neglect the plight of human beings. One strategy for managing these kinds of complexities is to bracket them at the outset and focus initially on what is involved in simply bringing out that animals have moral standing. Different approaches to this basic task encode different assumptions about what ethical thought about animals is like as well as about the kinds of methods proper to it, and it is helpful to start by asking what view of these matters we should adopt.

Two Examples from the Lives of Chimpanzees

Consider, to begin with, the following anecdote that moral philosopher and activist Lori Gruen tells in a recent paper:

> One of the chimpanzees I know living in sanctuary is now over forty years old. Every time humans come around he makes a ridiculous facial expression—he pops both of his lips out and folds them back so the inside of his lips show, making him look clown-like, with a big pink mouth. This chimpanzee was used in the entertainment business and presumably making himself look absurd garnered laughs and attention. He was undoubtedly taught to do this when he was young either by rewarding him when he did or, more likely, punishing him when he didn't. (Gruen 2014, 231)

Suppose that we are inclined to say that Gruen's chimp acquaintance was wronged by the treatment he received. This by itself is already a nontrivial claim, since it presupposes that the chimp is a being with moral standing who merits specific forms of treatment and who can be mistreated. But it still leaves unanswered many questions about the nature of the wrong or wrongs at issue. Was the chimp wronged by being subjected to the pain of punishment-based training? Was he also wronged by being held in captivity? Does the fact that he was forced to make a spectacle of himself for human beings—and that he was left in a condition in which he compulsively reenacts the spectacle—represent, as Gruen suggests, a further and distinct wrong? How significant are these wrongs? And, lastly, what resources do we have in ethics for answering these types of questions?

When animal advocates take a stand on what justifies us in representing animals such as this chimp as having moral standing, they are at the same time at least tacitly addressing this last question. We can see this by turning to what might aptly be called *traditional approaches to animal ethics*. This label can be used to collectively pick out a family of views about animals' moral standing that, while in many respects divergent, resemble each other in accepting a metaphysical picture (i.e., a picture of what kinds of things there are) that is ingrained in our intellectual culture and that provides the framework within which most research in ethics is now pursued. The picture is one in which the empirical or observable world is as such devoid of moral values. Ethical positions that encode this picture have arresting implications for how we conceive the resources available to us for moral thought. Here there is no question of our needing the exercise of moral capacities such as moral imagination in order to get aspects of the empirical world in view in a manner relevant to ethics. That is, there is no question of our needing to look at things from alternative cultural or historical perspectives or to imaginatively enter the experience of individuals very different from us in order to see some things clearly. The image of the empirical world with which we operate in ethics is supposed to be something that, far from being produced through this kind of moral effort, is handed down to us from disciplines such as the natural sciences, where these disciplines are conceived as independent of ethics. Moreover, since the overall image of the empirical world with which we operate in ethics necessarily subsumes within it any image that we have of the worldly lives of animals, this means that the task of bringing the worldly lives of animals into focus in ethics turns out to be a task that, instead of belonging to ethics proper, gets outsourced to disciplines outside it.

Acceptance of this outlook has a notable effect on the way in which we approach ethical questions that come up in reference to Gruen's example of the chimp. It obliges us to regard the project of arriving at the kind of empirical understanding of this creature's life that is relevant to ethics as one that we can entrust to disciplines such as biology, where these are conceived as external to ethics. Now we have to say that there is no room for an engagement with morally saturated accounts of chimpanzee existence to contribute necessarily to this project. Among the things that we thus exclude as essentially irrelevant is engagement with the many charged accounts that the primatologist Jane Goodall has given of her encounters with individual chimps. In one essay, Goodall writes about Jojo, an adult male chimpanzee who was born in the African forest and who, when she met him, had spent ten years in a medical research laboratory in a five-foot-by-five-foot steel cage. Goodall describes what human beings have done to Jojo and other captive chimps in these terms:

> [We have] deprived them of freedom, stole from them the dim greens and browns; the soft gray light of that African forest, the peace of afternoon; when the sun flecks through the canopy and small creatures rustle and flit and creep among the leaves. Deprived them of the freedom to choose, each day, how they would spend their time, and where and with whom. Deprived them of the sounds of nature, the gurgling of streams, murmuring wind in the branches, of chimpanzee calls that ring out so clear, and rise up through the tree tops and drift away in the hills. Deprived them of their comforts, the soft leafy floor of the forest, the springy, leafy branches from which sleeping nests can be made. (Goodall 2000, xii)

Goodall is here trying to impress on us the magnitude of the harm done to Jojo and other chimpanzees in captivity by giving us an evocative account of the humble glory of chimp life in the forest. It would be possible to allow the kinds of responses invited by Goodall's words to shape our efforts to bring the worldly existence of Gruen's chimp into focus. But if we accept an outlook in which the world is as such bereft of moral value, we will be compelled to reject at the outset any nonneutral perspectives that Goodall's writing invites us to adopt as incapable of contributing internally to the empirical understanding of Gruen's chimp that we want in ethics.

Traditional Approaches to Animals and Ethics

This outlook shows up within many of the most well-known and widely discussed approaches to animal ethics in the form of the idea that the fabric of the world is free of moral values. This idea is, for instance, at play in Peter Singer's *Animal Liberation* (Singer 1975). Singer's strategy for showing that animals matter, here and elsewhere, through to the present day, starts from a form of utilitarianism, a view on which the right action in a particular context is the one that best promotes the interests of all creatures concerned, where a creature's "interests" are an expression of her capacity for pain or pleasure. Singer's work has had a major impact, providing inspiration for many subsequent interventions in animal ethics, and what has been most influential is not his utilitarian stance itself but one of its core presuppositions. It is a presupposition of Singer's stance that any consideration a—human or nonhuman—creature merits is a reflection not of its membership in any group (say, her membership in the group "human beings" or in the group "horses") but rather of her individual capacities (e.g., her individual capacities for pain and pleasure). This presupposition provides the unifying principle for the most prominent family of approaches to animal ethics. Many animal advocates follow Singer's lead in treating moral status as a function of individual capacities while at the same time disagreeing with him about which individual capacities are morally relevant (e.g., suggesting that the capacity not for pain but for, say, subjecthood is the mark of moral standing; Regan 1983). Further, although it is in principle possible to endorse a doctrine that grounds moral status in individual capacities without assuming that the world is morally neutral, animal advocates who incline toward such doctrines in fact overwhelmingly integrate an assumption along these lines into their outlooks. Some—notably, Singer—even make their attachment to the assumption explicit (Crary 2016). The result is that members of this set of thinkers treat the identification of the individual worldly capacities that, as they see it, are morally significant as a job that is properly assigned not to ethics itself but to disciplines beyond it. Their preferred approaches to animal ethics thus clearly count as "traditional" in the above sense.

Advocates of these traditional approaches to animal ethics are sometimes referred to collectively as *moral individualists* (Rachels 1990; McMahan 2005). What speaks for this terminology is the fact that these thinkers ground an animal's— nonhuman and human—moral standing in a certain kind of attention to her as an individual (specifically, in morally neutral attention to her individual capacities of mind). However, as will emerge below, there are other thinkers who take up questions of animals and ethics and who, while differing substantially from Singer et al. in their views, nevertheless resemble them in basing their conclusions about moral standing on attention to individual creatures. In light of this convergence, it makes sense to withhold the generic label "moral individualism" from the projects of Singer and likeminded others and to place their work instead under the heading of *traditional moral individualism*.

It is not hard to see how Singer and others use the tenets of traditional moral individualism to show that animals are proper objects of moral concern. Once we're equipped with the thought that moral standing is a function of neutrally available individual capacities, we need do little more than allow that any capacities we regard as morally relevant in human beings are similarly morally relevant in animals. To be sure, this position only seems to have direct implications for the treatment of animals if it is combined with the thought that some human beings and some animals are in fact equally well endowed with morally relevant capacities. Taking their cue from this observation, animal advocates who favor traditional moral individualism often take the further step of drawing attention to cases of human beings who (as a result of illness, injury, age, or some congenital condition) are severely cognitively disabled and who are no better equipped than some animals with what the thinkers in question see as "morally significant capacities." Sometimes these thinkers suggest, for instance, that a person who has suffered a serious brain injury may be no better mentally endowed than a dog or a pig. With reference to the—morally objectionable—idea that severely cognitively disabled individuals are "marginal" cases of humanity, this argument is sometimes referred to as the *argument from marginal cases* (Dombrowski 1997). Wanting to avoid this offensive terminology, some theorists now speak of the *argument from species overlap* (Horta 2014). But a powerful case can be made for thinking that without regard to whether it is *described* in morally problematic terms, this argument *is* morally problematic—indeed outrageous—for implying that severely cognitively impaired individuals merit diminished respect by virtue of their impairments (Kittay 2010; 2016). Setting aside this topic, the point is that this argument is employed by members of the high-profile group of animal protectionists who appeal to the principles of traditional moral individualism to support their view that some animals merit moral attention.

It is a premise of the so-called argument from species overlap that the mere fact of being human (i.e., without regard to the level of one's individual capacities) is morally insignificant. Animal advocates who run versions of the argument, and who thereby effectively contest the tendency to treat being human as by itself morally important, sometimes attack this tendency under the heading of *speciesism* (Ryder 1989). Speciesism is typically understood as unwarranted prejudice in favor of one's species. What traditional moral individualists add to this is the thought that we express such unwarranted prejudice if we suggest that the mere fact of being human matters morally. It would, however, be wrong to assume that champions of all traditional approaches to animal ethics regard as speciesist the idea that simply being human is morally important.

One notable traditional approach that proceeds along different lines is developed in the work of Christine Korsgaard, a moral philosopher who draws her main inspiration from Kant's moral theory. Kant's stated view of animals is that they are mere moral instruments, that is, things that cannot themselves be harmed and that are only subject to "harms" that are indirect reflections of wrongs to human beings (Kant 1996, 192; 1997, 212–13). Korsgaard sets out to show that, despite this expressed view of animals, Kant actually equips us to show that animals matter. She stresses that she thinks a substantial part of the appeal of Kant's core claims about ethics has to do with the fact that they are consistent with the "traditional" idea that the world is a place with no trace in it of moral value or, as she puts it, a "hard" place (Korsgaard 1996b, 4). One of Korsgaard's explicit goals is to mount a modified Kantian defense of animals' moral standing that is likewise consistent with this idea. Her strategy centers on an adaptation of Kant's approach to establishing the moral standing of human beings. She tells us, in recognizably Kantian style, that when we act there is no way for us to avoid placing value on the natures we have as beings capable of rational action and that we are therefore obliged, on pain of inconsistency, to respect the rational natures of all rational beings. She goes on to echo Kant in representing all human beings (and not only those individuals actually endowed with reason) as having rational natures. Her aim in making this further Kantian move is to affirm that all human beings, and not merely the mentally well endowed, have moral standing and that the plain fact of being human is morally important (see chap. 20). Admittedly, there is significant disagreement about whether it is possible, within the framework of Kant's moral theory, to successfully argue for the view that bare humanity is morally important (Kain 2009; McMahan 2008), and Korsgaard herself appears to have advocated different argumentative strategies at different times (Korsgaard 2004, 1996a). Without entering into the dispute about the success of particular attempts to defend in Kantian terms the idea that merely being human matters, it is fair to say that Korsgaard's position patently departs from that of traditional moral individualists who reject this idea as "speciesist."

It does not, however, follow that Korsgaard's enterprise is anti-animal. On the contrary, Korsgaard claims that a subtle adjustment of her Kantian argument for establishing the moral standing of human beings serves to establish the moral standing of animals. Her suggestion is that when we act, there is no way for us to avoid valuing our own animal natures and that we are thus obliged, on pain of inconsistency, to respect the animal natures of all animate beings. That is how Korsgaard makes the case that animals matter, and, as she herself stresses, she makes it without assuming that moral values are part of the furniture of the universe (Korsgaard

2004). Her work thus presents another clear case of a traditional approach to animal ethics.

At this point, it should be clear that the set of traditional approaches to animal ethics comprises different strategies for dealing with the sorts of questions about animals and ethics that we confront when we consider, say, the treatment of Gruen's chimp acquaintance. Whereas traditional moral individualisms invite us to approach ethical questions about the chimp's treatment by asking whether we are granting him the same consideration that we extend to those individual humans who are equivalently endowed with whichever capacities we take to be morally significant, Korsgaardian Kantianisms invite us to proceed by asking whether we are placing the same value on his animal nature that, in acting, we invariably place on our own animal natures. These different strategies may, in some cases, lead to different moral conclusions about the treatment of animals such as Gruen's chimp. Whereas traditional moral individualism, such as Singer's, that ground moral standing in the capacity for pain don't have room for the idea of harms to a creature that the creature himself does not register, and whereas these views thus encourage us to repudiate the idea—championed by Gruen—that forcing this chimp to make a spectacle of himself in front of humans represents a distinct injury, beyond the injuries he suffered in being caged and subjected to painful training methods, Korsgaardian Kantianism may well have room for this idea.

Alternative Approaches to Animals and Ethics

Despite differences that exist among the ways in which traditional approaches call on us to address ethical questions about the treatment of animals, striking similarities remain. These approaches agree that the empirical understanding of animals' lives that we are after in ethics is the business of disciplines independent of ethics and that moral capacities cannot play a necessary role in this understanding. This is worth underlining because the operative image of what is involved in bringing animals empirically into view in ethics, while dominant, has outspoken critics. There are contributors to animal ethics who directly contest the image specifically by defending the idea that the exercise of moral capacities such as moral imagination is internal to our ability in ethics to arrive at an undistorted view of the world and, by the same token, internal to our ability to arrive at an undistorted view of the worldly lives of animals. Insofar as they thus represent the exploration of ethically charged perspectives as capable of immediately informing our efforts in ethics to get a clear view of aspects of reality, the champions of these oppositional projects adopt a stance that is also characteristic of many Marxist, feminist, and black theories of knowledge or epistemologies, intellectual projects that are sometimes referred to collectively as *alternative epistemologies* (Mills 1998). What distinguishes the animal advocates who sound these themes—and who might aptly be described as fans of *alternative approaches to animal ethics*—from other alternative epistemologists is the conviction that the features of reality that are ethically charged, and that are such that moral effort is required to illuminate them, include, in addition to aspects of rational human social life, also aspects of not wholly rational, animate (i.e., animal as well as human) life.

The debate between traditional and alternative approaches to animal ethics reaches back to the early years of the contemporary animal protection movement—back to the 1970s—and represents an intellectual divide within the movement that is easily one of the most fundamental and one of the most fruitful to explore. Among the noteworthy pioneering alternative interventions in this debate are Mary Midgley's (1983) *Animals and Why They Matter* and some early papers of Cora Diamond's (Diamond 1991, chaps. 13, 14). Within Diamond's writings an alternative approach to animal ethics is especially clearly worked out.

Diamond is explicit about rejecting an understanding of the world in which moral concepts are "hard," that is, "given for, or given prior to, moral thought and life" (Diamond 2010, 56), and she thus repudiates the philosophical outlook distinctive of traditional approaches to animal ethics. She presents us with a view of the world of moral concern that is instead "cloudy and shifting," and on which we need the exercise of moral capacities such as moral imagination to bring it into focus. The features of the world that she thinks require moral exertion to get in view include both human beings and animals, and she attempts to show that they "are not given for [ethical] thought independently of . . . a mass of ways of thinking about and responding to them" (Diamond 1991, 327). Elaborating on this point, she argues that ethical thought about human beings and animals of different kinds is shaped by concepts of them that, far from being merely biological, are imaginative and ethically loaded, and she tells us that when in ethics we attempt to capture the worldly circumstances of an individual creature, such a concept is invariably internal to our attempts. This means that, as Diamond sees it, human beings and animals figure in moral thought as beings who merit respect and attention simply as the kinds of creatures they are. So for Diamond—and in this respect she is representative of fans of alternative approaches to animal ethics—there is no question of opposing as "speciesist" the idea that merely being human (or merely being an animal of some kind) matters.

This is not to say that there are no noteworthy differences between, on the one hand, Diamond and other champions of alternative approaches and, on the other, those champions of traditional approaches, such as Korsgaard, who similarly treat the plain fact of being human as morally important. For Diamond and other fans of alternative approaches, human beings and animals only show up for us as mattering insofar as we look at their worldly lives through ethical conceptions, that is, insofar as we use methods that Korsgaard and other champions of traditional approaches are committed to rejecting as distorting. One corollary of this alternative posture is that a mode of instruction that gives new shape to our imaginative sense of the lives of animals of a given kind may directly contribute to the kind of—genuine—understanding of those animals that we seek in ethics. Diamond's work is peppered with examples of bits of writing that help to make a kind of direct contribution of our understanding of human and animal existence. For example, she discusses a poem of Walter de la Mare's that encourages us to look on a titmouse not as a merely biological thing but as "a tiny son of life," and she tries to bring out how, by virtue of thus shaping our attitudes, the poem may be immediately relevant to efforts to arrive at the sort of empirical understanding that is the business of ethics (Diamond 1991, 473–74).

The last decades have witnessed the emergence of a significant variety of projects that, like Diamond's, count as alternative approaches to animal ethics. This includes, among others, projects that derive their inspiration in part from critical theory (Sanbonmatsu 2011), critical race theory (Kim 2015), and, perhaps most conspicuously, feminist theory and the ethics of care (Donovan and Adams 2007; Gruen 2015). The common thread running through these projects is the idea that bringing the worldly lives of animals into focus in ethics requires a type of moral exertion that involves not only arriving at a conception of the things that matter in the lives of the animals in question, but also surveying those animals through the lens of that conception. Brought to bear on the case of Gruen's chimp acquaintance, the idea is that if we are to be able to clearly perceive his worldly circumstances in a manner relevant to ethics, we need to look at him through the lens of an ethical conception of chimp life, say, the sort of conception we might arrive at by engaging with works, like some of Jane Goodall's, that aim to impart a sense of what flourishing chimp life is like. It's worth pausing to note that the point is not that every attempt to give us a conception of what matters in the lives of chimpanzees will contribute internally to our ability to bring them empirically into focus in ethics. On the contrary, any particular attempt to foster such an ethical conception (such as, say, an attempt by Jane Goodall) may turn out to be manipulative or sentimental or distorting in some way. Still, the act of critically exploring novel conceptions of what matters in chimp life is taken to be integral to legitimate, world-guided moral thought about chimps. For the idea is that we need to have reference to a reasonably sophisticated conception if, for instance, we are to be able to determine whether some treatment to which the chimp has been subjected is a painful prod rather than a playful interaction or whether a given allocation to him of space and resources represents a continuation of captivity rather than sanctuary.

Insofar as we do in fact look at Gruen's chimp acquaintance in the relevant ethically inflected way, he will show up for us as meriting certain forms of treatment simply by virtue of being a chimpanzee. Notice that here our conclusions about the moral standing of the chimp are licensed by a certain kind of—morally nonneutral—attention to him as an individual. So it would not be unreasonable to talk in this connection about a *nontraditional* or *alternative moral individualism*. Abstracting from the terminological question of whether we should in fact speak of "alternative moral individualisms," it is also worth noticing that views that call for the morally nonneutral modes of attention that are in question encounter no problem—of the sort encountered by traditional moral individualists such as Singer who ground moral standing in neutrally accessible individual capacities—about allowing that the chimp may have suffered injuries that he himself can't register. So there is no obstacle to saying that the chimp was wronged by being forced to pop his lips out in a ridiculous grimace and to thereby make a spectacle of himself in front of human beings.

Advocates of traditional approaches to animal ethics have not devoted much energy to responding to alternative approaches, for the most part simply dismissing them as lacking in argumentative rigor. Alternative approaches call on us to look at the worldly lives of animals from specific ethical perspectives and, to the extent that advocates of traditional approaches comment on the work of their alternative counterparts, they tend to charge them with thereby recommending distorting modes of thought. This is because advocates of traditional approaches assume that the empirical world is as such devoid of moral values, and, granted this assumption, ethically nonneutral perspectives appear to have a necessary tendency to block our view of how things are. But it doesn't follow that advocates of traditional approaches have somehow succeeded in formulating a satisfactory rejoinder to advocates of alternative approaches. The dismissive attitude of advocates of traditional approaches seems question begging given that alternative thinkers reject the very metaphysical assumption that is supposed to justify their approach, viz, the assumption that the world is morally neutral. Advocates of alternative approaches reject this assumption and lay claim to a contrasting view of reality as including moral values, and they also hold that, in opposition to what their traditional correlates maintain, the worldly texture of human and animal lives is suffused with such values. So it appears to them that we have to use methods that are morally loaded if we are to arrive at the sort of empirical understanding of animal life that is the business of ethics. The upshot is that there is in fact a substantive quarrel between advocates of traditional and alternative approaches to animal ethics, specifically, a quarrel that turns on questions about how to conceive ethically the worldly lives of animals and the demands of knowing them.

Further Complexities

Discussions about these topics are yet more complex than talk of a broad opposition between "traditional" and "alternative" approaches might seem to suggest. Alongside traditional approaches to animal ethics, which tell us that we are limited to neutral methods in our attempts to bring empirical animal existence into focus, and alternative approaches, which counter that we require morally nonneutral methods, there are also—to mention one further prominent set of views—*poststructuralist* approaches. These approaches resemble alternative approaches insofar as they claim that any empirical methods we use in ethics are invariably morally nonneutral but also differ from them in giving this claim a skeptical inflection, suggesting that it follows that none of our world-directed modes of thought can lay claim to objective authority (Derrida 2008; Haraway 2003; Wolfe 2003). What thus emerges is that there are further wrinkles and subtleties to conversations about how in ethics to construe the challenges of knowing the worldly lives of animals. Nevertheless, for all their intricacy and abstractness, there is a sense in which these conversations are of interest to anyone concerned with the ethical treatment of animals not least because we invariably position ourselves within them when we grapple with questions about how an animal such as, for example, Gruen's chimp acquaintance should be regarded and, where relevant, treated. In grappling with such questions, we cannot help but commit ourselves to a view about the kinds of resources that are available to us for thinking morally about animals.

The view we arrive at will have implications for how we assess social contexts in which human as well as animal interests are at play and in which wrongs to animals are structurally connected with wrongs to human beings. We encounter such complexity when, for instance, we consider the treatment of animals within the practices of members of socially subjected human groups, say, the treatment of gray whales in revived versions of the traditional hunt of the Native American Makah (Kim 2015) or the treatment of animals of various kinds in the sacrificial rituals of Florida-based practitioners of the Santeria religion (Casal 2003). We also encounter such complexity when we set out to think about the treatment of animals in society-wide, modern institutions such as the industrial slaughterhouse, where in the United States since roughly the late 1990s the people who are on "kill floors," where the actually killing and butchering is done, and who are likeliest to be accused if there are allegations of cruelty to animals, are disproportionately men who are either Hispanic immigrants or African Americans who have been driven by need and lack of better opportunities to take nonunionized jobs that are dangerous, physically demanding, and poorly remunerated (Schlosser 2002; Compa 2005; and Pachirat 2011). How should we approach issues having to do with the treatment of animals that arise in these sorts of complex cases? If we proceed in the style of traditional approaches to animal ethics, then we will see the task of arriving at an adequate empirical understanding of the relevant circumstances as one that falls outside the purview of ethics, and we will take it that the core job for ethical thought is to deploy whatever ethical theory we favor. If, instead, we proceed in the style of alternative approaches, we will see the task of arriving at an adequate empirical understanding of the relevant circumstances as one that requires nonneutral methods (such as, e.g., methods that involve carefully capturing and critically exploring any relevant historical and cultural perspectives), and we will take it that this task is a core project for ethical thought. But without regard to whether we choose a version of one of these two broad approaches—or whether instead we choose a version of some hybrid or wholly different approach— the approach we select will affect not only the way in which we proceed but also the moral conclusions we draw about the treatment that animals of different kinds merit and about the kinds of harms they can be—and are—made to suffer.

Suggestions for Further Reading

Cavell, Stanley, Cora Diamond, John McDowell, Ian Hacking, and Cary Wolfe. 2008. *Philosophy and Animal Life*. New York: Columbia University Press.

Coetzee, J. M. 1999. *The Lives of Animals*. Edited by Amy Gutmann. Princeton, NJ: Princeton University Press.

Gruen, Lori. 2011. *Ethics and Animals: An Introduction*. Cambridge: Cambridge University Press.

Hearne, Vicki. 2007. *Adam's Task: Calling Animals by Name*. New York: Skyhorse.

References

Casal, Paula. 2003. "Is Multiculturalism Bad for Animals?" *Journal of Political Philosophy* 11 (1): 1–22.

Compa, Lance. 2005. "Blood, Sweat, and Fear: Workers' Rights in U.S. Meat and Poultry Plants." Human Rights Watch, January 24. https://www.hrw.org/report/2005/01/24/blood-sweat-and-fear/workers-rights-us-meat-and-poultry-plants.

Crary, Alice. 2016. *Inside Ethics: On the Demands of Moral Thought*. Cambridge, MA: Harvard University Press.

Derrida, Jacques. 2008. *The Animal That Therefore I Am*. Edited by Mary-Louise Mallet and translated by David Wills. New York: Fordham University Press.

Diamond, Cora. 1991. *The Realistic Spirit: Wittgenstein, Philosophy, and the Mind*. Cambridge, MA: MIT Press.

———. 2010. "Murdoch the Explorer." *Philosophical Topics* 38 (1): 51–85.

Dombrowski, Daniel. 1997. *Babies and Beasts: The Argument from Marginal Cases*. Chicago: University of Illinois Press.

Donovan, Josephine, and Carol J. Adams, eds. 2007. *The Feminist Care Tradition in Animal Ethics*. New York: Columbia University Press.

Goodall, Jane. 2000. Forward to *Rattling the Cage: Toward Legal Rights for Animals*, by Steven M. Wise, ix–xiii. Cambridge: Perseus.

Gruen, Lori. 2014. "Dignity, Captivity, and an Ethics of Sight." In *The Ethics of Captivity*, edited by Lori Gruen, 231–47. Oxford: University of Oxford Press.

———. 2015. *Entangled Empathy: An Alternative Ethic for Our Relationships with Animals*. New York: Lantern Books.

Haraway, Donna. 2003. *The Companion Species Manifesto: Dogs, People, and Significant Otherness*. Chicago, Prickly Paradigm.

Horta, Oscar. 2014. "The Scope of the Argument from Species Overlap." *Journal of Applied Philosophy* 31 (2): 142–53.

Kain, Patrick. 2009. "Kant's Defense of Human Moral Status." *Journal of the History of Philosophy* 47 (1): 59–101.

Kant, Immanuel. 1996. *Metaphysics of Morals*. Translated and edited by Mary McGregor. Cambridge: Cambridge University Press.

———. 1997. *Lectures on Ethics*. Edited by J. B. Schneewind. Translated by Peter Heath. Cambridge: Cambridge University Press.

Kim, Claire Jean. 2015. *Dangerous Crossings: Race, Species, and Nature in a Multicultural Age*. Cambridge: Cambridge University Press.

Kittay, Eva Feder. 2010. "The Personal Is Philosophical Is Political: A Philosopher and Mother of a Cognitively Disabled Person Sends Notes from the Battlefield." In *Cognitive Disability and Its Challenge to Moral Philosophy*, edited by Licia Carlson and Eva Feder Kittay, 393–413. Oxford: Wiley-Blackwell.

Korsgaard, Christine. 1996a. *Creating the Kingdom of Ends*. Cambridge: Cambridge University Press.

———. 1996b. *The Sources of Normativity*. Cambridge: Cambridge University Press.

———. 2004. "Fellow Creatures: Kantian Ethics and Our Duties to Animals." Tanner Lecture on Human Values, University of Michigan, February 6. https://dash.harvard.edu/bitstream/handle/1/3198692/korsgaard_FellowCreatures.pdf?sequence=2.pdf.

McMahan, Jeff. 2005. "Our Fellow Creatures." *Journal of Ethics* (9): 353–80.

———. 2008. "Challenges to Human Equality." *Journal of Ethics* (12): 81–104.

Midgley, Mary. 1983. *Animals and Why They Matter*. Athens: University of Georgia Press.

Mills, Charles. 1998. "Alternative Epistemologies." In *Blackness Visible: Essays on Philosophy and Race*, 21–39. Ithaca, NY: Cornell University Press.

Oderberg, David S. 2000. "The Illusion of Animal Rights." *Human Rights Review* 37: 37–45.

Pachirat, Timothy. 2011. *Every Twelve Seconds: Industrialized Slaughter and the Politics of Sight*. New Haven, CT: Yale University Press.

Rachels, James. 1990. *Created from Animals: The Moral Implications of Darwinism*. Oxford: Oxford University Press.

Regan, Tom. 1983. *The Case for Animal Rights*. Berkeley: University of California Press.

Regan, Tom, and Singer, Peter. 1989. *Animal Rights and Human Obligations*. Englewood Cliffs, NJ: Prentice Hall.

Ryder, Richard D. 1989. *Animal Revolution: Changing Attitudes toward Speciesism*. Oxford: Basil Blackwell.

Sanbonmatsu, John, ed. 2011. *Critical Theory and Animal Liberation*. Lanham: Roman and Littlefield.

Schlosser, Eric. 2002. *Fast Food Nation: What the All-American Meal Is Doing to the World*. London: Penguin Books.

Singer, Peter. 1975. *Animal Liberation*. New York: Harper Collins.

Wolfe, Cary. 2003. *Animal Rites: American Culture, the Discourse of Species, and Post-humanist Theory*. Chicago: University of Chicago Press.

11 EXTINCTION

Thom van Dooren

It was roughly two hundred years ago that the French naturalist Georges Cuvier forcefully introduced the Western world to the notion that species might become extinct. Today, far from being unthinkable as it was then, biology—grounded in the insights of Darwin, Wallace, and subsequent generations—presents extinction as an integral facet of life. "In evolutionary terms, extinction is not a remarkable aberration—species disappear continually and new species appear throughout the record of . . . recorded life" (Benton 1986, 129). Life is fleeting for both individuals and species (albeit usually over millions of years), and the record of their deaths plays an important functional role in "normal" evolutionary processes. From this perspective, however, one might reasonably ask why extinction should matter to us at all? If, as the paleontologist George Gaylord Simpson famously put it, Earth is a "charnel house for species" (1953, 281) on which an estimated 99.9 percent of all species that have ever existed are now gone, why should we exert great effort, or indeed any effort at all, in holding onto the ones that happen to share our period of tenure on this revolving planetary tomb?

So extinction, it seems, is inevitable. And yet it is oddly impossible, too. In Timothy Morton's (2013) words, "so dire is the paradox of evolution that Darwin should have used some kind of wink emoticon, had one been available, and scare quotes: The "Origin" of "Species"). The punchline of Darwin's book is that there are no species and they have no origin" (29). At the heart of this paradox is the gradual, processual nature of evolution. Emerging out of some previous forms—plural, because internal variability within species is also an inherent component of evolution by natural selection—a species is itself always multiplicitous. Both in any given moment and over extended periods of time, the organisms that we conveniently gather together under a single Linnaean name are varied to the point that the taxonomists who make it their business to classify them often have great difficulty doing so (Ruse 1992; Kirksey 2015; Mayr 1996). A species is always changing, always becoming other than itself. In the midst of all this uncertainty, all this change and variability, perhaps there is yet another reason not to worry about extinction: species don't really exist in the first place (see chap. 26).

If "species" are conventions of classification, and unavoidably fleeting ones at that, what is to be done about their "extinction"? Recent scholarship raises the stakes even further, insisting that the taxonomic classifications of Western science—around which concern for extinction is so often framed, managed, and publically communicated—are specifically colonial (Mitchell 2016), or perhaps anthropocentric (Ingold 2013), impositions on the fundamental heterogeneity of our living world. As with all systems of classification, we are reminded that this particular mode of making cuts, of labeling, of relating, naturalizes a set of assumptions, understandings, and power relations (Bowker and Star 1999). The goals and concerns of those who might classify differently—human and not (Kirksey 2015; Rose and van Dooren 2016)—are excluded, or perhaps even undermined, when this particular comportment toward the unraveling of our worlds takes up a hegemonic position.

From within the context of this uncertainty, it is hard to know how we ought to make sense of our current period. Indeed, if extinction is a meaningful phenomenon in our world, then it is one that in many ways characterizes our present moment. According to many biologists, we are now either well within or fast approaching the sixth mass extinction event since complex life evolved on this planet (Barnosky et al. 2011). The things we call "species" are disappearing at a hundred or perhaps a thousand times the rate that they have ordinarily done so. With this context in mind, it seems particularly urgent for scholars concerned with animals, or with any form of life for that matter, to come to terms with extinction. Alongside the "animal industrial complex," (Noske 1989) with its myriad cruelties and contributions to climate change, extinction must surely be at the heart of an Anthropocene Animal Studies. Over the past decade in particular, scholarly interest in extinction from within the humanities and social sciences has grown significantly. Much of this work falls under the banner of Animal Studies, but the broader environmental relevance of extinction and the fact that it affects a range of taxa beyond the animal kingdom, have drawn this work into productive dialogue with the environmental humanities, multispecies studies, and a range of other fields. Increasingly, this work is case-study driven: rather than engaging with extinction as a general process, it asks about the ways in which extinction takes shape in particular contexts, drawing in a diversity of human and nonhuman others. Through these detailed explorations, this scholarship offers answers, of sorts, to pressing questions about what extinction is, what it means, and why it matters—now, perhaps differently, than ever before.

Biocultural Entanglement

Extinction is a profoundly "biocultural" phenomenon. This proposition is the foundation for much of the recent work in the humanities and social sciences on extinction. But this is an insight that requires considerable fleshing out. While the loss of species has often been allocated to the natural sciences for both explanation and remediation, it is today clear that such an approach is inadequate. In both its causes and its consequences, as well as in its daily lived experience, extinction draws in multispecies communities of life that always include diverse humans and nonhumans. It might affect indigenous kinship systems on traditional practices of hunting and subsistence; cause the breakdown of systems of veneration, respect, and aesthetic appreciation; or even lead to the loss of livelihood opportunities and the spread of disease (Rose 2011; Hatley 2017; Sodikoff 2012; Ryan 2013). In all of these contexts we see that extinctions ripple out into the world, cutting across any simplistic division between "nature" and "culture." But the point here is not that extinctions affect "humanity." Rather, again and again, they affect particular individuals and communities, in ways that are always unequal; for example, it is poorer people who will experience the increase in the incidence of various diseases that is thought to be associated with the decline of India's vultures, who once cleaned up environments through their prodigious scavenging (van Dooren 2014).

It is perhaps less surprising that people are caught up in extinctions as a principal causal factor. Indeed, biodiversity loss is frequently narrated in a way that leaves little doubt about the "scourge" that is humanity, about the ways in which our actions are driving untold numbers of species over the edge. But here, too, attention to the specificity of particular cases matters. Specific individuals and groups, alongside specific forms of economic, cultural, and political life, are at fault. If, for example, we are to lose many of the endangered species of albatrosses in the coming years, it will (primarily) be the result of specific industrial fishing practices and the modes of consumption that they enable. As Ashley Dawson (2016) has succinctly put it: "In order to respond adequately to this planetary crisis, we need to transgress the boundaries that tend to keep science, environmentalism, and radical politics separate. Indeed, extinction cannot be understood in isolation from a critique of capitalism and imperialism" (15).

These differences matter if we hope to better understand how diverse people experience extinction and so how and why they might be willing and able to respond to it. To this end scholars have explored the diverse narratives we tell and the wider "cultural meanings of extinction" (Heise 2016, 2003; Huggan 2015). As a growing body of work in the humanities and social sciences is insisting, attention to this kind of human specificity is key to addressing environmental issues in ways that are not only sustainable but democratic, creative, and just (Neimanis, Åsberg, and Hedrén 2015; Rose et al. 2012). But the effort to acknowledge these "human dimensions" of extinction cannot be at the expense of the broader more than human aspects. A focus on the biocultural is about the rejection of this simplistic division of labor—between nature and culture, the "ecological" and the "human" dimensions—in favor of more complex stories that acknowledge the ways in which humans, other animals, and a diversity of other species are woven together in coforming multispecies communities. From this perspective, extinction takes the form of an unraveling, a breakdown of existing patterns of relationship. Whether it be the "ecological" loss of a pollinator or seed disperser or the "cultural" breakdown of a funerary system—in a world that is inherently relational, absences cannot help but bring about unravelings.

This focus on unravelings that cut across the human and the nonhuman has been central to recent humanities scholarship on extinction. The work of Deborah Bird Rose (2012) stands out as central here: she writes about the "great unmaking of life" (128), about the way in which mass killing can ripple out into multispecies communities to undermine the life-giving potential that death ordinarily has and produce what she calls "double death" (2006), and about the way in which the gifts of nourishment and meaning that are life, that flow between generations and across species, might become compromised and ultimately come apart (2012). In other work I have explored this space as the "dull edge of extinction": the slow unraveling of relationships and the breakdown of ways of life that happens well before and continues long after the death of the "last individual of a kind," a death that is so often taken to mark the precise moment of an extinction (van Dooren 2014). Instead of a fixation on the last individual, the so-called endling (De Vos 2007; Jørgensen 2017), this focus on unraveling positions extinction as an ongoing process. In this way each extinction must be understood to be radically unique; the unraveling of particular relationships and possibilities. Along these lines, Michelle Bastian (2012) has written about the cascading breakdown of marine ecosystems: as fish and many other species dwindle and jellyfish populations boom, that has a negative impact on some fishing communities and even on nuclear power station cooling intakes (which can become clogged with gelatinous bodies). As is clear in this work, unravelings are also reravelings; unmakings are remakings. The breakdown of existing relationships always creates new opportunities for others to thrive. The question is what is lost in the process, and on what grounds ought we to value and make a stand for some forms of shared life and not others. I will return to this topic again below.

Any sense of human life as radically distinct from a wider world breaks down in these analyses as our notions of "biology" and "culture" become fundamentally blurred. What is revealed here is that the cultures that we find ourselves "suspended" within—to borrow a lovely turn of phrase from Clifford Geertz (1973, 5)—are not exclusively human products but rather the working out of complex multispecies relations of meaning making, nourishment, and more. As a result, it becomes impossible to understand human societies and cultures in a vacuum, as has perhaps been most clearly articulated by some anthropologists (Shepard 1996; Rose 1992; Ingold 1990). But this relational notion of the constitution of human ways of life also means that we are vulnerable to extinction in particular ways. If who we are is constituted together, then as Judith Butler (2004) asks in a very different context: "Who 'am' I, without you? . . . I think I have lost 'you' only to discover that 'I' have gone missing as well" (22).

Having said all this, however, last individuals—like Lonesome George, Martha the Passenger Pigeon, and Benjamin the Thylacine—continue to be central to the way in which popular commentary represents extinction. And so, it makes sense that another prominent vein of recent scholarship on extinction has explored the uptake of these individuals as powerful sites of mourning and cultural expression. To this end, Rick de Vos (2007) has argued that "the last of its kind, or endling, constitutes a liminal figure, both real and ideal, singular in body and in time, holding together the notions of species and specimen" (183). In this way, these last individuals come to stand in for their whole species and sometimes the broader fact and possibility of extinction, in a way that lends persuasive power, but also creates important historical and ethical distortions (on this topic see Jørgensen 2017).

Ways of Life

Importantly, however, humanity is not at the center of everything that we might label "cultural." In fact, as Dominique Lestel argues, evolution invites us to appreciate the animal origins of all forms of cultural life (Lestel 2003). What we often call "culture" is not a uniquely human attribute added on to our biology but rather something grounded in particular, evolved, cognitive and emotional competencies that are distinctive and yet shared across species in important ways too. With this in mind, the unraveling that is extinction cannot be understood as a situation in which the "biological" relationships at stake are all those of nonhumans, with the "cultural" dimensions all relating to human life. Recent scholarship has insisted that we must understand extinction as, in addition to everything else, the loss of distinctive "animal cultures." More than *biodiversity*, in the reductive way that this term is often deployed—as a static inventory of genetic materials—evolved and evolving nonhuman ways of life are at stake in extinction: ways of mating and reproducing, of teaching young, of building nests, perhaps even of mourning the dead (Buchanan 2017; van Dooren 2013; Rose 2011; Hatley 2017; Chrulew 2017b). Vinciane Despret (2017) insists that "it is an entire world that has disappeared" with the passing of the passenger pigeon:

> The world has lost the taste of dry and fleshy fruits, of seeds and insects, the raindrops that slide off feathers, the air that dances and that shapes the paths of heat and density, the music in the throbbing murmur of thousands of wings applauding the flight. . . . All of this is no more. Humanity mourns the passenger pigeons. . . . But it is the world that bursts with its absence. (Despret 2017, 221)

Eileen Crist (2013) has also written powerfully about this loss of ways of life. She notes that in recent decades two "momentous realizations" have dawned on (some of) humanity: that we are losing species and destroying the environment at an incredible rate, and that we (in the West at least) have "tended to deny or underestimate the mental life of animals" (45). More than a mere occurrence alongside one another, Crist argues that there is a fundamental connection between the denial or denigration of animal mind and the destruction and impoverishment of environments, each reinforcing and enabling the other.

Of course, the breakdown of these ways of life does not happen in an instant. It, too, is part of the drawn out processes of unraveling that take place at the dull edge of extinction. In this way, the distinctive form of life that characterizes a species might also be lost well before the last individual breathes her last breath. Arguably, this was the case for the passenger pigeon: Martha, living on alone in the Cincinnati Zoo, unable to partake in any of the ways of being that Despret draws our attention toward (van Dooren 2014). The fact that ways of life do not necessarily travel along with the fleshy bodies that produced them informs another rich vein of recent extinction scholarship. This work explores the capacity for endangered species conservation to hold onto living bodies—to prevent extinction in this most literal of senses—but still to lose much that is unique and precious about a species. Matthew Chrulew (2017a) has closely explored the often diminished lives of animals in zoos and captive breeding facilities. He argues that these places are efforts to "secur[e] life against living itself," to "protect life from that which essentially characterises living: its embodiment, relationship to the world, to the environment, to other creatures, to human beings, the passage of time, reproduction via generations" (299). As a result of this diminishment of their worlds, many of these animals are unable to survive if released (Chrulew 2017b). But even if not taken into captivity, intensive conservation processes often threaten to reduce the "wildness" of animals, resulting in a range of efforts to limit such impacts (Reinert 2013). Alongside the brute question of survival, this situation raises a number of more philosophical questions about the goals of our efforts to stave off extinction: just how similar in behavior ought animals to be to their ancestors in order for "conservation" to have taken place? What if these behaviors are themselves part of what is placing the species at risk of extinction (van Dooren 2016)?

This focus on "ways of life" offers a kind of response to the ambiguity inherent in the notion of "species" mentioned at the outset. A way of life need not map neatly onto what taxonomists label as a species. Elsewhere, Deborah Bird Rose and I have used the notion of an ethos to think about what is at stake here (Rose and van Dooren 2016). This work has sought to question what will count as a distinct "way of life" for whom: how might valuable and disappearing forms of life cut across or lie within those groups that are ordinarily classified as "endangered species"? How might we pay attention to these kinds of losses too? In general, in contrast to the (fascinating, albeit highly technical) discussions of philosophers of biology that attempt to define what and whether a species is once and for all, work in Animal Studies and related fields has focused on the various ways in which species are being imagined, practiced, and enacted in the context of conservation and extinction. Eben Kirksey (2015) has documented some of the numerous ways that biologists make sense of species as well as the very real consequences for the species themselves, such as the *Leptolalax* frogs in Vietnam that, once identified and classified as endangered, might have appeared on the radar of people who would collect and illegally export rare animals for the pet trade. Meanwhile, the practice of "collecting" specimens for taxonomic research continues to contribute to the decline of some species (Kirksey 2015; De Vos 2017). In this research we also see that whether or not a group of organisms is deemed to be a species, or perhaps a subspecies or a hybrid, has very real consequences for conservation (Shrader-Frechette and McCoy 1994; Hinchliffe 2008) but also for the ways in which histories of "loss" are understood and narrated (De Vos 2017). In focusing on unique ways of life in this way, this approach might be understood as an effort to keep open the question of how and why we classify life while holding onto the pressing demand that we find some way to reckon with the contemporary extinguishing of myriad forms of being, insisting that in each instance something important, however uncertain and ephemeral, is being lost (Rose and van Dooren 2016).

Counterextinction Projects

While extinction has always been a more entangled and pervasive phenomenon than is often appreciated, in our current period this situation has been dramatically expanded. This is the case because of the scale at which both the loss of species and efforts to stem that loss are now taking place. All over the world, diverse "counter-extinction" projects (Chrulew 2017b) work to conserve endangered species and increasingly to bring back some of those already gone. In this way, the dull edge of extinction cannot be limited to the effects of an absence; it must also be understood to include the effects of the many and varied attempts to redress that absence. In taking up this topic, recent work on extinction has been required to engage with—and bring into conversation—two bodies of literature that have until recently remained quite separate. The first emerges primarily from animal ethics and is concerned with the impacts of conservation on all of those species caught up in it: from the "culling" of potential predators or competitors to the practices of captivity that sustain some endangered species. This is a large body of literature, much of it centered on the treatment of "introduced" species and the sacrificing of individual welfare for the good of the species or ecosystem (see Palmer 2010; Gruen 2011; Bekoff 2014). The second body of scholarship is primarily in anthropology, geography, and political ecology and has sought to understand how conservation remakes the world in the interests of some (human) communities and not others (Tsing 2005; West 2006; Lowe 2006). Here we see that contemporary extinction rhetorics—themselves emerging out of the modern environmental movement and the subsequent global uptake of "biodiversity conservation" as a central analytical and practical framework for these concerns and values—play a key role in animating, justifying, and funding the particular remaking of the world that we today call global conservation, with all of its consequences, good and ill. Bringing these two literatures into conversation, recent work on extinction has explored a broad space of competing understandings and ethical and political obligations (Thompson 2002; Heise 2016; van Dooren 2015; Kirksey 2012; Lorimer 2015; Münster 2014).

In addition to these efforts to hold onto disappearing species, scholarship in Animal Studies has increasingly addressed the growth of enthusiasm—or at least hype—surrounding efforts to resurrect (or "de-extinct") those that are already gone. This work has explored the ethical, imaginative, and ecological contexts of such projects with a particular focus on the way in which environments and human-animal relations are often (mis)conceived and appropriated by them. Support for de-extinction is frequently grounded in a similar logic to the one that reduces extinction to the death of the last of a kind. This logic is similar in two key ways: first, in its ability to condense the fate of a species into a single individual, but in place of the endling we have a (potential) progenitor, the beginning of a whole new line; and second, both of these fixations on individuals—last or first—demonstrate de-extinctionists' inability to think relationally, to think bioculturally, and so to grapple with the fact that an individual is not a species. A species is a way of life woven into a wider world, a situation that drastically increases the difficulties associated with "resurrection" (Turner 2007; Chrulew 2011; Jørgensen 2013; Friese 2013; van Dooren and Rose 2017).

Emerging out of this recent work in Animal Studies on extinction is a clear sense that there is no singular extinction phenomenon, no absolute answer to the question of what is wrong with extinction, or perhaps even to whether or not we ought always to work to prevent it. This situating of extinction allows us to set aside grand perspectives. While all life—including species—may ultimately be fleeting, there is no reason to believe that this "zoomed out," disembodied perspective has any special authority. Viewed from the position of eternity, most events in the history of this small planet seem inconsequential. But this vantage point tells us nothing about what matters for the mortal creatures—human and non—that happen to inhabit this planet at this time. It does not change the fact that each extinction unravels particular lives and possibilities. In each instance extinction seems to require us to ask what will this loss mean and for whom? What would be the costs of stemming it, and again, for whom? And so, ultimately, extinction asks us to consider what kind of relationships we want to cultivate in this place at this time.

Suggestions for Further Reading

Chrulew, Matthew, and Rick De Vos. 2018. "Extinction." In *The Edinburgh Companion to Animal Studies*, edited by Lynn Turner, Ron Broglio, and Undine Sellbach, 181–97. Edinburgh: Edinburgh University Press: 181–197.

Rose, Deborah Bird. 2011. *Wild Dog Dreaming: Love and Extinction*. Charlottesville: University of Virginia Press.

Rose, Deborah Bird, Thom van Dooren, and Matthew Chrulew. 2017. *Extinction Studies: Stories of Time, Death and Generations*. Edited by Deborah Bird Rose, Thom van Dooren, and Matthew Chrulew. New York: Columbia University Press.

Sodikoff, Genese Marie. 2012. *The Anthropology of Extinction: Essays on Culture and Species Death*. Bloomington: Indiana University Press.

van Dooren, Thom. 2014. *Flight Ways: Life and Loss at the Edge of Extinction*. New York: Columbia University Press.

References

Barnosky, Anthony D., Nicholas Matzke, Susumu Tomiya, Guinevere O. U. Wogan, Brian Swartz, Tiago B. Quental, Charles Marshall, et al. 2011. "Has the Earth's Sixth Mass Extinction Already Arrived?" *Nature* 471: 51–57.

Bastian, Michelle. 2012. "Fatally Confused: Telling the Time in the Midst of Ecological Crises." *Environmental Philosophy* 9 (1): 23–48.

Bekoff, Marc. 2014. *Rewilding Our Hearts: Building Pathways of Compassion and Coexistence*. Novato, CA: New World Library.

Benton, Michael J. 1986. "The Evolutionary Significance of Mass Extinctions." *Trends in Ecology and Evolution* 1 (5): 127–30.

Bowker, Geoffrey C., and Susan Leigh Star. 1999. *Sorting Things Out: Classification and Its Consequences.* Cambridge, MA: MIT Press.

Buchanan, Brett. 2017. "Bear Down: Resilience in Multispecies Cohabitation." In *The Routledge Companion to the Environmental Humanities,* edited by Ursula K. Heise, John Christensen, and Michelle Niemann, 289–98. London: Routledge.

Butler, Judith. 2004. *Precarious Life: The Powers of Mourning and Violence.* London: Verso.

Chrulew, Matthew. 2011. "Reversing Extinction: Restoration and Ressurection in the Pleistocene Rewilding Projects." *Humanimalia* 2 (2): 4–27.

———. 2017a. "Freezing the Ark: The Cryopolitics of Endangered Species Preservation." In *Cryopolitics: Frozen Life in a Melting World,* edited by Joanna Radin and Emma Kowal, 283–306. Cambridge, MA: MIT Press.

———. 2017b. "Saving the Golden Lion Tamarin." In *Extinction Studies: Stories of Time, Death and Generations,* edited by Deborah Bird Rose, Thom van Dooren, and Matthew Chrulew, 49–87. New York: Columbia University Press.

Crist, Eileen. 2013. "Ecocide and the Extinction of Animal Minds." In *Ignoring Nature No More: The Case for Compassionate Conservation,* edited by Marc Bekoff, 45–62. Chicago: University of Chicago Press.

Dawson, Ashley. 2016. *Extinction: A Radical History.* New York: OR Books.

Despret, Vinciane. 2017. "'It Is an Entire World That Has Disappeared.'" In *Extinction Studies: Stories of Time, Death and Generations,* edited by Deborah Bird Rose, Thom van Dooren, and Matthew Chrulew, 217–22. New York: Columbia University Press.

De Vos, Rick. 2007. "Extinction Stories: Performing Absence(s)." In *Knowing Animals,* edited by Laurence Simmons and Philip Armstrong, 183–95. Leiden: Brill.

———. 2017. "Extinction in a Distant Land: The Question of Elliot's Bird of Paradise." In *Extinction Studies: Stories of Time, Death and Generations,* edited by Deborah Bird Rose, Thom van Dooren, and Matthew Chrulew, 89–115. New York: Columbia University Press.

Friese, Carrie. 2013. *Cloning Wild Life: Zoos, Captivity, and the Future of Endangered Animals.* New York: New York University Press.

Geertz, Clifford. 1973. "Thick Description: Toward an Interpretive Theory of Culture." In *The Interpretation of Cultures,* 3–30. New York: Basic Books.

Gruen, Lori. 2011. *Ethics and Animals.* Cambridge: Cambridge University Press.

Hatley, James. 2017. "Walking with Ōkami, the Large-Mouthed Pure God." In *Extinction Studies: Stories of Time, Death and Generations,* edited by Deborah Bird Rose, Thom van Dooren, and Matthew Chrulew, 19–47. New York: Columbia University Press.

Heise, Ursula K. 2003. "From Extinction to Electronics: Dead Frogs, Live Dinosaurs, and Electric Sheep." In *Zoontologies: The Question of the Animal,* edited by Cary Wolfe, 59–82. Mineapolis: University of Minnesota Press.

———. 2016. *Imagining Extinction: The Cultural Meanings of Endangered Species.* Chicago: University of Chicago Press.

Hinchliffe, Steven. 2008. "Reconstituting Nature Conservation: Towards a Careful Political Ecology." *Geoforum* 39: 88–97.

Huggan, Graham. 2015. "Last Whales: Eschatology, Extinction and the Cetacean Imaginary in Winton and Pash." *Journal of Commonwealth Literature* 52 (2): 382–96.

Ingold, Tim. 1990. "An Anthropologist Looks at Biology." *Man*, n.s., 25 (2): 208–29.

———. 2013. "Anthropology beyond Humanity." *Suomen Anthropologi* 38 (3): 5–23.

Jørgensen, Dolly. 2013. "Reintroduction and De-extinction." *Bioscience* 63 (9): 719–20.

———. 2017. "Endling, the Mystique of the Last in an Extinction-Prone World." *Environmental Philosophy* 14 (1): 119–38.

Kirksey, Eben. 2012. "Living with Parasites in Palo Verde National Park." *Environmental Humanities* 1: 23–55.

———. 2015. "Species: A Praxiographic Study." *Journal of the Royal Anthropological Institute* 21 (4): 758–80.

Lestel, Dominique. 2003. *Les origines animales de la culture*. Paris: Flammarion.

———. 2013. "The Withering of Shared Life through the Loss of Biodiversity." *Social Science Information* 52 (2): 307–25.

Lorimer, Jamie. 2015. *Wildlife in the Anthropocene: Conservation after Nature*. Mineapolis: University of Minnesota Press.

Lowe, Celia. 2006. *Wild Profusion: Biodiversity Conservation in an Indonesian Archipelago*. Princeton, NJ: Princeton University Press.

Mayr, Ernst. 1996. "What Is a Species, and What Is Not?" *Philosophy of Science* 63: 262–77.

Mitchell, Audra. 2016. "Indigenous Visions of the Global Extinction Crisis."*Worldly*, May 27. https://worldlyir.wordpress.com/2016/05/27/indigenous-visions-of-the-global-extinction-crisis/.

Morton, Timothy. 2013. *Realist Magic: Objects, Ontology, Causality*. Ann Arbor, MI: Open Humanities Press.

Münster, Ursula. 2014. "Working for the Forest: The Ambivalent Intimacies of Human-Elephant Collaboration in South Indian Wildlife Conservation." *Ethnos: Journal of Anthropology* 81 (3): 425–47.

Neimanis, Astrida, Cecilia Åsberg, and Johan Hedrén. 2015. "Four Problems, Four Directions for Environmental Humanities: Toward Critical Posthumanities for the Anthropocene." *Ethics and the Environment* 20 (1): 67–97.

Noske, Barbara. 1989. *Humans and Other Animals: Beyond the Boundaries of Anthropology*. London: Pluto.

Palmer, Clare. 2010. *Animal Ethics in Context*. New York: Columbia University Press.

Reinert, Hugo. 2013. "The Care of Migrants: Telemetry and the Fragile Wild." *Environmental Humanities* 3 (1): 1–24.

Rose, Deborah Bird. 1992. *Dingo Makes Us Human: Life and Land in an Aboriginal Australian Culture*. Cambridge: Cambridge University Press.

———. 2006. "What If the Angel of History Were a Dog?" *Cultural Studies Review* 12 (1): 67–78.

———. 2011. "Flying Fox: Kin, Keystone, Kontaminant." *Australian Humanities Review* 50: 119–36.

———. 2012. "Multispecies Knots of Ethical Time." *Environmental Philosophy* 9 (1): 127–40.

Rose, Deborah Bird, and Thom van Dooren. 2016. "Encountering a More-than-Human World: Ethos and the Arts of Witness." In *The Routledge Companion to the Environmental Humanities*, edited by Ursula K. Heise, Jon Cristensen, and Michelle Niemann, 120–28. London: Routledge.

Rose, Deborah Bird, Thom van Dooren, Matthew Chrulew, Stuart Cooke, Matthew Kearnes, and Emily O'Gorman. 2012. "Thinking Through the Environment, Unsettling the Humanities." *Environmental Humanities* 1: 1–5.

Ruse, Michael. 1992. "Biological Species: Natural Kinds, Individuals, or What?" In *The Units of Evolution: Essays on the Nature of Species*, edited by Marc Ereshefsky, 343–62. Cambridge, MA: MIT Press.

Ryan, John. 2013. "Botanical Memory: Exploring Emotional Recollections of Native Flora in the Southwest of Western Australia." *Emotion, Space and Society* 8: 27–38.

Shepard, Paul. 1996. *The Others: How Animals Made Us Human*. Washington DC: Island.

Shrader-Frechette, K. S., and E. D. McCoy. 1994. "Biodiversity, Biological Uncertainty, and Setting Conservation Priorities." *Biology and Philosophy* 9: 167–95.

Simpson, George Gaylord. 1953. *The Major Features of Evolution*. New York: Columbia University Press.

Sodikoff, Genese Marie. 2012. *The Anthropology of Extinction: Essays on Culture and Species Death*. Bloomington: Indiana University Press.

Thompson, Charis. 2002. "When Elephants Stand for Competing Philosophies of Nature: Amboseli National Park, Kenya." In *Complexities: Social Studies of Knowledge Practices*, edited by John Law and Annemarie Mol, 166–90. Durham, NC: Duke University Press.

Tsing, Anna Lowenhaupt. 2005. *Friction: An Ethnography of Global Connection*. Princeton, NJ: Princeton University Press.

Turner, Stephanie S. 2007. "Open-Ended Stories: Extinction Narratives in Genome Time." *Literature and Medicine* 26 (1): 55–82.

van Dooren, Thom. 2013. "Mourning Crows: Grief and Extinction in a Shared World." In *Routledge Handbook of Human-Animal Studies*, edited by Garry Marvin and Susan McHugh, 275–89. London: Routledge.

———. 2014. *Flight Ways: Life and Loss at the Edge of Extinction*. New York: Columbia University Press.

———. 2015. "A Day with Crows: Rarity, Nativity and the Violent-Care of Conservation." *Animal Studies Journal* 4 (2): 1–28.

———. 2016. "Authentic Crows: Identity, Captivity and Emergent Forms of Life." *Theory, Culture and Society* 33 (2): 29–52.

van Dooren, Thom, and Deborah Bird Rose. 2017. "Keeping Faith with the Dead: Mourning and De-extinction." *Australian Zoologist* 38 (3): 375–78.

West, Paige. 2006. *Conservation Is Our Government Now: The Politics of Ecology in Papua New Guinea*. Durham, NC: Duke University Press.

12 KINSHIP

Agustín Fuentes and Natalie Porter

There is nothing more important than kin. Kinship defines our place in the world and our relationships with those around us. Kinship is the core of evolutionary biology and lies at the center of human societies. In biological, historical, and social realms, making kin, being kin, and aiding kin are the fundamental features of life.

As an object of analysis, kinship has long provided an entry point for viewing the diversity of human social organization across time and space, and as a concept, kinship has animated theoretical debates about nature-culture, self-other divisions and categories of being. Historically, anthropologists have posited a distinction between biological and social kinship and have interrogated the extent to which kinship is shaped both by a pregiven "natural" order as well as by human engagement (Schneider 1980). The writer Armistead Maupin tells us that we live in a world filled with both biological kin and logical kin. He refers to those kinship relations that we are born into as biological and those that we choose, construct, and nourish as logical.

More recently, however, the analytical distinctions that drive kinship studies have been unsettled by anthropological accounts that expose the simultaneous configuration of biological and social, natural and cultural kin relations. Humans make kin in part via biology but more profusely via creativity and comingling, processes whose effects are felt within and beyond biology. Here we refuse to limit ourselves to conceptual distinctions between biological and social kin and instead expand the reach of biologies and logics of relations to illustrate why and how kinship is a multispecies process. By exploring different forms of selective bonding across species divisions, we intend not only to interrogate traditional biological kinship but also to problematize biological-logical (natural-cultural) binaries in social analysis (Sahlins 2013). Instead of offering one definition—*kinship* is culture and not biology, or *kinship* is both biological and logical—we suggest that it is more fruitful to explore how, within particular multispecies relations, kinship troubles and transcends these analytical distinctions.

If *kin* are those closest to us in space, time, and flesh, then *kinship*, by definition, is a multispecies endeavor. In this essay we will illustrate the utility of "kinship" as a concept for navigating multispecies relationships, therein making the term *kinship* relevant for more than human life forms. We will demonstrate that humans and others are entangled in a myriad of kinship bonds and will offer a context and set of perspectives to place kinship at the center of investigations in Animal Studies.

More than Human Kinship

In the evolutionary context, replication is the name of the game. Delivering your particular genetic content into the next generation is considered the most basic way in which evolutionary success is achieved. Natural selection via ecological pressures, the contest between organisms and their environment, sets up a filter through which only organisms with particular clusters of traits, or certain variants, reproduce with regularity or ease. Over time those variants that do better in a given environment become more common in subsequent populations while those that did less well through the filter dwindle. These more successful variants (or rather, clusters of variants) all share certain heritable biological traits tying them to one another: they are kin. So kinship, in the most basic evolutionary sense, is the fact of shared commonalities between these successful biological clusters.

In humans these clusters, or kin, are made via biological amalgamations, melding bodies, or parts thereof, creating genetic legacies via inheritances and distributions of shared strings of DNA. We classify these kinds of kin as "blood" kin, flesh of our flesh. And for many, this is the basic human biological definition: kinship is the result of shared biological legacies. Humans produce other humans through biological exchanges and inheritances, which create the evolutionary kin noted above.

But even in this most basic definition of *kin*, humans are never alone; our biological clusters are multispecies communities. Human biological reproduction does not simply produce other humans; it reproduces whole communities of nonhumans. Human bodies are seething sites of multispecies kinship networks. *Demodex* mites, *Candida* fungi, and a horde of bacteria move from one human body to the next as soon as that next body emerges from the first, some even before that! Whole communities of commensal flora and fauna infuse the human being and are biologically passed from parent to offspring, replicated and shared, inherited and spread. The microbiomes of skin, gut, mouth, and genitals are part and parcel of the inheritances we share, the clusters we pass from generation to generation. That most iconic image of kin making, a mother nursing her infant, is the perfect exemplar of this process. In nursing, the breast milk passes not just sugars and fats, nutrition to grow kin, but a plethora of microbes, bacteria, and living communities, the components to make kin. This biological entanglement between mother and child facilitates the creation and (re)-creation of the multispecies kinship community that we call human.

Kinship is more than biology, more than culture, and more than species specific, and thus it is not necessarily bound to single bodies or lineages. As Donna Haraway explains, making kin entails bioculturally, biotechnically, biopolitically, and historically situated people acting in relation to and combining with other species assemblages as well as biotic and abiotic forces (Haraway 2015). Given that kinship is never a simple or single process or pattern, thinking with and about the term *kinship* forces us to recognize that it is a multivalent landscape. Kin are relatives, but relative to what and to whom? How relations are formed and the processes and embodied products of those relations become the central locus for our considerations.

The term *kinship* itself has power in academic and popular usage because of its tight association with biological inheritance, lineage unity, and its role as a central organizing factor in human social systems. However, when we utilize the term in the context of multispecies relationships, we deprioritize the exclusive nature of human kinship and redefine the core of relationships as one of humans and others becoming together. Such usages enable new kinds of relations to emerge from alliances and symbiotic attachments expanding beyond relationships structured by patrilineal descent or filiation. This is kinship in its rhizomatic form: multispecies relationships grow out without clear foundations or borders, start or endpoints (Deleuze and Guattari 1987).

Approaching kinship as a multispecies process moves the concept beyond the enactment of inheritance and into the realm of potentialities of exchange and the construction of relations. This is what Stefan Helmreich

(2003) identifies as a kinship organized less around practices of sex and more around a biopolitics of transfer. Material, visceral, bodily exchanges become patterns of association. Reciprocity, touching, and infection are pathways to familiarity. This view of kinship has implications for the articulation of similarities and differences as we navigate the myriad of relationships between humans and others in the world. In the following sections we illustrate how such kinships are, were, and might be construed.

Primordial Kin: The Case for Viruses

Viruses are arguably humans' most primordial partners. Viruses form part of the human metagenome, and within our bodies their numbers exceed both human and bacterial cells alike (Lowe 2014). Virus remnants in the human genome act as evolutionary agents guiding the mutation of human DNA and ultimately scripting life processes (MacPhail 2004). The viral gene, syncytin, for instance, codes for proteins in the placenta that transmit nutrients from mother to fetus. Through these and other life-giving acts, viruses have enabled the evolutionary emergence and continuation of humankind (Lowe 2014).

Here we depart from usual understandings of viral infection as a foreign agent invading its host and instead narrate the same relational process as a host inviting a virus into the fold of its relations: making kin. For what are virus attachment sites if not loci of recognition, accommodation, and *familiarity* between kith?

Viruses, we know, need hosts to survive. But not just any host. Viruses are discriminating entities; they create alliances with different species in accordance with biological and ecological conditions. Influenza viruses, for instance, bind to acid receptors on host cells that vary from species to species. Such species-specific compatibilities are expressed in the classification of kinds: avian flu, swine flu, and human flu, for instance. Further, what constitutes a host one season might not in the next. As they interact with each other and their environments, animals continually generate antibodies that discourage connections with particular viruses and encourage connections with others. Host-virus relations are processual and contextual; they rely on entities reading and responding to one another over time through selective bonding.

We also know that viruses cannot reproduce without a host. Viral reproduction is asexual, which means that the host does not contribute genetic material to the virus "offspring." But it is somewhat disingenuous to characterize virus reproduction as replication since many viruses constantly mutate through interactions with their hosts' antibodies, or acquired immunities. Phylogenetic analyses of virus evolution capture how hosts leave genetic marks on viruses in mutations that allow viruses to exploit different cellular environments, expand their relations across populations, and further drive their own evolution. Hosts and their off-spring, too, change in the process of viral infection and evolution. In addition to causing disease, virus-host exchanges leave traces that can then be passed on through the host's reproductive practices as inherited infections and immunities. HIV viruses leave their genetic material in host cells, and Ebola can propagate through semen. For both virus and host, existence and longevity is a multispecies relationship.

This relationship is not always a wholly antagonistic one. Replacing the metaphor of invasion with one of invitation and coalescence allows us to imagine other forms and practices of kinship. There are many ways of making kin, choice, love, and care being just a few. Though arguably the epitome of the unloved and unwanted other, anyone who has been vaccinated has chosen to meet and mingle with viruses (Greenhough 2012). Creating alliances with infectious agents boosts a host's immune system and allows it to form more discerning associations with other viruses and virus hosts. This is a process of familiarization and accommodation, selective incorporation and distinction that shape our everyday experience: colds, flus, and other mundane forms of companionability between kinds. Creating kinships with viruses affects our biology and our relations with other, nonviral, kin as well.

As emerging and reemerging viral infections become more and more common, the task of rethinking our relations with the often unloved others that live within and among us becomes more and more urgent (Kirksey 2014). Viruses cannot survive or reproduce without a host, but neither can humans survive or flourish without viruses or the myriad other creatures that occupy our bodies and drive our development. Viruses blur the biological-social boundary; they both incite and illustrate the cascading connections that shape life in multispecies communities. Let's suspend notions about the integrity or autonomy of bodies and species and instead appreciate the fact that none of us exists alone. We might start by more fully embracing our bonds (kinship?) with viruses and by thinking critically about how we might engage in practices of choice, care, and responsibility that better reflect our shared biological inheritances and histories and that enact more promising collective futures and potentials.

Biocultural Kinships

Multispecies kinship does not stop with the biological transmission, connections, and mutualities of the viral-human interface; it is also a core component of human cultural and ecological realities. Humans create kinship with other species though a myriad of methods, manners, and imaginings. The creation of kinship involves the reshaping of perceptions and behaviors (and yes, bodies) and is even more diverse, entangled, and generative than the replication and dissemination of those microbiota living in and around us. Dogs, chickens, and monkeys offer three excellent examples.

Kinship with Canines

More than 12,400 years ago at the site of Uyun al-Hammam (today Jordan), a group of humans dug a grave and laid two bodies into it, one on top of the other. One body was that of an adult woman, and the other was an adult red fox. They added a few tools and some red pigment, then covered the bodies (Maher et al. 2011). At about the same time at the Natufian site of Ain Mallaha (today Israel), another group stood in a grove of oaks and pistachio trees and laid a young woman to rest in a shallow grave, placing a deceased puppy next to her head. They laid her hand on the puppy before covering them both and sealing the grave (Davis and Valla 1978). The domestication of humans and dogs is the story of creating kinship and was driven by both species.

As much as 25,000–30,000 years ago there were two species of mammals spreading across the northern hemisphere: gray wolves and humans. Both were highly social animals with complex, interconnected social and ecological lives. Both had a strong sense of loyalty to their social groups, communal care of their young, and keen hunting abilities. Humans had big brains, thumbs, tools, fire, language, and in some places (such as the Levant and other parts of central Eurasia) they had figured out how to live more or less in one place. There were numerous times when human communities and wolf packs came into conflict, especially over kills. When humans moved into an area, the wolves would have noticed that a new top predator was in town and that they'd be second. Wolves started following human hunting groups seeking to scavenge as best they could, sometimes maybe even challenging the humans (and eating one or two) if they caught a few of them alone. But most of the time the wolves avoided the sharp spears and arrows and fire that the humans wielded (Shipman 2011). In some areas such encounters became more common, and in a few instances the relations between the two species began to change.

The longer the packs hung around humans, the more likely it was that a few, or all, of a pack started to stick close to the humans for easy food, garbage and waste, kill scraps, and the protection of the nightly human fire. The humans would have noticed this and initially driven the wolves away with weapons and shouts, but over time the wolves remained persistent, just out of reach but always nearby. Changing their hunting patterns, the wolves began to follow the humans as they moved about or stuck around the camps if they settled in for a season or two. Slowly, over many generations of wolves and humans, the wolves' changes affected the humans, who began to tolerate their presence. These humans noted that there were a few benefits to having the wolves around. They made noises if other predators or large animals approached the camps at night; they also shadowed the humans when they hunted, sometimes driving out smaller prey that the humans could capture and eat as they moved on to find larger game. Finally, there were times when pups or young wolves ended up in human camps, abandoned or injured, and on occasion the humans tended them rather than killing them. And the kinship began. Humans and wolves started to rely on one another as safety nets, as hunting partners, and as friends, as kin (Olmert 2009; Shipman 2011).

The wolves who grew up in the human communities started to change as human creativity got into the mix. Humans noticed differences in the personalities of pups and the eventual adults. These personality variants made a difference in how well the wolves got along with the human community, how adept they were at communicating with the humans and following human cues, and even how they interacted with the human children. Some showed more attachment to humans and more responsiveness to human signals and were less likely to fight with humans over kills—even bonding tightly with certain men, women, or children (kin?), following them around and almost keeping watch over them. These wolves were transferring their wolf-pack allegiance, their recognition of kinship, to the humans, and the humans offered theirs in return. Before long humans recognized that mutual interactions, behaving as family, with pups right away produced the best companions. As humans began selectively to spend time with the best and most human-friendly pups, the behavior and bodies of the wolves changed, they became dogs, and together we created kin. By approximately 15,000–20,000 years ago we can find bones (Larsen and Fuller 2014; Zedder 2012) that are wolflike but show signs of domestication—being smaller, less angular, and a bit more puppylike—in association with human archeological remains. It is likely that this is the first indication that wolves (*Canis lupus*) had changed enough to be called dogs (*Canis familiaris*). And they were shaping us at the same time.

Recent research demonstrates that dogs have tapped into human hormonal and psychoneuroendocrine pathways and co-opted the human capacity for communally caring for young and each other. As humans were shaping dogs, those dogs also shaped the human responses to them: wolfs/dogs inserted themselves into human communities and began to elicit the same kinds of compassionate and physiological responses that human community members elicited from their human kin. Work by multiple research teams demonstrates that dogs tap into the human oxytocin response system and have shaped it to respond to their actions. As humans mastered the dog, the dog also mastered a bit of us. The relationship flourished. Some have even suggested that it was this relationship that made our direct human ancestors better hunters and maybe gave us the extra social and ecological support we needed to succeed across all the habitats that we encountered. Other groups of humans who did not contribute extensively to the modern human lineage (such as the Neanderthals) never entered into this kinship relationship, and they are extinct (Shipman 2011). One could argue that the establishment of kinship with dogs set the stage for the dramatic expansion of human kinship to a multitude of other species through domestication processes.

Kinship with Chickens

The chicken rivals the dog in the multiplicity of its entanglements with human communities: there are more chickens in the world than any other species of bird or domestic animal. The origins of the domestic fowl (*Gallus domesticus*) are difficult to pinpoint in time and space (Kanginakudru et al. 2008), but archaeological evidence suggests that chicken domestication began with the red jungle fowl (*Gallus gallus*) somewhere in Asia around seven thousand years ago.[1]

There was much to recommend *Gallus gallus* for incorporation into

---

1. Three other species of jungle fowl have been shown to contribute DNA to *Gallus domesticus*: the Java, the Ceylon, and the jungle grey (Smith and Daniel 2000). There is also evidence for indigenous South American chickens, or at least pre-Columbian chickens brought over by Polynesian sailors or Egyptians (Kockelman 2016).

human communities. The birds are striking. The males' heads are painted with streaks of red wattles and combs; their feathered bodies, legs, and tails flicker a brilliant green blue; and their voices carry in early morning calls. The males also strut and pose, and they use their foot spurs to jump and strike, angling their airborne bodies against those who would steal their mates. Such an intrepid spirit appealed to humans, and the cockfight became one of the earliest spectator sports in history. Though more muted, females, too, had much to offer. Humans learned that removing females from tree nests could induce them to lay additional eggs. Soon enough, people were capturing and raising birds for protein and play. This was a relatively easy process. Red jungle fowl have heavy bodies and short wing-spans that compromise their ability to fly or range far distances. Such were the biosocial conditions for creating fowl familiars.

The process of making chicken familiars has entailed multifaceted attunements to animal life. The Romans began selecting different breeds for eggs, flesh, and sport and experimented with methods to fatten the creatures and encourage egg production. Drawing both on observational evidence and philosophical maxims, they built egg nests, henhouses, and feeding troughs that catered to chicken comportment (Squier 2012). Ancient Egyptians relied on their sensory flesh to calibrate the temperature in clay brick enclosures, hatching thousands of chicks to feed complex urban societies with great labor forces. Ritual and symbolic processes have also brought humans and chickens into often-otherworldly relation. Ancient Greeks offered chickens to Aesculapius, the god of medicine, and maintained them in the temples of Hercules and Hebe. The name of the last sovereign Inca ruler, Atahualpa, is the Quechua name for chicken. Chicken oracles and auguries utilized by the hill tribes of Asia found their match in Greece and Rome and in the famed poison oracles of African Nuer (Evans-Pritchard 1940). Poultry ontologies among Quechan-speaking Maya include affine hens and consanguine cocks (Kockelman 2016).

Chickens' potency, biological plasticity, short developmental cycle, and fecundity have made them available to a degree of manipulation unrivaled among domestic animals. The majority of chickens today are bred to be fleshy, proportional, and fast-growing creatures amenable to large-scale, high-volume, and standardized industrial production. These patterns of cohabitation and coevolution have further transformed both species. Those of us implicated in industrial farming systems tend to eat too many animals; our food is overly processed, overly preserved, overly medicated, and overly contaminated. Foodborne outbreaks are on the rise, along with overweight and obesity rates linked to heart disease, cancer, stress, and diabetes. Humans who labor alongside chickens in industrial ecologies face further risks, working long hours for indecent wages and under perilous conditions without recourse to basic health care, all to create chicken defined more by their biological productivity and commercial function than by their lineage within a broader family of fowl (Horowitz 2004).

The fact that many chickens today are often unable to stand, walk, or flap their wings has prompted some to argue that industrial chickens are no longer chickens (Smith and Daniel 2000), that the process of making chicken "products" has rendered humans and chickens strangers to ourselves and each other. Chickens, however, will not let us cut our ties so easily. The evolution of zoonoses such as the highly pathogenic avian flu reveals that as humans evolve alongside chickens, we also generate commensal viral communities that thrive in our shared spaces and substances. However unrecognizable or unfamiliar, we are tied to chickens in a mutual struggle to survive the contagions we have cocreated.

But lest we understand these developments as a story of interspecies competition and survival of the fittest, we should remember the lessons that chickens have taught us about evolution. Since Aristotle, researchers have turned to the chicken embryo as a means to understand how organisms unfold themselves in conversation with the actors and entities around them. It was chicken embryo experiments that led C. H. Waddington to propose the field of epigenetics, and developmental biologists today continue to draw on chicken embryology to posit an evolution based on collaboration rather than competition and a cooperative genome guided by interactive, *un*selfish genes (Squier 2012; Weiss and Buchanan 2009). Viewing human-chicken bonds from the perspective of the chicken embryo should discourage estrangement from our fowl familiars and instead encourage us to recognize that chickens, humans, and viruses inhabit a generative world of entangled flesh and shared substance. Control, containment, culling, and competition may be fools errands in these mutual ecologies. (Selective) cooperation and tolerance, (discerning) affinities and attachments—these are the ties that bind and sustain us.

Kinship with Monkeys

Humans are primates, and thus we share a very specific biological kinship with other primates: phylogeny. Phylogeny is the evolutionary relationships between groups of organisms. But the story doesn't end there. In many human groups, relationships with certain other primates represent deep connections, but not in simple or linear terms. There is fluidity, biological and cultural, in these relations that include other primates as pets, food, sacred figures, and persons. Often the nonhuman primates can occupy more than one of these roles in relation to humans, even roles that appear contradictory in nature (e.g., food-persons, sacred figures-pests). Even the briefest survey of the variety of human-other primate interfaces illustrates our assertions about the multispecies reality of kinship relations.

For the Guajá of Brazil, monkeys are simultaneously kin and food (Cormier 2003). In the world of the Guajá, monkeys play social, religious, and nutritional roles. Here both monkeys and humans share aspects that constitute personhood. For the Guajá, endocannibalism is a central aspect of their worldview and spiritual practice and is achieved via consuming their kin, howler monkeys. For the Guajá, both primates share more than biology in common; they are siblings of the same origin and thus partner in the ritual cycle of spirit and body mutual consumption.[2] Multiple species of monkeys are hunted by the Guajá. Often hunters kill a mother, and if she has a young infant, they keep the infant alive and bring her into their social group. Orphaned monkeys enter the Guajá formal kinship system and are often given names, breastfed, treated as child-kin of the humans, and raised with their human siblings. Social status for mothers can be tied to the total number of offspring, human and monkey, she successfully nurtures.

Another example of the myriad of kinship relations between humans and primates are those of the macaque-human interfaces across Asia. Ranging from prey to pests to pathogen sharers to central characters in mythical narratives and economic processes, macaques weave in and out of kinships with humans. Macaques are hunted as food across Asia, but equally (or even more so) they constitute neighbors sharing space, food, and water with humans at Hindu and Buddhist temples across South and Southeast Asia. Increasingly, macaques and humans are messmates in urban landscapes from Delhi to Singapore to Jakarta and a myriad of cities and towns in between. Macaque populations in urban settings may today

2. While in this world the monkeys do not consume humans, in the spirit and past worlds the relations can reverse and shift; thus, the Guajá see this relationship as involving mutual consumption.

outweigh the numbers of macaques in forested environments. In such contexts, humans and their macaque kin do as kin often do: they conflict, fight, and consider one another a nuisance. But at the same time, humans offer food to the monkeys, and the monkeys serve as sources of income for the humans via coconut picking, performances, and tourist attractions (Fuentes 2012).

Macaques play central roles in the Ramayana, the Journey to the West, and a number of other central narratives that construct human ideologies, practices and beliefs. Human constructed and shaped landscapes structure the ecology, behavior, and even genetic patterns of macaque populations. Macaques and humans exchange pathogens on a regular basis, ranging from malarias to viruses to gastrointestinal parasites. Recent work demonstrates that in Southeast Asia (and probably many other locales) the macaque-human relationship entails a very fluid co-biology, a three-way kinship between two primates and a range of viral participants (Jones-Engel et al. 2008). Mutual influences shaping one another's behavior, biology, and perceptions, macaques and humans cocreate each other as kin again and again.

Expansive Kinship

In 2003, Janet Carsten published *After Kinship*, a cross-cultural comparison that unsettled received notions of kinship and refused to circumscribe human relations to strictly biological or cultural domains. Indebted to feminist cultural analysis, this provocation continues to be a fruitful one in anthropology, particularly in studies of new reproductive technologies and gender relations that explore the mutual configuration of human biologies and sociality. Here we have extended the comparison across species and posited a more expansive kinship that reflects the myriad of relationships, patterns, and processes that occur between humans and others on the planet. The multispecies entanglements described here reveal nonhuman entities as humans' primordial progenitors, contemporary companions, eco- and cosmological cohabitants, pathogen sharers, and ritual partners. Kin.

The term *kinship* is critical for multispecies analyses. Animal Studies needs an expansive approach to kinship in order to best reflect the myriad of relationships, patterns, and processes between humans and others on the planet. In bringing the concept of kinship to the multispecies arena, we have also enriched understandings of kinship in social analysis. Our aim is not to argue that kinship is more cultural (or logical) than biological or vice versa; rather, we have offered a new avenue for challenging an analytical tradition that holds these conceptual categories apart. We have shown that in addition to novel technologies, scholars can also look to much more sedimented (yet still dynamic) relationships between beings as a means to destabilize nature-culture and bio-social dichotomies. Considering how humans remake bodies, selves, and relations *with and alongside* nonhuman familiars prompts a conceptualization of kinship, and multispecies relations, that recognizes how culture and biology inflect on and merge into one another without discrete patterns of cause and effect. Asking whether nature follows from culture (or vice versa) or whether kinship is biological or social, human or other, is rather like asking whether the chicken preceded the egg. The more pressing task at hand is to illuminate the multiple processes and outcomes of kin relations that challenge staid categories of life and that generate new possibilities for living, being, and becoming with (Haraway 2007).

Suggestions for Further Reading

Fuentes, A., and L. D. Wolfe. 2002. *Primates Face to Face: The Conservation Implications of Human and Nonhuman Primate Interconnections.* Cambridge: Cambridge University Press.

Haraway, D. 2003. *The Companion Species Manifesto: Dogs, People, and Significant Otherness.* Chicago: Prickly Paradigm.

Kirksey, E., and S. Helmreich. 2010. "The Emergence of Multispecies Ethnography." Special issue, *Cultural Anthropology* 25 (4).

References

Carsten, J. 2003. *After Kinship.* Cambridge: Cambridge University Press.

Cormier, L. 2003. *Kinship with Monkeys: The Guajá Foragers of Eastern Amazonia.* New York: Columbia University Press.

Davis, S. J. M., and F. Valla. 1978. Evidence for Domestication of the Dog 12,000 Years Ago in the Natufian of Israel. *Nature* 276: 608–10.

Deleuze, G., and F. Guattari. 1987. "1730: Becoming-Intense, Becoming-Animal, Becoming-Imperceptible." In *A Thousand Plateaus: Capitalism and Schizophrenia,* translated by Brian Massumi, 232–309 (Minneapolis: University of Minnesota Press, 1987).

Evans-Pritchard, E. E. 1940. *The Nuer.* Oxford: Oxford University Press.

Fuentes, A. 2012. "Ethnoprimatology and the Anthropology of the Human-Primate Interface." *Annual Review of Anthropology* 41:101–17.

Greenhough, B. 2012. "Where Species Meet and Mingle: Endemic Human-Virus Relations, Embodied Communication and More-than-Human Agency at the Common Cold Unit 1946–90." *Cultural Geographies* 19 (3): 281–301.

Haraway, D. 2007. *When Species Meet.* Minneapolis: University of Minnesota Press.

———. 2015. "Anthropocene, Capitalocene, Plantationocene, Chthulucene: Making Kin." Environmental Humanities 6: 159–65.

Helmreich, S. 2003. "Trees and Seas of Information: Alien Kinship and the Biopolitics of Gene Transfer in Marine Biology and Biotechnology." *American Ethnologist* 30 (3): 340–58.

Horowitz, R. 2004. "Making the Chicken of Tomorrow: Reworking Poultry as Commodities and as Creatures, 1945–1990." In *Industrialising Organisms: Introducing Evolutionary History*, edited by S. Schrepfer and P. Scranton, 215–36. New York: Routledge.

Jones-Engel, L., C. C. May, G. A. Engel, K. A. Steinkraus, M. A. Schillaci, A. Fuentes, A. Rompis, et al. 2008. "Diverse Contexts of Zoonotic Transmission of Simian Foamy Viruses in Asia." *Emerging Infectious Diseases* 14 (8):1200–1208.

Kanginakudru, S., M. Muralidhar, R. D. Jakati, and J. Nagaraju. 2008. "Genetic Evidence from Indian Red Jungle Fowl Corroborates Multiple Domestication of Modern Day Chicken." *BMC Evolutionary Biology* 8: 174.

Kirksey, E. ed. 2014. *The Multispecies Salon.* Durham, NC: Duke University Press.

Kockelman, P. 2016. "The Chicken and the Quetzal: Incommensurate Ontologies and Portable Values in Guatemala's Cloud Forest." Durham, NC: Duke University Press.

Larson. G., and D. Fuller. 2014. "The Evolution of Animal Domestication." *Annual Review of Ecology, Evolution, and Systematics* 45: 115–36.

Lowe, C. 2014. "Infection: Living Lexicon for the Environmental Humanities." *Environmental Humanities* 5: 301–5.

MacPhail, T. 2004. "The Viral Gene: An Undead Metaphor Recoding Life." *Science as Culture* 13 (3): 325–45.

Maher, L. A., J. T. Stock, S. Finney, J. J. N. Heywood, P. T. Miracle, and E. B. Banning. 2011. "A Unique Human-Fox Burial from a Pre-Natufian Cemetery in the Levant (Jordan)." *PLoS ONE* 6 (1): e15815.

Maupin, Armistead. 2017. *Logical Family: A Memoir.* New York: Harper Collins.

Olmert, M. 2009. *Made for Each Other: The Biology of the Human-Animal Bond.* Philadelphia: Da Capo.

Sahlins, M. 2013. *What Kinship Is—And Is Not.* Chicago: University of Chicago Press.

Schneider, D. 1980. *American Kinship.* Chicago: University of Chicago Press, 1980.

Shipman, P. 2011. *The Animal Connection: A New Perspective on What Makes Us Human.* New York: W. W. Norton.

Smith, P., and C. Daniel. 2000. *The Chicken Book.* Athens: University of Georgia Press.

Squier, S. 2012. *Poultry Science, Chicken Culture: A Partial Alphabet.* New Brunswick, NJ: Rutgers University Press.

Weiss, K. M., and A. V. Buchanan. 2009. "The Cooperative Genome: Organisms as Social Contracts." *International Journal of Developmental Biology* 53 (5/6): 753–63.

Zedder, M. A. 2012. "The Domestication of Animals." *Journal of Anthropological Research* 68: 161–90.

13 LAW

Kristen Stilt

Studies of the behavior of nonhuman animals have taught us that some animals have expectations of justice. Sarah Boysen, a primatologist, says that the notion of justice among chimpanzees is strong: "Deviations from their code of conduct are dealt with swiftly and succinctly, and then everybody moves on," she explains (Gellene 2007). Expectations of justice are seen in species beyond primates. According to Robert Solomon, "Wolves have a keen sense of how things ought to be among them" and "pay close attention to one another's needs and to the needs of the group in general. They follow a fairly strict meritocracy, balanced by considerations of need and respect for each other's 'possessions,' usually a piece of meat" (Solomon 1990, 141). As interest in justice in animals grows, so does what we learn.

This research shows that communities of animals have developed behavioral expectations of their members with potential consequences for transgressors. If we adopt a minimalist definition of law, such that it simply prevents conflict and perhaps even provides enforceable social norms to regulate a particular society, we could conclude that indeed, communities of animals have their own laws. Those laws are not articulated in a way humans can easily understand, and traditionally the lack of language in the human sense has been used as a justification to deny animals even moral consideration. But under a particular conception of law, whether animals have a language that humans deem adequate for certain purposes becomes irrelevant.

And so we might begin our discussion of "law" at this point and investigate the many ways that communities of animals regulate themselves.

But that is not where this conversation typically begins in the field of law.[1] Publications of "animal law" are not about customs, practices, norms, or rules of behavior of animals (as humans might understand them) but rather about the laws of humans as we create and apply them to animals. So ubiquitous is this approach that it is almost never discussed as a choice.

Animals did not ask us, as humans, to make laws that apply to them. Nor did we ask them if they wanted our laws. Under an alternative framework, we might relate to animals as differing self-governing societies relate to one another.[2] Instead, we impose our jurisdiction and our laws on animals. We use law to put animals into categories of our own choosing and to control them, both conceptually and physically. We use law not to recognize, embrace, cultivate, and enable their own innate characteristics and abilities but rather to position them in a way that is convenient and conducive to our wants and needs. Animals do not have the ability to challenge us on our own terms because they are not formally part of any society's lawmaking process. They are not citizens.

There is not, of course, a single human "we." Laws are a reflection of who is in power among the citizenry. As with any area of lawmaking, special interests have great influence over the content of the laws. And there is an important difference between the law on the books and the law as applied. But the stark line that humans have drawn between "us" and "them" — when we are all animals — is firmly entrenched and widely accepted.

Animals as Property

Most fundamentally, and across virtually all legal traditions, we categorize animals as legal property. We have adhered to the property category even as both ethological studies and understandings of our own relationships with animals defy that categorization. In a recent Texas Supreme Court decision, the justices confronted the case of a family dog, named Avery, who escaped from the home of the Medlens during a thunderstorm and ended up in the local shelter. When the Medlens went to bring Avery home, they did not have the necessary funds to release him, and so a "hold for

1. One notable exception is Catherine MacKinnon, who has criticized the "lack of serious inquiry into animal government" (MacKinnon 2004, 270).

2. For the idea that wild animals form "sovereign communities, whose relations to sovereign human communities should be regulated by norms of international justice," see Donaldson and Kymlicka (2011, 157).

owner" tag was placed on Avery's cage. They returned ready to pay the fees only to find that the shelter had made a mistake and Avery had been euthanized. The court followed long-standing precedent to declare that "pets are property in the eyes of the law," and as property, Avery had virtually no value at all.[3] Despite the common phrase and legal fact of "ownership," very few "owners" say that their pets are property in the commonsense meaning of that term.[4] The Texas Supreme Court justices admitted that it did not sound quite right, adding that "a beloved companion dog is not a fungible, inanimate object like, say, a toaster," but in the eyes of the law, the property status is clear.[5]

In the "eyes of the law," animals are property, and so the starting point is that we, as humans, can do what we want: we can buy, sell, modify, manipulate, and destroy animals. We are entitled to the fruits of their labor and are entitled to take their offspring and reproduce our use of them. But of course the law does not have eyes of its own; we create law, and it is contingent on time and place. Nevertheless, there is one remarkable consistency: animals are virtually always considered property, and even in the few places in which the label of "property" has been modified, they are still property in substance.[6]

Animals are not exactly like a toaster, as the Texas Supreme Court admitted, and are not like every other kind of property. For example, no rules protect toasters from mistreatment by humans. You may destroy your toaster however you want, consistent with local laws that are concerned with human health and safety. Animals do receive some protection, but the framework for that protection is whatever we choose to give them. This is true even when the basis for "what they get" is religion, since ultimately it is humans who interpret religious texts and decide how to apply them in the world.

3. Strickland v. Medlen, 397 S.W.3d 184 (Tex. 2013), 185.

4. According to the most recent statistics from the American Veterinary Medical Association, only 1 percent of pet owners they surveyed consider their animals to be "property," whereas the other 99 percent consider them to be "family members," "pets," or "companions" (American Veterinary Medical Association 2012, 5).

5. Strickland v. Medlen, 397 S.W.3d 184 (Tex. 2013), 185–86.

6. France, for example, changed its Civil Code so that animals are sentient living beings rather than property, but it does not change the way that animals are treated under French law. Nicolas Boring, "France: Animals Granted New Legal Status," Library of Congress, Global Legal Monitor, April 25, 2014, http://www.loc.gov/law/foreign-news/article/france-animals-granted -new-legal-status/.

Animal Protection Law

Sometimes we use the law to protect animals from humans, although the reasons that we restrain ourselves are not always clear. Criminal law is an area of substantial growth worldwide in the area of animals for reasons and with implications that have not yet been explored fully, and many if not most jurisdictions around the world have criminal law protections for animals.[7] Criminal law provides the main source of protection and in some cases the only source of protection for animals, but the level of enforcement differs from jurisdiction to jurisdiction and largely depends on the priorities of prosecutors. In the United States, for example, all states and the District of Columbia have at least one anticruelty law that is a felony (Waisman, Frasch, and Wagman 2014).

We have chosen to use the criminal law to protect some animals in some circumstances, often for our own benefit. Criminal law in its earliest formulations tended to protect animals as the property of another human and so often only applied when an animal was harmed by someone other than his or her owner (Livingston 2001–2002; Waisman, Frasch, and Wagman 2014). Today, more commonly criminal law protects animals regardless of ownership, at least for affirmative harms. But the main reason for criminal law is not necessarily to recognize and protect the animal's own interests. Rather, a common rationale is the belief that cruelty to animals is socially unacceptable or more specifically hardens individuals to the prospect of causing harm to humans (Livingston 2001–2002; Waisman, Frasch, and Wagman 2004).

In the case of Egypt, for example, animal advocates worked to persuade constitutional drafters to include animal protections in the new constitution following the 2011 revolution that ousted authoritarian president Mubarak. The most outspoken advocate garnered support for her cause by linking violence against animals to violence against humans, and the spokesperson for the constitutional drafting commission, when announcing that a clause protecting animals had been included in the constitution, praised that argument: "They say that violence in society—and as we are now combatting violence and terrorism—begins with treating animals with cruelty. For people become used to cruelty this way. If people get used to treating animals kindly, they will be kind to humans also" (Stilt 2017).

7. See the survey on animal legislation conducted by the Global Animal Law Project, Global AnimalLaw.org, https://www.globalanimallaw.org/database/national/index.html.

As a result, the Egyptian constitution provides in Article 45 that "the state will provide for . . . the kind treatment of animals [al-rifq bil-hayawan], all according to law."

In a criminal law example from the United States, a defendant was charged with killing the pet goldfish of the children of his girlfriend. In a fit of rage, the defendant threw the fish tank across the room and then killed the goldfish by stomping on them in view of the children. One issue in the case was whether killing the goldfish in this manner constituted aggravated cruelty, which would raise the charge from a misdemeanor to a felony. The defendant argued that he caused the instantaneous death of the fish, and so the fish did not feel the extreme physical pain associated with an act of aggravated cruelty. In determining that aggravated cruelty was indeed present, the court made clear why the criminal law is concerned with animal protection: "the legislative history of the statute indicates that the crime was established in recognition of the correlation between violence against animals and subsequent violence against human beings."[8] The concern is thus the defendant's state of mind: this defendant exhibited aggravated cruelty in his behavior, and it is that behavior that we are concerned about for what it might—and has—done to other humans.

In some cases, criminal law does focus directly on the experiences of animals, recognizing that animals have intrinsic value and that their protection is not just a means to human benefits. For example, police found emaciated and deceased animals, mainly horses and goats, on a farm in Oregon. The legal issue was whether each of the twenty animals suffering from severe neglect counted as an individual victim or whether they were the collective victim of a single crime. Merging the twenty convictions into one would have resulted in a lower penalty for the defendant. Advocating for an instrumental justification for concern for animals, the defendant argued that the victim of the abuse is not the animal but rather the public at large, with the result that the public counts as one victim. The court rejected this argument and found that each animal was indeed an individual victim: "it appears that the legislature's primary concern was to protect individual animals as sentient beings, rather than to vindicate a more generalized public interest in their welfare."[9]

More generally, we use the law to provide animals with the level of protection we want them to have. For example, we deem some animals

8. People v. Garcia, 29 A.D.3d 255 (2006), 261.
9. State of Oregon v. Nix, 283 P.3d 449 (Or.App. 2012).

as more deserving of protection than others and some kinds of pain and suffering as more worthy of prevention than others. Historically, animals with economic and utility value were the main ones protected from harm caused by others (Livingston 2001–2002). But there has been a slow and major change in the criminal law, and large agricultural interests have played influential roles in the law's development. In the United States, for example, the traditional family farm is largely gone, and now "farm animals" means the animals farmed, or more precisely mass produced, for food in factory farms. These animals are no longer at the center of criminal law protection. In fact, to a great extent, they have been pushed out of it by the large agricultural interests that own the animals and do not want *their* treatment of *their* animals regulated—echoing criminal law's original purpose of protecting animals owned by others. Farm animals receive no federal protection while raised, and state criminal laws often contain broad categorical exemptions, including "animals and specific practices used in agricultural industries" (Waisman, Frasch, and Wagman 2014, 74).[10]

Legal protections for animals reflect not just human values generally but, specifically, local values and political dynamics. In India, cows, calves, and other milch and draught cattle are legally protected (which, in practice, is very limited) because of a human determination based on Hindu religious beliefs (Eisen and Stilt 2017). In Egypt, dogs face terrible treatment as a matter of practice because of widely held religious beliefs that dogs are impure, and it is impermissible to keep or care for them (Stilt 2017). Local practices define the experiences of animals, and there is very little protection for animals in general that comes from an international or supranational source (Peters 2016).[11]

The way in which we impose our priorities through the law is seen clearly in the US Animal Welfare Act (AWA). Originally enacted as the Laboratory Animal Welfare Act in 1966, its prompt was the concern that stolen pets were ending up in research facilities, and it covered dogs, cats, nonhuman primates, guinea pigs, hamsters, and rabbits intended for use in research. As expanded over the decades, by its title it now purports to cover animal

10. The full range of common exemptions includes "traditional veterinary practices; animals used for medical, educational or scientific research; hunting, fishing and trapping; animals and specific practices used in agricultural industries; pest control; animals and practices used in entertainment, such as films and television, rodeos, circuses and other public exhibitions; and certain animal training methods" (Waisman, Frasch, and Wagman 2014, 74).

11. The Universal Declaration of Animal Welfare and other similar attempts have not been effective to date.

welfare but actually only protects some warm-blooded animals and only when they are intended for use in research, exhibition, the wholesale pet trade, and when they are transported in connection with these activities. It also covers animal fighting. Significantly, the AWA does not cover animals raised for food, which are 98 percent of all warm-blooded animals raised or used by humans in the United States. This means, for example, that farm animals in petting zoos are covered by the AWA, but the billions of farm animals raised for food every year are not. All fish, even those raised for the pet trade or for exhibition, are excluded, as are reptiles. While this constrained definition has been criticized as a shortcoming of the AWA or as an indication of inconsistencies in the law, it actually is simply an expression of human concerns and priorities.

As many as 94 percent of Americans agree with the statement that "animals raised for food on farms deserve to be free from abuse and cruelty."[12] However, the vast lobbying power of the agricultural industry has often prevented the development of farmed animal protections through the legislative process at both the federal and state levels. Instead, advocates increasingly have turned to public ballot initiatives in the states that allow them such that voters may directly enact state laws regarding the manner in which farmed animals are raised and in particular the conditions of their confinement in cages. The ballot initiative has become a primary vehicle of legal change in the area of farmed animals, and Massachusetts has become the most recent state to adopt laws protecting some farmed animals in this manner.

The Independent Interests of Animals

Do we ever use the law to recognize animals for *who* they are and attempt to protect their needs as they express them? Under the AWA, only primates and marine mammals receive mention of their social needs. The Secretary of Agriculture is directed to promulgate standards that include "minimum requirements . . . for a physical environment adequate to promote the psychological well-being of primates."[13] The resulting regulations require facilities with primates to develop and follow an environmental

12. These results are from a survey announced in a February 17, 2012, press release by the American Society for the Prevention of Cruelty to Animals (http://www.aspca.org/about-us /press-releases/aspca-research-shows-americans-overwhelmingly-support-investigations -expose).

13. Animal Welfare Act, 7 U.S.C. § 2143(a)(1)-(2).

enhancement plan that addresses, among other things, social needs. But exceptions allow for primates to continue to be held singly even though primatologists have determined that living in social groups is considered essential to the psychological well-being of many, if not most, primate species.[14] Some of most remarkable language on the issue of recognizing animals for who they are comes from the Supreme Court of India in a case involving the rehoming of Asiatic lions. The Indian Supreme Court said that "our approach should be ecocentric and not anthropocentric and we must apply the 'species best interest standard,' that is the best interest of the Asiatic lions."[15] Even here, though, the focus is on the level of the species and not individual animals.

In 2009, the Treaty of Lisbon entered into force, adopting the following article from the Treaty on the Functioning of the European Union:

> In formulating and implementing the Union's agriculture, fisheries, transport, internal market, research and technological development and space policies, the Union and the Member States shall, *since animals are sentient beings, pay full regard to the welfare requirements of animals*, while respecting the legislative or administrative provisions and customs of the Member States relating in particular to religious rites, cultural traditions and regional heritage. (emphasis added)[16]

Paying "full regard to the welfare requirements of animals" does not mean that they receive some inherent protection; it does not challenge the basic framework that humans determine what animals receive. Legislative and administrative provisions and customs of the member states are respected—and in particular religious rites, cultural traditions, and regional heritage. The recognition of "religious rites" highlights that the law applied to animals is not just the law adopted in official legal processes; humans also impose on animals rites, traditions, and heritages, in the EU and beyond and use the law to protect the interests of humans in carrying them out.

For example, religious slaughter, and specifically kosher and *halal* slaughter, is becoming a contentious issue around the world. In general,

14. Animal Welfare Act Regulations, 9 C.F.R. § 3.81(a).

15. Centre for Environmental Law, WWF-I v. Union of India (UOI) & Ors (2013) 8 SCC 234, 256.

16. https://ec.europa.eu/food/animals/welfare_en.

nonreligious, or secular, slaughter involves the stunning of animals, which is intended to render the animals insensible to pain right before they are killed. Stunning is not a panacea, and it can fail, causing animals to be slaughtered while fully conscious, although the intention is to render them insensible to pain first. In contrast, kosher and many forms of *halal* slaughter require that the animals are fully conscious when their throats are cut.

In the United States, religious slaughter is accepted in the law. The Humane Methods of Slaughter Act declares as policy "that the slaughtering of livestock and the handling of livestock in connection with slaughter shall be carried out only by humane methods." The Act provides for two humane methods of slaughtering and handling: (1) "all animals are rendered insensible to pain by a single blow or gunshot or an electrical, chemical or other means that is rapid and effective, before being shackled, hoisted, thrown, cast, or cut," and (2) "by slaughtering in accordance with the ritual requirements of the Jewish faith or any other religious faith that prescribes a method of slaughter whereby the animal suffers loss of consciousness by anemia of the brain caused by the simultaneous and instantaneous severance of the carotid arteries with a sharp instrument."[17] Because the act deems religious slaughter a per se "humane method," in the United States, there is no consideration of whether ritual slaughter is humane in any commonsense or scientific meaning of the word; it is protected under a particular US understanding of religious freedom.

In Europe, EU regulations do not require stunning before slaughter but allow member states to impose stricter rules if they want, including requiring stunning with the effect of preventing religious slaughter. Some nations, including Sweden, Norway, Switzerland, and Iceland, have effectively banned religious slaughter (Needham 2012). Critics of bans on religious slaughter claim that the rules are not about protecting animal interests but rather about using the cause of animals to carry out an anti-Semitic or anti-Muslim agenda. Seeing the issue as a human conflict between groups, for which animals are merely the object involved, flows naturally from the way that humans have created laws for animals. It is not possible to separate the human element from the resulting law because humans have made that law. This is true when animals lose protections, such as when interest groups seek to have their practice exempted from an

17. Humane Methods of Slaughter Act, 7 U.S.C. 1902.

anticruelty law, and it is also true when animals might benefit, as with the laws on religious slaughter.

When humans are recognized as having a right to do something involving animals, the balance tips even further in the direction of human interests. For example, German animal advocates viewed the absence of a constitutional protection for animals as an obstacle to their aims because animal interests were consistently trumped by constitutional rights of humans, such as freedom of religion, freedom of artistic expression, and freedom of research. The Christian Democrat Union and Christian Social Union coalition initially prevented movement toward a constitutional amendment to add some protections for animals but then supported it after the Federal Constitutional Court controversially exempted a *halal* butcher from the anticruelty laws and allowed him to slaughter animals without stunning (Evans 2010).

The phrase "and the animals" was added to an existing environmental protection provision in the German Basic Law, Germany's constitution, to read

> Mindful also of its responsibility toward future generations, the State shall protect within the framework of the constitutional order the natural bases of life and the animals by legislation and, in accordance with law and justice, by executive and judicial power.

As a state objective, the provision does not establish justiciable rights and does not require any state action in terms of implementation. While German courts, legislatures, and the executive are all formally bound by the objective, they have significant discretion to determine laws and policies to meet its requirements. Further, in any legal analysis involving competing constitutional claims, the German courts conduct a proportionality analysis, weighing the constitutional human rights at stake alongside the animal interests. The Federal Administrative Court held that while animal protection has greater constitutional weight as a result of the amendment, the exemption from the anticruelty act for religious slaughter still stands unless amended by the legislature. The Federal Constitutional Court has upheld a zoophilia prohibition even though Germany recognizes a human right to sexual autonomy. Relying in part on the new constitutional protection of animals, the court deemed the prohibition a reasonable limitation (Eisen and Stilt 2017).

Constitutional recognition can be valuable, and eight countries now have a prescriptive constitutional animal protection provision.[18] "Protection" of animals, "prevention of cruelty" to animals, "compassion" for living creatures, and "kindness" to animals are the most common phrasings, and all indicate that animals should be protected from some harms as determined by humans without questioning the basic framework that animals receive what we choose to give them (Eisen and Stilt, 2017).

A New Framework for Animal Law?

What would it look like to release our grip, even slightly, on this current legal framework, whereby we use the law to exercise human preferences for the lives of billions of animals? What would it look like to allow animals, or some animals, to step outside this framework and stand on an equal playing field with the interests of humans? What would it look like to recognize some interests of animals as so important, so fundamental, that they escape the legal hierarchy we have created? What would it look like to allow some animals to access some of *our* law—not the law we have written for them, but the law we apply to ourselves?

We can see a glimpse of answers to these questions, as well as the ways the status quo makes it difficult to even ask them, in the oral argument in a New York state court on the habeas corpus claim filed by attorney Steven Wise on behalf of his client, a chimpanzee named Tommy, who, at the time of the suit, was held alone in a small cage. In the oral argument, the judges focused on Tommy's condition—and perhaps improving it—in his current state of confinement. Wise focused on something different: using habeas corpus, a remedy traditionally used by humans, to release Tommy from his current confinement.

JUDGE GARRY. So can we safely assume that the goal of this proceeding is to promote the well-being of the chimpanzee?

STEVEN WISE. No, there's only one goal for the proceeding because it's a common law habeas corpus proceeding to discharge the chimpanzee. . . .

18. Eisen and Stilt (2017 §§ 1–2) define "animal protection provisions" as provisions that "treat individual animals as legitimate beneficiaries of constitutional protection—rather than merely as resources or symbols." Animal protection provisions include "(1) any federal jurisdictional assignments that reference 'protection,' 'welfare,' 'cruelty,' or other language suggesting concern for individual animals ('jurisdictional provisions'); and (2) any duty, mandate, or goal—placed on the state or on private citizens—to protect or consider the well-being of individual animals ('prescriptive provisions')."

JUDGE GARRY. So are you saying that you're not interested in promoting the chimpanzee's well-being?

STEVEN WISE. That is not the purpose of our suit.[19]

In stating that he is not interested in the well-being of Tommy, Wise initially may sound like an attorney who does not care about his client. But his concern is on a different, fundamental level, and to get there, he has to bypass, even reject, the legal framework we have set up for animals. He was not seeking to improve the conditions of Tommy's current confinement under the laws applicable to animals, even though some of those laws might have been violated. Instead, he sought to end the current confinement and transfer Tommy to a sanctuary that Wise identified using laws that have not traditionally applied to animals. A sanctuary is not freedom from confinement in the absolute sense, but it was considered by Wise to be as close to freedom as possible given Tommy's own capabilities and needs.

Wise is essentially saying, I am not here to discuss the welfare conditions of his confinement, under the Animal Welfare Act or other laws that we have written for animals. Instead, let Tommy out of the legal framework we have imposed on him that accepts his confinement because he is the property of his owner. Give him access to *our* legal remedy of habeas corpus to challenge the sheer fact of his confinement and, because of his capabilities, release him. Allow him access to *our* laws, to *our* legal system, in a way not yet seen, but in a way that you, as judges relying on the common law, have the power to do.

Even if we grant animals access to some of our laws—and so far Wise has not won his case—humans will still be the lawmakers, advocates, and adjudicators because it is our legal system, not theirs. Even if we decide that we will use the law to provide animals what they want or need, we will still be the ones deciding what we think they want or need. With this inevitable responsibility, we return to the study of animal behavior, or ethology, where this chapter began, and the obligation to learn as much about as many species of nonhuman animals as possible on their own terms.

Suggestions for Further Reading

Bekoff, Marc 2007. *The Emotional Lives of Animals*. Novata, CA: New World Library.
Donaldson, Sue, and Will Kymlicka. 2011. *Zoopolis*. Oxford: Oxford University Press.

19. The exchange can be seen in the documentary about the NonHuman Rights Project, *Unlocking the Cage.*

Eisen, Jessica, and Kristen Stilt. 2017. "Animals, Protection and Status of." In *Max Planck Encyclopedia of Comparative Constitutional Law*. Oxford: Oxford University Press.

Francione, Gary L. 2000. *Introduction to Animal Rights: Your Child or the Dog?* Philadelphia: Temple University Press.

Garner, Robert. 2013. *A Theory of Justice for Animals: Animal Rights in a Nonideal World*. New York: Oxford University Press.

Sunstein, Cass, and Martha Nussbaum, eds. 2004. *Animal Rights*. New York: Oxford University Press.

References

American Veterinary Medical Association. 2012. *U.S. Pet Ownership and Demographics Sourcebook*. Schaumburg, IL: Center for Information Management, American Veterinary Medical Association.

Donaldson, Sue, and Will Kymlicka. 2011. *Zoopolis*. Oxford: Oxford University Press.

Eisen, Jessica, and Kristen Stilt. 2017. "Animals, Protection and Status of." In *Max Planck Encyclopedia of Comparative Constitutional Law*. Oxford: Oxford University Press.

Evans, Erin. 2010. "Constitutional Inclusion of Animal Rights in Germany and Switzerland: How Did Animal Protection Become an Issue of National Importance?" *Society and Animals* 18: 231–50.

Gellene, Denise. 2007. "A Sense of Fair Play Is Only Human, Researchers Find." *Los Angeles Times*, October 5. http://articles.latimes.com/2007/oct/05/science/sci -unfair5.

Livingston, Margit. 2001–2002. "Desecrating the Ark: Animal Abuse and the Law's Role in Prevention." *Iowa Law Review* 87: 1–73.

MacKinnon, Catherine. 2004. "Of Mice and Men." In *Animal Rights*, edited by Cass Sunstein and Martha Nussbaum, 263–76. New York: Oxford University Press.

Needham, Christopher. 2012. "Religious Slaughter of Animals in the EU." Library Briefing, November 15, Library of the European Parliament. http://www.europarl.europa .eu/RegData/bibliotheque/briefing/2012/120375/LDM_BRI(2012)120375_REV2_EN .pdf.

Peters, Anne. 2016. "Global Animal Law: What It Is and Why We Need It." *Transnational Environmental Law* 5 (1): 9–23.

Solomon, Robert. 1990. *A Passion for Justice*. Reading, MA: Addison-Wesley.

Stilt, Kristen. 2017. "Constitutional Innovation and Animal Protection in Egypt." *Law and Social Inquiry*.

Waisman, Sonia S., Pamela D. Frasch, and Bruce A. Wagman, eds. 2014. *Animal Law: Cases and Materials*. Durham, NC: Carolina Academic Press.

14 LIFE

Eduardo Kohn

An ongoing challenge for Animal Studies is to conceptualize how to live with other kinds of living beings. Deepening our understanding of what that life that unites us is can help us find ways to relate better to those beings with whom we share our fragile home in a time of planetary ecological crisis.

This endeavor, however, is still hampered by a dualistic metaphysics that limits much of our thinking about life by placing humans, with our minds, gods, and cultures on one side of a great divide, and animals, with their bodies, biologies, and materialities on the other. Although there are many approaches that seek to overcome this divide, when the solution is to give animals humanlike minds or to find our unity in a shared embodied nature or to merge minds and bodies through a conceptual hyphen (e.g., the natures-cultures of Science and Technology Studies), we end up reproducing the very dualism we seek to overcome.

Dualism is not just a scholarly problem getting in the way of our ability to make conceptual sense of the life we share with other kinds of beings, for it is also an urgent political problem. The so-called Anthropocene, the proposed name for this geological epoch in which we all now live (see Chakrabarty 2016), is characterized by the ways in which humans, increasingly separate from the world of nature, have become a veritable force of nature. Understanding what makes us humans separate from the world and what continuities we nonetheless share with it despite that separation is a first step toward developing an ethical orientation for the Anthropocene.

Life in Mind

The way to do this is to rethink what life itself is. What, in general terms—
those that may account for life elsewhere in the universe—is life? Life,
first off, is a dynamic in which thinking takes place. A living dynamic is the
locus, wherever it may be found, of the production of new thoughts. One
could say that life is mind.

By mind I mean any entity capable of learning by experience. Consider
the emergence over evolutionary time of an adaptation such as wings in
the lineage of birds. Although at first it may sound strange, one can think
of a wing as a growing thought about the world. Through the "experience"
of many lineages of protobirds over vast periods of time, wings increasingly came to resemble something about the currents of air birds can now
"grasp" in order to fly. Wings re-present air; they are an idea of it that has
helped birds learn something about what air is.

Let me draw out some implications from this simple example. First,
living dynamics are representational, or semiotic, dynamics. That is, they
involve signs that represent something about the world to somebody in
some way or another (see Peirce 1955). In this example, the sign would be
the wing; the part of the world they represent, with increasing accuracy,
air; and that somebody or self for whom they represent it, the future generations of birds who will interpret that sign (the wing) in the eventual
constitution of their own bodies. Signs are the stuff of thoughts. Minds
in fact are the products of signs and not the other way around. And signs
are the universal currency of minds wherever in the universe these are to
be found.

Thinking about representation, or signs, in this way can seem counterintuitive for those of us thinking from within the canon of social theory.
This is because as social theorists, we tend to conflate the sign with the
symbol, a kind of sign that, on this planet at least, is uniquely human.
Symbols form the basis for human language and everything that makes
human thought and human life distinctive. But symbolic reference is a
special kind of sign process that is, in fact, made up of the more basic kinds
of sign processes intrinsic to the rest of life (Deacon 1997).

A symbol is a kind of sign that refers to its object of reference indirectly
by virtue of a prior relationship it has to a circular system of other such
symbols that constitutes it as a sign. For example, the English word *dog*
only refers to the furry four-legged creature in question by virtue of the
fact that its meaning as a sign is determined by the ways in which it is held

by a linguistic context of similarly arbitrary and conventional signs. This representational modality has certain important properties. Language, culture, consciousness, and fiction are all products of distinctively human forms of symbolic thought insofar as they all depend on the ways in which symbolic reference creates a relatively separate system of thought that can then become a domain from which to think back on the world from which it emerged (Deacon 1997).

Distinctively human emotional experiences also depend on symbolic reference insofar as they are the products of displacing the self across temporalities—a decidedly symbolic feat. Nostalgia, for example, involves the special feeling of pain that arises when one is unable to fully integrate past and present selves. Anxiety involves the ways in which a symbolic mind can populate the future with an array of possible dangerous scenarios (Deacon 2003).

Symbolic thought allows us humans to think and live differently in the world. It is what makes us so special. Symbols, because they refer to the world out there indirectly, by virtue of the prior relationships they have to the set of symbols that form their interpretative contexts, are relatively more separate from the world. That is, although all signs are involved in some sort of mediated relation to the world (a wing is not the air it represents), symbols are so to a degree that is heretofore unprecedented in the world of living minds.

How we think about ethics in relation to other kinds of living beings is also affected by this way of thinking about life semiotically. One can think of ethics as the ability to think about the good of an other that is not us. This makes ethical thinking a quintessentially symbolic and human endeavor. For symbols allow us to think with sufficient abstraction so that we can see a good that is not necessarily good for us. But this kind of ethical thinking is based on something more fundamental—that there exists an other for which there *is* a good.

That we can think of life in terms of value in this way is the product of the fact that value emerges with life. That is, although we can attribute value to anything, it is only in the realm of life where value emerges as a constitutive element. A living self is shaped by its search for an environment that is good for it. Wings—those embodied thoughts about air—exist because they are good for those organisms equipped with them. Although it would be incorrect to look for a one-to-one functional correspondence between any semiotic feature and its value—play and indeterminacy are central to evolution and to life—the upshot of this way of thinking about life for Animal Studies is that it provides a framework through which one can approach how to think ethically about other kinds of living beings. Living beings can be the object of our thinking about the good because, as subjects, their orientation toward a good constitutes who they are.

On this view, that which would guide our ethical thinking concerning other kinds would not always necessarily involve the question of suffering, for suffering here is just one index of something that is potentially bad for the being in question. There may well be many things that are bad for many kinds of beings that don't necessarily cause suffering to those beings.

Recognizing the ways in which ethics emerges from the value intrinsic to the realm of life does not provide any easy blueprint for making ethical decisions. And in thinking of ethics in relation to life, one has to remember that death is a central part of life. With Kaja Silverman (2009) I hold that the ethical injunction to recognize our psychic limits means that for others to live, we must die. And for us to live, others must die as well. With Donna Haraway (2008) I hold that our ways of living must also incorporate and take responsibility for the killing that we necessarily do to live.

Killing well–something that can follow no easy formula–is, as strange as it may sound, for Haraway an important part of an ethical life. This changes how we think about a host of things related to animal life. One may well take a "pro-choice" stand on abortion, but this will be based on the conscious decision that one is taking some sort of a life. One may well choose to eat meat, but the flesh one eats is that of a person. A "pro-life" stand and veganism, respectively, cannot fully shield us from the problems these practices raise (see chap. 27). For all living involves killing: our forebears must die to make room for us to live; we have survived thanks to the deaths of others; and all of our food, even our vegetables, are living selves.

The separation that symbolic reference confers is the reason why we humans so often think dualistically (it is further reinforced by the fact that we, like the vast majority of animals, happen to exhibit bilateral symmetry). A symbolic mind can see itself as separate from the physical universe because to a certain extent it is relatively more separate from material and energetic processes as compared to other forms of thinking.

But symbolic reference is the emergent product of a set of more basic sign processes on which it is wholly dependent. And these processes don't create the dualisms we so often associate with human thought. These other signs are "icons," which in and of themselves exhibit some of the qualities of the entity they represent, and "indices," which point to or have some sort of correlation with the entities they represent (Peirce 1955). Insofar as a symbol, for it to function as a sign, must ultimately have some sort of a *resemblance* to the object it *points to* in the world, it, too, must partake of icons and indices for it to be meaningful. And insofar as a symbol, for it to be alive, must have some sort of an effect on the lives of those who interpret it, it must also ultimately be connected to the material and energetic processes from which it, like all other sign processes, emerges.

Life as Signs

The representational theories most often used in the humanities and social sciences, grounded as they are in the work of Ferdinand de Saussure (1959), treat the human linguistic sign as the prime example of what a sign is. This has the effect of projecting symbolic properties onto all sign processes, obscuring, in the process, many other distinctive properties of nonlinguistic semiosis. And because Saussurean semiology does not theorize the ways in which the symbolic sign, despite its unique properties, is nonetheless grounded in more basic sign processes, it comes to see humans as living, via representation, in a dualistic prison house with no idea of how to get back in the world without jettisoning the sign altogether.

Within the context of this way of thinking about representation, if we want to understand our relation to animals, we must either attribute symbolic capacities to them (making them into little humans who inhabit the kinds of foreign mutually unintelligible worlds that different languages give us—witness Wittgenstein's aphorism that if lions could talk we would not understand them), or we are forced to relate to animals based on a supposedly prerepresentational shared embodiment (see chap. 21).

Conflating representation with language makes it hard for us to understand how we can think outside of or beyond language. It makes it appear as if our socially constructed or innate categories of thought constrain all of our thinking. Although it is obviously important to recognize the historically contingent nature of human systems of meaning, dependent as these are on symbolic thought—and working with this recognition has been the humanistic and social sciences' great contribution to scholarship—we have to keep in mind that we can and do constantly think beyond the circular system of language that holds so many of our thoughts.

This also applies to how we can think about life itself. Contra Michel Foucault (1970), we can think about "life itself" independently—to a certain degree—of the human social, cultural, political, and intellectual contexts that make so much of our thinking possible, and we can think with animals in ways that are not always necessarily bounded by the historically contingent ways we think we should think with animals.

In short, we humans are not the only ones who think, and not all thinking is of the human sort. Animals (and plants as well) have minds and think. They are selves (perhaps persons). This is because they are alive. Life is a sign process, and it is one in which we partake even if we sometimes forget how to listen for its distinctive nonsymbolic manifestations. Thinking better in the Anthropocene will involve learning to listen better for and think with these kinds of thoughts as they emerge and become manifest through our relations with other kinds of selves, not all of whom are human.

The example of a wing-as-thought points to other features of this way of thinking about life. The first involves future. Signs are alive insofar as they extend into a possible future. The bird's wing is an idea about air to be interpreted by a future bird as that future bird builds its body. Life and signs are never mere products of their pasts, for they always involve this kind of telic orientation. In fact, one could say that what is unique to life is this way in which, via the mediation of signs, the future comes to have an effect on the present. For a cat to successfully pounce on a moving ball, for example, she must jump to where the ball *will* be. The potential future position of the ball as she re-presents it affects the cat's actions in the present. Life always traffics with time in this way. We always have one foot, or paw, as the case may be, in the future.

Another important attribute of life when understood in these semiotic terms is its special relation to absence. This is sometimes difficult to appreciate given the centrality of presence to our metaphysical frameworks. We ground ourselves in materiality, in existence—in things. But what makes life so distinctive is its multiple orientations to what is absent (Deacon 2011).

All signs re-present what is not present. This is one absence. They do so in regard to an absent future event (e.g., that future position of the ball as *re*-presented by the cat), and they do so for an absent self. This absent self could be the future bird for which the wing-as-sign will stand for air, or it could be you, the reader of this essay, to whom I hope to make myself understood. In the realm of living thoughts, there is another absence as well—the dead. In an evolutionary dynamic, these dead would be all the variants that were selected out, leaving the living self in a lineage of thought extending to the future but held by and thus indebted to all the absent dead that it is not. The life of human symbolic thought involves its own kind of relation to absence because of the ways in which the symbolic signs through which it thinks are wholly dependent on the absent context of other signs for their reference. However, all living signs, and not just symbols, work through absences.

Another important attribute of this way of thinking about life in semiotic terms involves generality. Our metaphysical framework privileges the specific as something more real than the general. We tend to adopt a nominalist position whereby the sole source of generalities, such as abstractions and categories, is the human mind. In this view the world is made of discrete entities, and we humans, with our innate cognitive capacities and our cultural categories, create the abstract structures that give it order and meaning.

But seeing life semiotically encourages us to appreciate the ways in which generality emerges in the world and is not just the product of human minds. Much of this involves the counterintuitive workings of icons. Iconicity is a representational process in which not noticing difference has its own generative productivity. Take the example of the tick, as made famous by the ethologist Jakob von Uexküll (1982). The tick is, for some, "world-poor" (see Agamben 2004) in that it doesn't distinguish among the many kinds of beings it parasitizes; any mammal radiating heat and the odor of butyric acid will do. But this kind of cognitive confusion or poverty—treating what might otherwise be considered to be different species as being of a kind—is productive, for it can create new *kinds*. In the woodlands and suburbs of the North American northeast, this new kind is the class of beings through which Lyme disease can pass. This is a new general, not reducible to any one kind of animal, nor is it imposed by the human mind. It is real, even if its existence cannot be precisely located. Biological entities such as lineages and species exhibit this kind of real generality as well. It may well be that the individual animal is the existent manifestation of the lineage or species, but this does not in any way make the lineage or species any less real as a kind. We need to be careful not to conflate existence with reality. The absential dynamic that is so central to life depends also on the reality of emergent generals that exceed the existence of the individuals through whom they become manifest.

To be sure, there are risks associated with focusing on life, as I do, as a separate domain of reality with qualities that distinguish it from nonlife. Aren't I simply replacing one dualism with another? Elizabeth Povinelli (2016), for example, would associate this way of thinking of life with what she critically refers to as the "carbon imaginary," a historically given discursive formation that encourages us to police the boundaries between life and nonlife, informed by a heteronormative injunction to accord ultimate value to reproduction. Donna Haraway has sought to hold open a space for political possibility through the figure of the cyborg. A cyborg is both living and machinic. It is the product of both science and "fabulation," and as such it eludes any easy definition or categorization that would restrict its ability to do conceptual work in the world (Haraway 1991).

In my insistence on getting right what life "is" I have taken a different tack. Part of giving potential to a concept is delimiting what it is not. My goal is to harness some of the properties of life as a way to think better for our times. And this potentiation must necessarily involve creating conceptual constraints. If we don't do this, whenever we look for animacy beyond humans we will end up narcissistically projecting on to other kinds what, from a human and humanist perspective, we understand life to be.

Mine is a general view of life (sexual reproduction is just one kind of re-presentation that could count as life). It holds that we need to be able to say something specific, not just about what, in general, life is, but what generality itself is. By a "general" I mean any dynamic whose own reality exceeds the specific existent manifestations it will also necessarily come to have.

Life creates and propagates generals. To be provocative, life creates not only kinds but also souls, spirits, and even gods. All of these exhibit their own kind of general reality, a reality that extends beyond and yet is nonetheless continuous with the individual lives from which it emerges. In emphasizing the spirit quality of life, that which makes the animal animate, I am simply recognizing the reality of all of those absential general qualities that sustain the life of the individual of any living kind.

Animism

In this regard, the view of life I am proposing here bears much in common with the "new" animism. *Animism* is a nineteenth-century anthropological term that, in its traditional meaning, distinguished so-called primitive people from moderns by virtue of the fact that they (mistakenly) attribute animate properties to the world beyond humans (Tylor [1871] 1920). This concept has been revitalized in a nonpejorative fashion as a way to critique our modern assumptions about humans and our place in the world.

In Philippe Descola's (2013) revamping, animism can be contrasted with naturalism, the ontological orientation that characterizes modernity.

Naturalism holds that what we humans share with the rest of the world is our exteriority—that is, our bodies, our materiality. Interiority, in this view, by contrast, is a specific quality reserved for humans. Only we, according to the naturalist, have minds, souls, and culture.

Animism, by contrast, which is associated with many different groups throughout the world but is especially prevalent in the Amazon, where both Descola and I conduct anthropological research, inverts this formula. What we humans share with all other beings is our interiority. The animist has no problem recognizing that animals have souls and partake in spirit life—these insubstantial features confer on them the quality of persons, which they share with us humans. What distinguishes one kind of being from another is the different kind of body it possesses.

It is no coincidence that the Amazon rainforest is home to so many of the world's animist societies, for it is an extraordinarily "animate" place. Because it houses so many kinds of life forms, it is also home to so many more generals, absences, even spirits. When a hunter dreams of a spirit master of a peccary and then, based on this dream, goes out and hunts a peccary the following day, he is connecting that real general spirit life of the forest to the kinds of fleshy concrete existence from which it emerged. Amazonian animism makes manifest the animate, spirited—in a word, enchanted—nature of all life. When we immerse ourselves in its dense ecologies we cannot but appreciate this. And in appreciating this we can come to see the ways in which it is a constitutive feature of all life, not just tropical life.

Animism is not, to reiterate, a mistaken belief. Rather, it constitutes an opening onto the animistic potential in the world. And, as Ghassan Hage (2012), following Eduardo Viveiros de Castro (Viveiros de Castro 1998), has noted, for this reason it may serve as a space from which to imagine an "alter-politics" for our times.

This, to my mind, is why thinking animistically can provide us with an orientation for thinking in this time of global ecological crisis. The Anthropocene is characterized by the fact that we humans, as a life form, have become increasingly separate from other forms of life to such an extent that our culture has now become a separable force of nature with deleterious effects for life on Earth as we know it. Animism may provide a position from which to critically reflect on this separation and to find another way to live in this world.

This separation, which is now posing a threat to the animate world, can be seen to have many origins. One can locate it with the emergence of symbolic thought as a distinctive representational modality. Symbolic reference, recall, acquires a circular life of its own, which is relatively divorced from the world from which it grows and to which it refers. As such, it has the effect of separating us from the world. When we build our conceptual theories out of this fact, as the humanities and social sciences often do, without recognizing the semiotic and physical connections that nonetheless ground us in the world we share with the rest of life, we further accentuate this separation.

We humans have become increasingly separated from the world through practices engendered by our distinctive manner of thinking. Agriculture, based as it is largely on the domestication of weeds, depends on the simplification of ecosystems in ways that privilege early successional species over dense ecologies of climax associations. This creates one separation. Animal domestication, which allowed us to remove our sources of animal protein from the dense ecologies we once needed to navigate to procure these, results in another separation. These developments enabled the rise of cities, further severing our built environments from nonhuman ones. The industrial revolution, which replaced human and animal labor with the work of machines, and more recently, the increasing global circulation of abstract systems involving money and information, have further divorced human life from a "nature" that is increasingly treated as a material resource whose sole function is now to fuel the increasingly separate domain of human life.

All of these ways in which we have come to be divorced from the rest of the living world make us who we are as humans, and it would not be particularly productive to think that we can or should slough them off altogether. But the alter-politics for the Anthropocene that I believe we need would involve recognizing and revitalizing the connections we have with the rest of life. Remembering that we are not the only kind of *we*, and that in this recognition, there is the possibility of forging a larger sense of self that involves other kinds of creatures (Haraway 2008), the grounds for this connection would involve reconnecting with that part of mind and spirit that is emergent with life.

What is so unique to life is that life itself thinks. What would it be like to learn to listen again to the kind of thinking we share with other living beings? This would require cultivating practices that involve thinking less in our distinctively human way and more like the rest of the living, thinking world, practices that might involve working and especially playing with companion species, the cultivation of dream interpretation, gardening, drawing, picking blueberries in the forest, meditation, bird-watching, and even hunting. All of these activities break our symbolic habits of thought. They encourage us to cultivate something I call *sylvan thinking*–a kind of thinking that is wild, untamed, one that we share with the rest of life (Kohn 2013). This is the real *pensée sauvage* (Lévi-Strauss 1966) that we need to learn again to cultivate.

Thinking again with and like life may help us cultivate another kind of political and ethical orientation for the Anthropocene. For, if we can learn to listen anew for it, the life with which we can thus think can provide its own thoughts on how to keep alive this kind of thinking that is also ours. Thinking about life in this way, as the domain in which absences, generals, and kinds grow, is a reason why the ecological politics we need to cultivate for our times will also perforce involve an important spiritual reorientation as well. This is the animism we need to rediscover as we search for better ways to live among the many selves that make us.

Suggestions for Further Reading

Deacon, Terrence W. 2011. *Incomplete Nature: How Mind Emerged from Matter*. New York: W. W. Norton.

Kohn, Eduardo. 2013. *How Forests Think: Toward an Anthropology beyond the Human*. Berkeley: University of California Press.

Peirce, Charles S. 1955 "Logic as Semiotic: The Theory of Signs." In *Philosophical Writings of Peirce*, edited by Justus Buchler, 98–119. New York: Dover.

References

Agamben, Giorgio. 2004. *The Open: Man and Animal*. Stanford, CA: Stanford University Press.

Chakrabarty, Dipesh. 2016. "The Human Significance of the Anthropocene." In *Reset Modernity*, edited by B. Latour, 189–99. Karlsruhe: ZKM Center for Art and Media; Cambridge, MA: MIT Press.

Deacon, Terrence W. 1997. *The Symbolic Species: The Co-evolution of Language and the Brain*. New York: W. W. Norton.

———. 2003. "The Hierarchic Logic of Emergence: Untangling the Interdependence of Evolution and Self-Organization." In *Evolution and Learning: The Baldwin Effect Reconsidered*, edited by B. Weber and D. Depew, 273–308. Cambridge, MA: MIT Press.

———. 2012. *Incomplete Nature: How Mind Emerged from Matter*. New York: W. W. Norton.

Descola, Philippe. 2013. *Beyond Nature and Culture*. Chicago: University of Chicago Press.

Foucault, Michel. 1970. *The Order of Things: An Archaeology of the Human Sciences*. London: Tavistock.

Hage, Ghassan. 2012. "Critical Anthropological Thought and the Radical Political Imaginary Today." *Critique of Anthropology* 32 (3): 285–308.

Haraway, Donna. 1991. *Simians, Cyborgs, and Women: The Reinvention of Nature*. New York: Routledge.

———. 2008. *When Species Meet*. Minneapolis: University of Minnesota Press.

Kohn, Eduardo. 2013. *How Forests Think: Toward an Anthropology beyond the Human*. Berkeley: University of California Press.

Lévi-Strauss, Claude 1966. *The Savage Mind*. Chicago: University of Chicago Press.

Peirce, Charles S. 1955 "Logic as Semiotic: The Theory of Signs." In *Philosophical Writings of Peirce*, edited by Justus Buchler, 98–119. New York: Dover.

Povinelli, Elizabeth. 2016. *Geontologies: A Requiem to Late Liberalism*. Durham, NC: Duke University Press.

Saussure, Ferdinand de. 1959. *Course in General Linguistics*. New York: Philosophical Library.

Silverman, Kaja. 2009. *Flesh of My Flesh*. Stanford, CA: Stanford University Press.

Tylor, Edward B. (1871) 1920. *Primitive Culture*. London: John Murray.

Uexküll, Jakob von. 1982. "The Theory of Meaning." *Semiotica* 42 (1): 25–82.

Viveiros de Castro, Eduardo. 1998. "Cosmological Deixis and Amerindian Perspectivism." *Journal of the Royal Anthropological Institute*, n.s., 4: 469–88.

15 MATTER

James K. Stanescu

French philosopher Rene Descartes, in his *Discourse on Method*, famously compared nonhuman animals to "automatons" (Descartes 1985, 139). Descartes further explains his position in a 1649 letter to British philosopher Henry More—"it is more probable that worms, flies, caterpillars, and other animals move like machines than they have immortal souls. . . . Since art copies nature, and people make various automatons which move without thought, it seems reasonable that nature should produce its own automatons, which are even more splendid than artificial ones—namely the animals" (Descartes 1991, 366). While this stance is well known, I am not sure the import of this Cartesian dualism is equally understood. As soon as Descartes is done with this discussion of animals as machines in *Discourse on Method*, he immediately argues that "I described the rational soul, and showed that, unlike the others things of which I have spoken, it cannot be derived in any way *from the potentiality of matter*" (Descartes 1985, 141, emphasis added). So the Cartesian dualism sets on one side the soul, consciousness, language, and even sentience itself. On this side reside beings toward whom we have moral and political duties. On the other side are all the beings that come from just the potentiality of matter itself, and this group includes other animals, who might as well be machines. And while many people would not subscribe to a view as extreme as Descartes', it is this trick by which matter ceases to matter that I will address in this chapter. Matter matters for animal scholars because matter, broadly understood, stands in for the nonhuman, which is pushed to the side by anthropocentrism.

To think about matter we may go back to the concept's beginning, perhaps starting with Aristotle's *Physics* and *Metaphysics* and then moving through millennia of iterations, including the materialism of Feuerbach and Marx. Or we could focus on the new feminist materialism of Karen Barad and others who are developing philosophies of things, objects, and stuff. There is another sense of matter—one captured by Judith Butler's title *Bodies That Matter* that suggests that matter isn't just all the substance of the world; it is also a term we use to signify import, even hierarchy. This valence between *matter* the noun, the materialness of things, and *matter* the verb, how we decide what we care about and what gets to count, will be the force that drives this chapter.

Matter against Anthropocentrism

One of the major projects of Animal Studies is figuring out how to escape the confines of anthropocentrism (see chap. 3). This is a task that is easier said than done; especially because virtually everything about our society, and our thinking, is built around human exceptionalism. Thinkers are much more likely to say something makes us uniquely human than they are to say something makes us like other animals. But anthropocentrism is not simply the ways that human beings might be unique or morally exceptional. Anthropocentrism, taken to its extreme, is a denial of the world mattering outside of human acknowledgment and intervention. In this strong form, the world is viewed as built fundamentally on human representation, or human language, or human action, or at the very least, human interaction. This view that reality is produced and built through human interaction is sometimes called "social construction," and it places the human as the central protagonist in the story of the world. Animals, vegetables, and minerals (not to mention atoms, gravitational waves, viruses, and the like) are at best a supporting cast, fundamentally unimportant, unless humans are on the scene to interpret, intervene, and give meaning to the events unfolding. But, as Karen Barad puts it

> Language has been granted too much power. The linguistic turn, the semiotic turn, the interpretative turn, the cultural turn: it seems that at every turn lately every "thing"—even materiality—is turned into a matter of language or some other form of cultural representation. . . . Language matters. Discourse matters. Culture matters. There is an important sense in which the only thing that doesn't seem to matter anymore is matter. (Barad 2007, 132)

I would note that everything that is important in those turns Barad talks about are the very things that we so often assume belong uniquely to the human. So, if we are looking at the most recent resurgence of thinking about matter in the social and human sciences, it is in many ways a response to the anthropocentrism of social construction.

There have been many different and overlapping ways theorists have taken matter seriously in the recent years, from Jane Bennett's vibrant matter to Donna Haraway's semiotic-material matrices to Karen Barad's agential realism to Bruno Latour's actor-network theory to Mel Chen's animacies to Manuel Delanda's assemblages to Stacy Alaimo's transcorporality, and many more. While there are important differences among these various approaches, one of their central commonalities is they all seek to undermine the actor-object dichotomy. According to anthropocentrism, humans are the only possible actors, the only agents of change. The rest of existence comprises objects on which we act. But according to these new materialisms, the nonhuman is as much an actor as the human. In other words, the object is also the actor.

At the end of *Vibrant Matter*, Jane Bennett writes, "I believe that encounters with lively matter can chasten my fantasies of human mastery, highlight the common materiality of all that is, expose a wider distribution of agency, and reshape the self and its interests" (Bennett 2010, 122). Challenging anthropocentrism by taking seriously the agency of nonhuman actors opens up new understandings of our place as one among many. Consider this example: the Anthropocene is the name given by some scientists and activists to the geological time period where human action has shaped and changed the world itself, mostly through the rise in greenhouse gases and the resulting global warming. And while this may seem like an extension of anthropocentrism, a quick analysis of global warming reveals something else. We have island nations that are disappearing, weird and extreme weather events that are becoming common, mass extinctions, and all of these events point to the end of a certain human mastery. The world itself has become an actor, and the matter of the world has come back like the return of the repressed, threatening to disrupt all of human life.

What we might learn from the example of the Anthropocene is that seeing matter as active and formative hurts our human narcissism. And probably the most classical form of that hurt is seeing humans as one form of biological matter subjected to the same forces and drives that all biological organisms are shaped by. Indeed, evolution has always fundamentally been a kind of materialism. Elizabeth Grosz wants us to turn to the work of Darwin and Bergson in order to understand the ways that matter and life interact with each other. As Grosz states,

> Life is, for Bergson, an extension and elaboration of matter through attenuating divergence or difference. Matter functions through the capacity of objects to be placed side by side, to be compared, contrasted, aligned, and returned to their previous states. Relations between objects are external, a reflection of their relative positions in space. Following Darwin, Bergson understands that life emerges from matter through the creation of an ever-broader gulf or discontinuity between cause and effect or stimulus and response. In its emergence, life brings new conditions to the material world, unexpected forces, forms of actualization that matter in itself, without its living attenuations, may not be able to engender. But with the emergence of life and its ever-divergent forms of elaboration, life comes to possess a difference in kind from matter. . . . Life is not some mysterious alternative force, an other to matter, but the elaboration and expansion of matter, the force of concentration, winding, or folding up that matter unwinds or unfolds. (Grosz 2011, 30–31)

Grosz's argument can be understood as saying that evolution is the process of a relationship between matter and life. Life, rather than being some outside force to matter, as with Descartes, comes from the potentialities of matter itself, but life then changes, shapes, and pushes certain forms onto matter. I point this out because the discourse of human exceptionalism is in opposition to the idea of a relationship between matter and life; human exceptionalism is a type of creationist thinking. Evolution is the name we give to the forces that help select for traits that allow one to reproduce and reach the age of reproduction in certain environments. Evolution is this dynamic between matter and life. Evolution doesn't really care who you are. As such, it tends to repeat adaptive traits across multiple species. To believe that humans are unique and exceptional, against all other forms of life, requires a certain, shall we say, faith. That there exists a group of beings with a certain cluster of traits, that this group has fuzzy boundaries, and that almost any individual trait can be found in another type of being makes sense. But belief in the existence of an indivisible line between animals and humans (one so strong that the phrase "human animal" sounds odd) requires a certain transcendental intervention into what it means to be a human. It should not, therefore, come as a surprise that the creationist advocacy organization, the Discovery Institute, has a program in human exceptionalism, and whose senior fellow, Wesley Smith, spends most of his time attacking vegans and animal rights' advocates (Smith 2010). To think the human as an exception from other animals is to think the human as an exception from evolution; in other words, to think as a creationist. To take seriously the dynamic of matter and its agency is to lessen the power of anthropocentrism.

Animal Performativity

So far we have touched on the importance of a new turn in social theory, one that takes seriously the agency of matter. But this materialist theory includes animals mostly because they had been excluded with all the other matter. Many materialist theorists are as likely to talk about metal or bacteria as they are to talk about other animals. So, the cat sitting in my lap now has agency, but so, too, does the can of soda sitting next to my laptop. So, what we still need is some analysis of the ways that things come to matter, come to be important. Obviously, that is far too big of a topic for an essay of this scope, but through an analysis of material and animal performativity, we can come to understand one of the ways we assign hierarchy to beings.

Drawing on J. L. Austin's famous analysis in *How to Do Things with Words*, a performative utterance is a speech act that performs the thing it speaks. Classical examples include a minister announcing people are married or an umpire calling a strike. The very statement of the action causes it to manifest. This concept was taken up by Judith Butler to articulate the idea of gender performativity in her books *Gender Trouble* and *Bodies That Matter*. For Butler, gender is something iterative, that we perform with our bodies over and over again. Take the notion that pink is a feminine color and that blue is a masculine color. Now clearly, there is nothing inherent in those colors that make them gendered. Rather, by acting as if pink is for girls and blue is for boys, we perform gender. We create norms within society, and those norms get disseminated and reiterated until the whole thing becomes naturalized. Thus, just as a minister can make someone married with a performative utterance, we make gender through the performances of our bodies and the norms of society. But that also means gender is a profoundly social experience. Performativity only works within a social field. I am not a minister, and I cannot run up to strangers on the street and declare them married and expect that my declarations make it so. In the same way, I cannot really have a private gender. Thus, for Butler, gender only makes sense within a social field.

Now, you may be asking, isn't this the same sort of social construction we began this chapter rejecting? Here is where Barad's work comes in, because she adapts Butler's understanding of gender performance and proposes a "posthumanist performative approach . . . that specifically acknowledges and takes account of matter's dynamism" (Barad 2007, 135). Performativity is not something one does in isolation, but as Barad shows, it is also not merely social. To perform an action requires you to engage with other beings, and the material reality of those you intra-act with pushes back against a merely social performance. As Barad argues, "matter does not refer to a fixed substance; rather, matter is a substance in its intra-active becoming—not a thing but a doing, a congealing of agency" (151, emphasis excluded). When we perform, our bodies and beings are put into a relationship with other bodies and beings. If you are a sculptor, you perform that role in relationship with the material you sculpt. But that relationship is true with all performances, which cause us to understand ourselves as embodied and undergoing processes of change with material actors outside of ourselves and outside of mere representation. I would hope that Barad's understanding of posthuman performativity can be seen as a way of beginning to undo the Cartesian dualism with which we began our discussion. Rather than seeing ourselves removed from the potentiality of matter, finding our embodied relationship with the world around us gives us a way to understand how we produce and co-constitute with the matter around us. Rather than separated out from the world, we are entirely embedded, entangled, and bound up with the matter around us. As I type, the caffeine in my system is helping to fuel the late night writing; the computer and books that are balanced precariously on my lap are reminding me of the force of gravity, and the cat who is trying to sleep on this keyboard is telling me that he would really like to be fed, please. Posthumanist performativity is an experience of embodiment that exceeds purely human sociality and language and undoes the dualism that excludes the nonhuman from our normative demands.

Because of the reality of matter, we experience posthuman performativity as a kind of resistance to our fantasies of isolated and independent action. Even for Butler, matter resists. As Butler points out,

What I would propose in place of these conceptions of construction is a return to the notion of matter . . . as a process of materialization that stabilizes over time to produce the effect of boundary, fixity, and surface we call matter. That matter is always materialized has, I think, to be thought in relation to the productive and, indeed, materializing effects of regulatory power in the Foucaultian sense. (Butler 1993, 9–10, emphasis excluded)

That last part is important, for one of the ways that matter enters the social field, for Butler, is as a regulatory norm. Take the concept of gendered colors again: just because pink's link to femininity is socially constructed does not protect someone who is seen as male and wears pink from being ridiculed, stigmatized, and even physically attacked. And that threat of violence and marginalization means that even if norms are socially constructed, they still can be successful at regulating performance. Butler is pointing out that the norms of gender are maintained through material disciplinary relationships. So, one of the ways we build and maintain hierarchies—in other words, one of the ways we determine which bodies and lives matter—is by how well those bodies perform within the norms that are placed upon them. And it is not just gender that is regulated; animals and animality are similarly regulated.

Lynda Birke, Mette Bryld, and Nina Lykke (2004) argue in their article "Animal Performances" that "the noun 'animal' is linked to a plethora of hegemonic discourses (philosophical, scientific, etc.), which rely on underlying assumptions about the essence or identity of 'animal' or 'human'. Their effect is to sustain the opposition of Human/cultural subject versus Animal/natural object" (169). What they suggest is that animals need to perform what humans understand as their very animality. If animals fail the norms we expect for those animals, they are met with the sort of violent regulatory power that Butler understood for those who fail the norms of gender. In order to support that point, you need only think of how humans react when wild animals come into urban areas, or when domesticated animals act "feral" and hurt a human, or the vast campaigns of extermination we take against so-called invasive species. We have a whole apparatus designed to materially discipline other animals so that they perform the norms we expect. But Birke, Bryld, and Lykke (2004) are not simply interested in the ways we violently demarcate the line between humans and animals; they are also interested in the ways we become undone as solitary humans in our relationships with other animals. "In particular, understanding 'what animals do' when the animals in question are living in close proximity to us (e.g., companion animals) means understanding how animals themselves participate. It also means understanding how both human and animal are engaged in mutual decision-making, to create a kind of choreography, a co-creation of behavior" (174). This multispecies choreography reminds us of Barad's intra-acting posthumanist performativity, in which our relationship to the material other is always a doing and a co-becoming.

At the same time, Birke brings forth one of Butler's key insights that she adapts from Foucault's model of resistance. If gender is performative, then that performance has to be done over and over again—it is iterative. And while there are real, material, and violent mechanisms making sure the norms are correctly performed, each performance is also a chance to resist, to perform otherwise and potentially break down the governing norms and regulations. In the same way that we can produce different performances of the human and the animal, we can engage in different kinds of co-becoming and co-belonging. Birke's specific analysis is of the lab rat. Rats are often made to linguistically and discursively disappear, such as "300 rats were used," or they are treated simply as another tool in the laboratory with little or no discussion of the living conditions of the rats (Birke, Bryld, and Lykke 2004, 172). But in order for the lab rats to disappear as mere objects in articles requires a whole apparatus of production. There are specialized cages and equipment that have to be used and different strains of rats that are chosen for different kinds of experiments. There is intensive training that human laboratory staff has to undergo in order to handle the rats appropriately for scientific protocols and safety. Furthermore, the rats themselves can become resistant subjects through biting, squealing, and running away. Laboratory rats are not mere tools or objects but are agents in an expansive process of intra-action that seeks to materially produce the performance of *laboratory rat*. Birke, Bryld, and Lykke (2004) hope that focusing on the materiality of the rodents will refuse the discursive trick that makes them disappear and instead help us to realize "we are all matter, and we all matter" (178).

I want to push this analysis in a slightly different direction. If so much of human-animal intra-action is about reproducing certain kinds of animal performances, then what sort of changes can we make to alter the iterations and produce ways of co-becoming? A farm sanctuary, for example, can change the norms of how we are to interact with other animals who were once considered livestock. Veganism can be seen as a kind of performance meant to change the norms by which we are to understand human relationships with animals (see chap. 27). Animal sanctuaries, instead of zoos, cause us to change what it means to respect other animals (see chap. 23). Creating alternative performances are necessary to produce new norms about humans and animals, and those norms are key to shifting how we think about what bodies and lives matter.

The Night Where All Cows Are Stone

So far we have been focused on how taking seriously the importance of matter allows us also to make other animals matter more. But there is a worry that focusing on matter, rather than extending an inclusion, could undermine the very normative grounds on which animal ethics stand. As Peter Gratton (2014) points out in his discussion of object-oriented ontology,

> To be a bit pithy, this risks an object-oriented political correctness, where difference accorded to human and animal others is dismissed normatively on ontological grounds. To paraphrase Arendt's dictum from a very different context, namely that where all are responsible . . . none are: where *everything* is due justice and respect, then nothing is. (Gratton 2014, 119, emphasis in original).

What Gratton is arguing is that there might be something uniquely morally important about humans and animal others but that our increased concern for things generally might actually undermine appeals to extend justice and respect to marginalized humans and other animals. One example of something like this can be seen in Bennett's *Vibrant Matter*. In a footnote, Bennett (2010) claims, "this fermentation seems to require some managing to ensure, for example, that all the ingredients are in the pot. It seems to require humans to exercise this 'executive' function" (150). This assertion seems to fall in conjunction with her claims that she cannot understand any public where human needs "would not take priority. . . . To put it bluntly, my conatus will not let me 'horizontalize' the world completely. I also identify with members of my species, insofar as they are bodies most similar to mine" (104). So, Bennett undermines a certain anthropocentrism with one move, but at the same time she brings forward claims for human exceptionalism and speciesism. To put it another way, while new materialism has done a remarkable job of challenging the anthropocentrism of ontology and charting new ways to conceive of a more egalitarian understanding of actors in this world, maybe it has only done so by increasing the normative hierarchy of humans. Let's take a more common example to understand how this works.

Those of us who have advocated for vegetarianism and veganism in our diets have heard again and again the question, "What about plants?" This question is seldom raised by anyone who is morally concerned about the eating of plants. Indeed, because eating animals who are fed plants causes more destruction of plants than a vegan diet, those few who do actually care about plants tend to advocate veganism (Hall 2011). Rather, the question is raised as a way of implying that if all food is morally problematic to eat, than all food must be allowed. The logic is something like this: if all things we eat have some sort of moral worth, than eating can never be innocent. And if we are all guilty when we eat, than the vegan and the meat eater are on the same ethical plane, and the only thing that matters again is arbitrary human exceptionalism. As Lori Gruen (2015) puts it, writing specifically about new material feminists, "recognizing life and its various entangled processes doesn't necessarily help us to respond to differences among kinds of fellow creatures" (69). And so the "what about plants?" objection can be seen as trying to reduce the differences between different kinds of life, different kinds of matter. Now, somehow, seaweed is the same as tofu, which is the same as oysters, which is the same as pigs, which is the same as, well, you get the idea. This example shows us how a supposed egalitarian move, to include plants, seeks to remove differences, and in so doing, ends up recentering the very anthropocentrism that some forms of materialism claims to be dismantling. The issue of plants is never meant to extend our moral calculations but rather to reduce them.

The specter that Gratton and Gruen raise, then, is very real. If all matter is active, lively, agential in this world, than why should I care more about our treatment of cows than our treatment of stones? If everything matters, than does anything really matter? Was the problem of Cartesian dualism never the dualism but simply that he excluded animals from counting as lives that mattered?

I won't pretend that these are easy questions. And it may be that, in the end, attending to matter is not helpful for animals after all. However, it should also not come as a surprise that our systems of morality and politics, shaped by an almost exclusively anthropocentric world, might not be immediately up to the challenge of understanding a posthumanist world. While the questions "What about plants?" and "What about stones?" are meant to reduce our ethical reasoning, maybe we should take the test seriously to bring our ethics and politics to stranger places. We can only do so by learning to balance the demands of egalitarianism with the respect and care of our differences. We might not be able to get to those places on our own. If our experiments with understanding matter have taught us anything, it is that we are not the independent protagonists of this more than human story. Maybe it is as we develop new performances, new co-becomings, new norms and relationships with other animals and our very humanity that we can ultimately produce an ethics outside of human exceptionalism. Maybe it is only together that we can finally figure out what *matters*.

Suggestions for Further Reading

Alaimo, Stacy. 2010. *Bodily Natures: Science, Environment, and the Material Self.* Bloomington: Indiana University Press.

Barad, Karen. 2007. *Meeting the Universe Halfway: Quantum Physics and the Entanglement of Matter and Meaning.* Durham, NC: Duke University Press.

Bennett, Jane. 2010. *Vibrant Matter: A Political Ecology of Things.* Durham, NC: Duke University Press.

Chen, Mel Y. 2012. *Animacies: Biopolitics, Racial Mattering, and Queer Affect.* Durham, NC: Duke University Press.

DeLanda, Manuel. 2011. *Philosophy and Simulation: The Emergence of Synthetic Reason.* New York: Continuum.

Grosz, Elizabeth. 2011. *Becoming Undone: Darwinian Reflections on Life, Politics, and Art.* Durham, NC: Duke University Press.

Haraway, Donna. 2008. *When Species Meet.* Minneapolis: University of Minnesota Press.

References

Barad, Karen. 2007. *Meeting the Universe Halfway: Quantum Physics and the Entanglement of Matter and Meaning.* Durham, NC: Duke University Press, 2007.

Bennett, Jane. 2010. *Vibrant Matter: A Political Ecology of Things.* Durham, NC: Duke University Press, 2010.

Birke, Lynda, Mette Bryld, and Nina Lykke. 2004. "Animal Performances: An Exploration of Intersections between Feminist Science Studies and Studies of Human/Animal Relationships." *Feminist Theory* 5 (2): 167–83.

Butler, Judith. 1993. *Bodies That Matter: On the Discursive Limits of "Sex."* New York: Routledge.

Descartes, René. 1984. *The Philosophical Writings of Descartes.* Vol. 1. Cambridge: Cambridge University Press.

———. 1991. *The Philosophical Writings of Descartes.* Vol. 3. Cambridge: Cambridge University Press.

Gratton, Peter. 2014. *Speculative Realism: Problems and Prospects.* New York: Bloomsbury Academic.

Grosz, Elizabeth. 2011. *Becoming Undone: Darwinian Reflections on Life, Politics, and Art*. Durham, NC: Duke University Press.

Gruen, Lori. 2015. *Entangled Empathy: An Alternative Ethic for our Relationships with Animals*. New York: Lantern Books.

Hall, Matthew. 2011. *Plants as Persons: A Philosophical Botany*. Albany, NY: SUNY Press.

Smith, Wesley. 2010. *A Rat Is a Pig Is a Dog Is a Boy: The Human Cost of the Animal Rights Movement*. New York: Encounter Books.

16 MIND

Kristin Andrews

Mind is what makes us so characteristically ourselves and what is of utmost significance to us. The mind is the atheist's soul. It is the thing that will remain as long as I am me. And while my mind *is* me, it is also what *allows* me to experience the world and myself, and it is the repository of memory and the source of phenomenal experience. It is what is trapped in the limbless, disfigured body of a World War I soldier in *Johnny Got His Gun*, what Helen Keller was able to finally share once Anne Sullivan tapped a representation of water into her wet palm.

Something like this reflects the contemporary dominant Western opinion about the nature of mind. The mind is a physical object, a system, or a process that is the seat of consciousness, self, rationality, cognition, emotion, perception, and sensation. But because it can express so much, the term *mind* can be easily misunderstood.

Functioning as either a noun, verb, or adjective, *mind* is often associated with a host of other words that are equally poorly understood. Associated with mind are emotions, thoughts, sensory experiences, mental imagery, understanding things (language, other people, math, counterfactuals, biology, and so forth), dreaming, will, action, consciousness, and the self. These aspects of mind are not fully unified; while we might agree that my cat has sensory experiences, we might also doubt that she understands how to host a dinner party or that she considers what she might do tomorrow. We might even doubt that she dreams or is a self. The ability to even ask these questions is enough to suggest that there are various kinds of minds—and that when examining mind we shouldn't look for one monolithic type (Dennett 1996).

For example, the dolphin mind probably involves echolocating inside objects, perhaps even observing the speed of their companions' beating hearts. Using sonar to "see" organs and fetuses inside of other dolphins might facilitate a kind of telepathy such that feelings are transparent. Dolphins might further use echolocation to develop a group mind (White 2007).

The octopus mind allows octopuses to engage in multiple activities at once, as the distribution of neurons in the legs and the central body appears to permit simultaneous divergent control. Central neurons control gross movement while local neurons control local movement, and in some cases these sources of control may be in conflict—the typical functioning of the octopus may involve a sense of cooperation between two or more agents living within one body (Godfrey-Smith 2016b).

We can also find variation among human minds. Neurodiversity, cultural differences (Nisbett 2003), and even linguistic differences (Boroditsky 2003) can shape the way people see the world. For example, some argue that autistic minds are more sensitive to sensory stimuli than neurotypical minds (Pentzell 2013).

The existence of different types of minds has both an upside and a downside. The diversity of minds can be helpful if we are interested in properties that are necessary for mind; candidate properties would have to be evidenced in all the exemplars, while the differences would be interesting variations. Given vast differences in the minds we have already identified, what is shared may turn out to be very shallow, or evolutionarily old. Knowing what is shared can in turn help us identify yet more types of minds on our planet.

However, the diversity of minds also raises some methodological and epistemic concerns about how and whether anyone can know a mind that is a different type from one's own. The philosopher Thomas Nagel answered his question "What is it like to be a bat?" by saying "We cannot know" (Nagel 1979). He thought that since the minds of bats are so different from the minds of humans, we cannot imagine being a bat (who flies, uses sonar, has bat social structures, etc.) but that the best we could do is to imagine what it would be like for *us* to be bat-like. That there are so many different and potentially unknowable minds on this planet adds to the complexity of providing a clear analysis of mind.

René Descartes, in his *Discourse on Method*, appears sympathetic to Nagel's position insofar as Descartes's cogito argument concludes that we can only know our own mind. He says I can doubt many things—that there is an external world, that the laws of math are true—but the one thing I cannot doubt is that I exist as a minded thing because the necessary condition for doubting is to be a doubting thing, and doubting things are minded. I might be a brain in a vat, a ghost, or stuck in the *Matrix*, but nothing is more certain than the existence of my experience.

However, while I may have intimate knowledge of my sensations—hunger, heat, euphoria—it doesn't follow that I have intimate knowledge of my *mind* if mind is an entity, system, or process that permits those sensations. To understand any mind, we need to do more than look inward. One useful strategy to understand mind better is to start with an investigation into what mind lets us do—what the *functions* of mind appear to be. If we can understand the purpose of having a mind and the benefits mind provides to organisms, then we can begin to identify different kinds of minds. And once we have an array of different kinds of minds identified, we can better investigate what all those minds have in common.

Functions of Mind and Structures of Mind

What are the sorts of things minds do? Minds experience, they feel things. Conscious awareness (or qualia—how things feel) is among the most apparent aspect of our minded selves. Minds (or minded organisms) smell the perfume of night-blooming jasmine, feel the stickiness of the humid summer evening, taste the sour wedge of lime. Minds (or minded organisms) feel hungry, tired, playful. Minds (or minded organisms) also think about things. We can consider how to solve a problem, where to go to get food, who is an ally or foe, when the fruit in this tree will be ripe, why a friend is banging two rocks together. Some minds can also think about what they think or feel; we can consider what to do to resolve hunger or whether we enjoy the scent of jasmine. Minds think and feel in order to act, to move about in the world. Hunger causes one to approach food, and thinking the fruit is now ripe will lead one to the correct tree. Feeling playful and seeing an ally can lead one to solicit social play from the right individual.

That minds do these things and that many species of animal do many of these things should result in a default position that many other animals have minds. However, that only humans have minds has been a familiar position in the Western world. This may be due in part to the Abrahamic tradition's notion of an immaterial soul that can preserve the self after bodily death. This view is less common in some Eastern traditions that do not create a dualism between nonhumans and humans (Asquith 1991).

Once philosophers and scientists started challenging the idea that there is an immaterial soul, physicalist approaches to mind started to gain traction, and the idea of nonhuman consciousness became more readily accepted. In 2012 a group of scientists signed the Cambridge Declaration of Consciousness in Non-human Animals, which states that

> convergent evidence indicates that non-human animals have the neuro-anatomical, neurochemical, and neurophysiological substrates of conscious states along with the capacity to exhibit intentional behaviors. Consequently, the weight of evidence indicates that humans are not unique in possessing the neurological substrates that generate consciousness. Nonhuman animals, including all mammals and birds, and many other creatures, including octopuses, also possess these neurological substrates. (Low et al. 2012, 2)

The document also reports that because neuroscientific research routinely uses animals as research subjects, most studies assume homologous brain structures that would imply nonhuman subjects are also conscious. If they were studying consciousness in nonconscious subjects, then their research wouldn't have gone nearly as well as it has.

Nonhuman animal thinking is another matter. When we ask whether animals think, we might be wondering whether they solve problems, engage in logical reasoning, or whether they predict what others are going to do by thinking about what others believe and want. These abilities may be interrelated. Consider the thought experiment introduced by the Stoic philosopher Chrysippus: a dog is tracking a rabbit's scent along a path when he arrives at a three-way crossroad. The dog quickly sniffs each of the first two paths, then, not finding the scent in either of the first two locations, runs down the third path without smelling first. Chrysippus's dog appears to have made a logical inference based on the premise that the rabbit had to run down one of the three paths, the rabbit didn't take path A or path B, and therefore the rabbit must have taken path C.

While Chrysippus's story may be apocryphal, we know today that various species can solve similar problems. Psychological research into exclusion reasoning (along the lines of A or B, not A, therefore B) found the ability to reason in this way in different species—including birds, such as the kea (O'Hara et al. 2016), dogs (Erdőhegyi et al. 2007), and nonhuman great apes (Call 2004, 2006; Marsh and MacDonald 2012).

If we look at the behavior, we can say that the animals are able to solve logical problems. However, one might think that behavior isn't definitive evidence and that we need to know the structure of the mind or brain in order to know whether the animal thinks logically. For example, logical reasoning might be understood as the ability to represent information as propositions (the meaning of sentences) and draw inferences from sets of propositions. This intuitive position, advocated by the philosopher Jerry Fodor, takes thought to have a unique vehicle—we think in something like language. The *language of thought hypothesis* proposes that we think in a language-like representational medium, implicitly relying on rules and symbols that permit rational thought needed to solve logical problems (Fodor 1975).

Chrysippus's dog may be reasoning logically if dogs think in the language of thought, and if this dog represents the following information: the rabbit can only run down path A, B, or C; the rabbit did not run down path A or B; therefore, the rabbit must have run down path C. Given the premises, there is no other possible conclusion—the conclusion cannot be false so long as the premises are true. While we describe our own thoughts and the thoughts of others in terms of propositions and the attitudes individuals take toward propositions, that may be more of an interpretative move than reflective of the structure of thinking. We may say that the dog *believed* the rabbit took path C, and that the dog *desired* to catch the rabbit, and that belief coupled with that desire caused the dog to run down path C. On the language of thought hypothesis, there is literally an internal language that performs this operation.

There is another possibility; the dog's cognitive structure may be organized differently but still permit the same patterns of behavior.

Though a painting can convey information, communicating that information to another will require using linguistic representations that are not existent in the painting itself. That communicated information can be true. Similarly, our attributions of thoughts to humans and other animals can be true without requiring that those individuals think in a language of thought. Perhaps the dog thinks in terms akin to a diagram or map—a representation that still has parts much like a language but doesn't have the full expressive power of a language (Camp 2007). A map can be compositional, since there are parts to a map that can be organized differently— the rabbit may have had two paths to choose from rather than one. And such a map can express different things, such as the possibility that the rabbit ran down one of these two paths, and so maps can be productive and express new meanings. Or perhaps the cognitive structure of the dog is probabilistic in nature. Michael Rescorla suggests that the dog could be using a Bayesian reasoning system that allows one to form and update probability formulas given changes in perceptual information without any awareness of this operation (Rescorla 2009). On this account, the dog implicitly understands before sniffing any road that there is a 33.3 percent likelihood that the rabbit ran down any one of the three roads, and after sniffing the first road, the dog updates the probability for the first road to 0 percent and the probabilities for the other two roads to 50 percent. After sniffing the second road, the dog again updates the probabilities such that there is a 0 percent likelihood the rabbit ran down the first two roads and a 100 percent likelihood the rabbit ran down the third road, so the dog immediately takes off down the third road without needing to gather any additional information.

Rationality is just one function of mind, but one that allows organisms to make choices about how best to fulfill their goals given a complex and potentially dangerous environment. Chrysippus's dog story is an illustration about this aspect of mind. But it would be a mistake to see rationality as separate from other aspects of mind. The dog has emotions motivating the chase, the dog has perceptual capacities that aid in predation, and the dog may even plan what to do when he finally catches the rabbit. Various functions of mind can work together, and when investigating any one function, it is important to determine which other functions may be coming along for the ride.

Contents of Minds

If the dog does not think in language, can we still use language to describe what the dog is thinking? Can Fido believe that there is a bone buried in the backyard? One worry about an affirmative answer to this question is that we will misrepresent Fido's beliefs and desires if we use language, and our concepts, to describe them. If Fido doesn't have the concept of *bone*, along with knowledge of different kinds of bones (including ear bones, whale bones, etc.) and knowledge of the function of bones and skeletons, then it would be misleading to say that Fido has bone beliefs at all (Stich 1979). We run into morally suspect territory when we ascribe beliefs and intentions to individuals who cannot have those mental states, particularly when the situation is not morally neutral. Differences in gender, culture, development, ability, and so forth, can make it difficult to accurately capture what someone is thinking and, correspondingly, what it is like to be someone other than oneself. When we are dealing with other species, and particularly with species dissimilar from humans in sensory, cognitive, and physical capacities, we run the risk of anthropomorphizing when we ascribe mental states. This is because we are describing animal actions in human terms, and so we are taking them to be members of our human community, with human interests and human concerns. If anthropomorphism is always wrong, then ascribing content to animals in this will lead us astray.

A neutral definition of *anthropomorphism* is the attribution of human traits to nonhumans. From this definition it cannot always be wrong to anthropomorphize, because humans actually share many traits with other organisms. We are physical objects. We are biological entities; we have eyes, hearts, live births, and so forth. A normative definition of *anthropomorphism* is the attribution of uniquely human traits to nonhumans. Given this definition, any anthropomorphic attribution to a nonhuman would be a mistake, because the trait wouldn't be uniquely human if a nonhuman also had it. The scientific study of human and nonhuman mind includes the attempt to find out what properties humans and other animals do indeed share, and until we know what if anything makes humans unique, the charge of normative anthropomorphism will be premature.

While worries characterized as normative anthropomorphism may be premature, we can still worry about bias in attributing properties to other animals, just as we worry about bias in attributing properties to other humans, and in particular humans who somehow differ from oneself. When attributing content to other humans, we must be careful to avoid egocentrically thinking that others are just like us. On the other hand, we have to be careful not to xenophobically think that others are completely different from us. Similarly, when considering other animals, we need to avoid an unwarranted anthropectomy, or the rejection of human traits in nonhuman animals (Andrews and Huss 2014).

Instead of attributing content that reflects our standpoints or biases, we can carefully ascribe content to nonlinguistic creatures just as we do for humans who are other—to humans from other cultures who speak languages other than one's own or to human infants who do not yet use any language. The anthropomorphism critique suggests we ought not apply concepts to animals that the animals cannot apply to themselves. "We" should not apply our concepts to "them." But this position is too strong, as it ignores that there are many different human standpoints from which we attribute content to many different kinds of individuals.

Consider infant research. For adults, infants are other. Nonetheless, human parents and caregivers develop a kind of expertise regarding the minds of human infants. Perhaps part of this expertise comes from being members of the same species, but there are also significant differences in physiology between adult and infant humans, since infants are just developing their sensory systems, digestive systems, and cognitive-affective systems. A significant condition for developing expertise in the minds of human infants comes from spending lots of time observing them and interacting with them. From their quotidian interaction with infants and from their need to coordinate with infants to soothe them, feed them, and entertain them, parents and caregivers have come to interpret infant behaviors as having certain meanings. This prescientific knowledge base has allowed scientists to construct methods for studying human infants. For example, in order to determine whether infants think two things are the same, or go together, researchers will show infants some stimulus over and over again until the infant gets bored. The researcher *interprets* that the infant is bored because she stops looking at the image. At that point the infant is shown a new stimulus, and if she looks longer at the new stimulus, the researcher *interprets* this as evidence that the infant sees the new image as something different from the first. There is no problem with the researcher interpreting infant behavior in order to draw these conclusions; interpretation is needed to get the research off the ground. It is also needed to get communication off the ground. However, we have to remain vigilant and acknowledge that the interpretation might be wrong, particularly when we find additional behavior that doesn't fit neatly into the current picture.

Being minded can involve having beliefs, and if beliefs involve acting in certain interpretable ways, then we can ascribe beliefs to individuals who lack language without getting things wrong. The philosopher Daniel Dennett argues that beliefs are not identical with brain states or processes. Investigating the brain isn't going to help us figure out what someone really believes. We must instead analyze patterns of behavior. According to Dennett's intentional systems theory, any system whose behavior can be reliably and voluminously predicted by attributing folk psychological terms about belief and desire will be an intentional system (Dennett 2009). Dennett's "intentional stance" is an interpretive lens through which we can see behavior as being caused by beliefs and desires. These mental states together lead to action. Folk psychology is the approach to understanding intentional behavior in these terms.

To illustrate, when playing chess, we might consider the beliefs and desires of our competitor. As I threaten the rook, I suppose the competitor thinks the rook is threatened, and I suppose that they want to save the rook. This will lead me to look for ways my competitor will try to achieve the goal of saving the rook. But just believing the rook is threatened isn't enough to try to save the rook, because the competitor might want to sacrifice the rook in order to move me into a vulnerable position. This method isn't infallible, and it doesn't always lead to accurate prediction. But after the fact, we can explain the belief and desire the competitor had at that point by considering both what they did and what their goal was. Beliefs and desires together cause action, on the folk psychological model, and can only be analyzed given a large body of behavior.

Dennett would say that being minded is just acting in such a way that behavior can be accurately predicted and explained through the attributions of beliefs and desires. There is nothing more to having a belief than to act reliably as if you have that belief. Thus, as in this case of understanding what beliefs a monkey might have, to determine whether any animal has a belief, Dennett suggests we take the following steps:

> First, observe their behavior for a while and make a tentative catalogue of their needs—their immediate biological needs as well as their derivative, *informational* needs—what they *need to know* about the world they live in. Then adopt what I call the *intentional stance*: treat the monkeys as if they were—as they may well turn out to be—rational agents with the "right" beliefs and desires. Frame hypotheses about what they believe and desire by figuring out what they *ought* to believe and desire, given their circumstances, and then test these hypotheses by assuming that they are rational enough to do what they ought to do, given those beliefs and desires. The method yields predictions of behavior under various conditions; if the predictions are falsified, something has to give in the set of tentative hypotheses and further tests will sift out what should give. (Dennett 1998, 207)

Here we have a picture of mind not as some object that interacts with the body in some way but as the sorts of thing some bodies do.

How Do We Know Other Minds?

As previously noted, for some time in the West the connection between having a soul and having a mind led some to deny mind to nonhuman animals. But the worry about whether something has a mind need not be related to any religious commitment. Some skeptical philosophers even wondered whether other humans have minds—solipsism is the philosophical position that the only thing that exists is one's own mind and that all one sees is a creation of one's mind. Descartes methodological skepticism demonstrated how we can doubt other minds along with the external world. While Descartes himself avoids the solipsistic consequences of his argument by appealing to the existence of a benevolent god, that move was not universally compelling. Thus, Descartes raised the problem of other minds: *how do we know that there are other minds?*

John Stuart Mill suggested that we can solve this problem by arguing from analogy. By considering our own characteristics and noticing that other humans also share those characteristics, we can conclude that other humans have minds. The more characteristics we have in common, the more likely it is that the characteristic we are investigating is also shared by the target. Consider two oranges. You peel one and eat the fruit; it is sweet. You have another orange, unpeeled, next to the first. You know many properties shared by both fruits, but what you don't know is whether the second is going to be sweet. You know that they both come from the same tree, that they both look the same, that they have the same weight in your hand. As you peel the second orange, you gain further information that the flesh looks just as plump and juicy as the first orange. As you gain knowledge, your certainty that the second orange will also be sweet increases because the more the two pieces of fruit have in common, the more reason you have for inferring that the unknown property will also be shared.

Similarly, the argument from analogy for other human minds works by noting similarities between oneself and all other humans. The argument goes like this.: I have a mind and I have some set of properties M. Other humans also have the set of properties M. Therefore, other humans probably have a mind. One problem arises when we ask which properties are M properties. I know I have a mind because I have phenomenological experience of wondering whether others have minds, and that experience is sufficient for having mind. But if we have direct access to our own minds but not the minds of others, those properties will never be apparent to us in the other. If we choose M properties that are irrelevant, such as shaped like me, or having the same skin color as me, then our inference is unjustified. By looking just at oneself, it is hard to know which properties would be relevant. Having a brain or being alive are thought to be associated with mind not because we generalize from merely our own case but because we take it for granted that all humans are minded.

Starting with the assumption that humans have minds (so long as they are old enough and not functioning atypically—issues that raise their own set of questions), we can apply the argument from analogy to determine whether other species have minds. So, one might reason as follows. All humans who have minds have some set of properties M. Individuals of species A have the set of properties M. Therefore, individuals of species A probably have minds. Once again we have the question of what the properties are, but here we can use the scientific method to investigate what all those humans who are minded (call this set B) have in common and which of those things are also had by humans who we think are not minded (call this set C). Those properties in set B that are not also in set C will give us the set of properties M.

However, this method presupposes we have a clear sense of what it is to have a mind. How is it that we decide what is minded? We cannot assume all humans are minded, since clumps of human DNA lack brains and do nothing that appears to be minded, and dead humans do not appear to be minded. So what is the appearance of mindedness? Again, the answer is that we look toward patterns of behavior that we can interpret in a meaningful way; we turn back to the functions of mind. What seems to be central to appearing as minded is to be comprehensible in the sense of folk psychology such that the behavior makes sense to us. We see mind when having mind is the best explanation for what the individual does.

Folk psychology, that commonsense understanding of people as being moved by their desires, goals, feelings, perceptions, and so forth, is impossible to do without. It is our way of seeing certain kinds of movement in the world. Worries concerning whether there are other minds and whether they are knowable are symptoms of a kind of philosophical sickness; they are just as fantastic as worries about the existence of the external world. They force us to deny what we see clearly in front of our face. Rather than figuring out whether other individuals are minded, we might directly see minds, and the worry that we do not stems from the metaphysical commitments of the Abrahamic religions, which take as given a species of Cartesian dualism between mind stuff and body stuff. As David Hume put it, "Next to the ridicule of denying an evident truth, is that of taking much pains to defend it; and no truth appears to me more evident than that beasts are endowed with thought and reason as well as man. The arguments are in this case so obvious, that they never escape the most stupid and ignorant" (Hume, *Treatise of Human Nature*, 176). We experience others as minded, and so they are minded. To think otherwise would be to deny the obvious. We directly perceive minds in humans and in other animals when we are interacting with them (Searle 1994; Jamieson 1998).

A worry arises when we consider that humans may see minded action where there isn't any. We too often project what we hope to find. In a famous experiment, psychologists showed humans a film of three geometrical objects moving around a screen. Participants were asked to describe what was going on, and in many cases they described movement as intentional. For example, when asked to describe the situation, some participants created a story: the two triangles are men who are fighting over the circle—a woman (Heider and Simmell 1944). Attributions of folk psychology commonly extend beyond minded individuals. Humans commonly see facial expressions in clouds, cars, and buildings. We may be fooled by computer programs like Siri or ELIZA into thinking they have beliefs and desires. Androids are being developed to mimic human facial expressions. Our seeing minded action doesn't mean that minds are in fact causing action.

One possible reply is to show how the worry relies on a commitment to Cartesian dualism. The assumption here is that there is a thing—the mind, which has causal powers over the body, and which is affected by effects on the body. This commitment is a thread that runs through most worries that arise about the existence of animal minds. If the commitment could be put to rest, many of the arguments against animals being minded creatures would be as well. One set of worries still remains. The human perspective on the world is naturally constrained in terms of environment, size, and timescale. For individuals who live a lifetime in a day, it may be difficult for us to notice subtle movements that, were they slowed down so we could notice them, would look like intentional minded action. For individuals who move in ways different from us because their bodies are morphologically very different from our own—fish, spiders, bees, whales—we may fail to pick out the meaningful patterns that reflect the mindedness of these species. And for animals who live in the air, underground, or in the water, we would similarly have difficulties seeing minds.

Panpsychism, Biopsychism, Zoopsychism

So far we've been looking at minds in individuals and the evidence for mindedness in interpretable behavior. Other arguments for mind come from considerations about the nature of mind—and in particular consciousness—itself. We can briefly look at three views that have different implications about what sorts of things have conscious experience. Panpsychism is the view that mentality is a part of the fabric of the world, not something limited to animals or living organisms. Everything is minded to a certain extent. Biopsychism is the view that everything that is alive is minded—animals, plants, and the biological building blocks of animals and plants, including cells. "Where there is life there is mind, and mind in its most articulated forms belongs to life" (Thompson 2007, 1). Zoopsychism is the view that all animals are minded.

Panpsychism has a long history and comes in a variety of flavors. Idealists are natural panpsychists because of their commitment to mind as the only thing that exists. Physicalists, who think that the world consists only of concrete, spatially, and temporally located physical phenomena, can also be panpsychists. A physicalist argument for panpsychism can be made from the notion *ex nihilo nihil fit*—that nothing comes from nothing. Since there is mentality in us, there must be mentality in what we are made of. But since we are made from material substance, mentality must be a property or part of that material substance, and the complex mentality humans enjoy is a matter of being built up from the bits of mentality found in the matter that composes us. Galen Strawson (2006) argues that a real commitment to physicalism requires a commitment to panpsychism, since any other option involves some kind of magical emergence of mind from nothing.

Biopsychism is aligned with the idea that mind is not something one has but rather something one does. In Evan Thompson's view (2007), living beings are autonomous agents that actively generate and maintain themselves, creating their own cognitive environments. The nervous system of animals would be an example of an autonomous dynamical system, but so would the self-sustaining processes of a cell that is actively involved in a causal loop that self-organizes the cell and preserves its boundaries via metabolic processes. Living cells exist in relation to an environment that they shape into an *Umwelt*—an environment of norms and meaning. This leads Thompson to commit to the idea that cells are only fully understood in terms of their norms in that cells modify their environments so as to fulfill their goals (Thompson 2007). In this view, mind is a kind of know-how.

Zoopsychists accept a similar argument but see these actions as only applicable to animals—organisms who can sense their environment and actively cope with it. A zoopsychist might point to a set of properties necessary for mind similarly to those who made arguments for the existence of other minds by analogy. Peter Godfrey-Smith suggests that there may be a distinctive set of animal properties that are essential for subjective experience consisting of sensory-motor capabilities that allow for behavior in real time, fine-grained processing of sensory information, learning and remembering (Godfrey-Smith 2016a). Since we don't see this set of properties at the cellular level, the biopsychists' claim that cells are minded doesn't hold. More cells would be needed, and information would have to flow between systems. However, multicellularity didn't evolve only once on this planet; animals and plants share a common ancestor that most likely was a single-celled organism that existed 1.6 billion years ago (Meyerowitz 2002). When we turn to look at nonanimal multicellular organisms, such as plants, there is interesting recent scientific research suggesting plants have the same set of distinctive properties as animals. As we continue to study plants, they continue to surprise us; recently scientists published a study showing evidence of associate learning in the pea plant (Gagliano et al. 2016). There may be no tenable way to hold that all animals are minded without also accepting the existence of plant minds.

Charles Darwin described plants in terms of their sensory-motor capacities in one of his last works, *The Power of Movement in Plants*:

> It is hardly an exaggeration to say that the tip of the radicle [root] thus endowed, and having the power of directing the movements of the adjoining parts, acts like the brain of one of the lower animals; the brain being seated within the anterior end of the body, receiving impressions from the sense-organs, and directing the several movements. (Darwin 1880, as quoted in Calvo Garzón and Keijzer 2011, 161)

With the "brain" of the plant at the root, it is in a location that made root behavior hard to study by early naturalists who didn't have transparent growing medium available. However, the field of plant neurobiology has grown in the early twenty-first century with the creation of the Society for Plant Neurobiology and the journal *Plant Signaling and Behavior*. While plants lack anything like an animal's neurons, we find neurotransmitters like glutamate, dopamine, and serotonin in plants and signaling via electrical activity (Baluška and Mancuso 2009).

Summarizing the current state of the science of what he calls "minimal intelligence," the philosopher Paco Calvo notes that "the list of plant competencies has been growing at a considerable pace in recent years. Plants can, not only learn and memorize, but also make decisions and solve complex problems. In a sense, plants can *see*, *smell*, *hear*, and *feel* (Calvo 2016, 1329). He evaluates a range of scientific studies, but we can simply turn to one case that Darwin was particularly interested in—climbing plants. Behavioral studies out of Mancuso's lab using time-lapse photography show two bean plants looking for a pole to climb up; they both orient to the same pole, and when one of the plants makes contact, wrapping its body around the pole, the other plant veers off in another direction[1] (as reported in Calvo 2016). Mancuso thinks that we have an example of sensory-motor operation through a form of echolocation between two agents. If any of this is approximately right, then we appear to have another case of minds evolving to respond to other minds, beyond animals.

The question of whether there are other minds is ultimately a question emerging from our human curiosity, loneliness, and our complex desires to be unique and at the same time connected.

Suggestions for Further Reading

Andrews, Kristin. 2015. *The Animal Mind: An Introduction to the Philosophy of Animal Cognition.* New York: Routledge,.

Andrews, Kristin, and Jacob Beck. 2017. *The Routledge Handbook of the Philosophy of Animal Minds.* New York: Routledge.

Barrett, L. 2011. *Beyond the Brain: How Body and Environment Shape Animal and Human Minds.* Princeton, NJ: Princeton University Press.

Godfrey-Smith, P. 2016. *Other Minds: The Octopus, the Sea, and the Deep Origins of Consciousness.* New York: Farrar, Straus and Giroux.

Thompson, E. 2007. *Mind in Life: Biology, Phenomenology, and the Sciences of Mind.* Cambridge, MA: Belknap.

References

Andrews, K., and B. Huss. 2014. "Anthropomorphism, Anthropectomy, and the Null Hypothesis." *Biology and Philosophy* 29 (5): 711–29.

Appel, H. M., and R. B. Cocroft. 2014. "Plants Respond to Leaf Vibrations Caused by Insect Herbivore Chewing." *Oecologia* 175 (4): 1257–66.

Asquith, P. J. 1991. "Primate Research Groups in Japan: Orientations and East-West Differences." In *The Monkeys of Arashiyam:. Thirty-Five Years of Research in Japan and the West,* edited by L. Fedigan and P. J. Asquith, 81–98. Albany, NY: SUNY Press.

Baldwin, I. T., R. Halitschke, A. Paschold, C. C. von Dahl, and C. A. Preston. 2006.

---

1. You can see the video narrated by author Michael Pollen: https://www.youtube.com/watch?v=MPql1VHbYl4.

"Volatile Signaling in Plant-Plant Interactions: 'Talking Trees' in the Genomics Era." *Science* 311 (5762): 812–15.

Baluška, F., and S. Mancuso. 2009. "Deep Evolutionary Origins of Neurobiology: Turning the Essence of 'Neural' Upside-Down." *Communicative and Integrative Biology* 2 (1): 60–65.

———. 2013. "Root Apex Transition Zone as Oscillatory Zone." *Frontiers in Plant Science* 4: 354.

Bastien, R., T. Bohr, B. Moulia, and S. Douady. 2013. "Unifying Model of Shoot Gravitropism Reveals Proprioception as a Central Feature of Posture Control in Plants." *Proceedings of the National Academy of Sciences* 110 (2): 755–60.

Boroditsky, L. 2003, "Linguistic Relativity." In *Encyclopedia of Cognitive Science*, edited by L. Nadel, 917–22. London: Macmillan.

Call, J. 2004. "Inferences about the Location of Food in the Great Apes (*Pan paniscus, Pan troglodytes, Gorilla gorilla,* and *Pongo pygmaeus*)." *Journal of Comparative Psychology* 118 (2): 232–41.

———. 2006. "Inferences by Exclusion in the Great Apes: The Effect of Age and Species." *Animal Cognition* 9 (4): 393–403.

Calvo, P. 2016. "The Philosophy of Plant Neurobiology: A Manifesto." *Synthese* 193 (5): 1323–43.

Calvo Calvo Garzón, P., and F. Keijzer. 2011. Plants: Adaptive Behavior, Root Brains and Minimal Cognition. *Adaptive Behavior* 19 (3): 155–71.

Camp, E. 2007. "Thinking with Maps." *Philosophical Perspectives* 21 (1): 145–82.

Chamovitz, D. 2012. *What a Plant Knows: A Field Guide to the Senses.* New York: Scientific American / Farrar, Straus and Giroux.

Darwin, C. 1880. *The Power of Movement in Plants.* New York: D. Appleton.

Dennett, D. 1998. *Brainchildren: Essays on Designing Minds.* Cambridge, MA: MIT Press.

———. 1996. *Kinds of Minds: Towards an Understanding of Consciousness.* New York: Basic Books.

———. 2009. "The Part of Cognitive Science That Is Philosophy." *Topics in Cognitive Science* 1 (2): 231–36.

Dicke, M., A. A. Agrawal, and J. Bruin. 2003. "Plants Talk, but Are They Deaf?" *Trends in Plant Science* 8 (9): 403–5.

Dumais, J. 2013. "Beyond the Sine Law of Plant Gravitropism. *Proceedings of the National Academy of Sciences* 110 (2): 391–92.

Erdőhegyi, Á., J. Topál, Z. Virányi, and Á. Miklósi. 2007. "Dog-Logic: Inferential Reasoning in a Two-Way Choice Task and Its Restricted Use." *Animal Behaviour* 74 (4): 725–37.

Fodor, J. A. 1975. *The Language of Thought.* New York: Thomas Y. Crowell.

Gagliano, M., S. Mancuso, and D. Robert. 2012. "Towards Understanding Plant Bioacoustics. *Trends in Plant Science* 17 (6): 323–25.

Gagliano, M., V. V. Vyazovskiy, A. A. Borbély, M. Grimonprez, and M. Depczynski. 2016. "Learning by Association in Plants." *Scientific Reports* 6: 38427.

Godfrey-Smith, P. 2016a. "Individuality, Subjectivity, and Minimal Cognition." *Biology and Philosophy* 31 (6): 775–96.

————. 2016b. *Other Minds: The Octopus, the Sea, and the Deep Origins of Consciousness*. New York: Farrar, Straus and Giroux.

Heider, F., and M. Simmel. 1944. "An Experimental Study of Apparent Behavior." *American Journal of Psychology* 57: 243–59.

Hodge, A. 2009. "Root Decisions." *Plant, Cell and Environment* 32 (6): 628–40.

Jamieson, D. 1998. "Science, Knowledge, and Animal Minds." *Proceedings of the Aristotelian Society*, n.s., 98: 79–102.

Low, P., J. Panksepp, D. Reiss, D. Edelman, B. Van Swinderen, P. Low, and C. Koch. 2012. "The Cambridge Declaration on Consciousness." Publicly proclaimed at the Francis Crick Memorial Conference on Consciousness in Human and non-Human Animals, Cambridge, July 7.

Marsh, H. L., and S. E. MacDonald. 2012. "Information Seeking by Orangutans: A Generalized Search Strategy?" *Animal Cognition* 15 (3): 293–304.

Meyerowitz, E. M. 2002. "Plants Compared to Animals: The Broadest Comparative Study of Development. *Science* 295 (5559): 1482–85.

Nagel, T. 1974. "What Is It Like to Be a Bat?" *Philosophical Review* 83 (4): 435–50

Nisbett, R. E. 2003. *The Geography of Thought: How Asians and Westerners Think Differently . . . and Why*. New York: Free Press.

O'Hara, M., R. Schwing, I. Federspiel, G. K. Gajdon, and L. Huber. 2016. "Reasoning by Exclusion in the Kea (*Nestor notabilis*)." *Animal Cognition* 19 (5): 965–75.

Pentzell, N. 2013. "I Think, Therefore I Am: I Am Verbal, Therefore I Live. In *The Philosophy of Autism*, edited by J. L. Anderson and S. Cushing, 103–8. Lanham, MD: Roman and Littlefield.

Rescorla, M. 2009. "Chrysippus' Dog as a Case Study in Non-Linguistic Cognition." In *The Philosophy of Animal Minds*, edited by R. Lurz, 52–71. New York: Cambridge University Press.

Searle, J. 1994. "Animal Minds." *Midwest Studies in Philosophy* 19 (1): 206–19.

Stich, Stephen P. 1979. "Do Animals Have Beliefs?" *Australasian Journal of Philosophy* 57: 15–28.

Strawson, G. 2006. "Realistic Monism: Why Physicalism Entails Panpsychism." In *Consciousness and Its Place in Nature: Does Physicalism Entail Panpsychism?*, edited by A. Freeman, 3–31. Exeter: Imprint Academic.

Thompson, E. 2007. *Mind in Life: Biology, Phenomenology, and the Sciences of Mind*. Cambridge, MA: Belknap.

Trewavas, A. 2009. "What Is Plant Behaviour? *Plant, Cell and Environment* 32 (6): 606–16.

White, T. 2007. *In Defense of Dolphins: The New Moral Frontier*. Malden, MA: Blackwell.

17 PAIN

Victoria A. Braithwaite

Pain is an enigma. We all know what it is, but only you can experience the pain you feel. Your pain, the subjective feelings it creates, and the hurt that it engenders, are entirely yours. As humans, we can at least use words to describe our aches and pains to one another; we can probably recognize certain aspects of the described symptoms, and we will most likely have some level of empathy and sympathy for someone describing a pain state. But what do we know about the pain of those that cannot talk to us? Here we move into a much murkier realm; what do we know about the pain of an infant or an animal?

In newborn human babies, for example, the sharp lance that pierces the skin of the heel to obtain a tiny drop of blood for tests or the needle of a syringe used to deliver vaccines prompt reactions that look and sound as though the baby is in pain, but is she really? This question tested the ingenuity of neuroscientists and medical researchers, and until relatively recently, several procedures were done on preterm and new born babies with little to no use of analgesics (Qiu 2006). In part, this was because safe, pain-relieving drugs for such small infants had yet to be developed, but it also stemmed from the belief of some clinicians that such young babies did not have a brain that was sufficiently well developed to experience pain—or if they did, they would not remember procedures done to them at that early stage in life, so analgesics were not considered necessary (Fitzgerald 2005, Bartocci et al. 2006; Slater et al. 2006). Today, a

I thank Paula Droege, Dan Weary, and Adam Shriver for their many stimulating discussions in the run-up to writing this chapter and to the Institute for Advanced Studies, Berlin, Germany, for funding a sabbatical year and providing us with the opportunity to have these discussions as part of a focus group on "Pain."

growing understanding of pain mechanisms in young infants is helping the development of procedures and drugs that aim to minimize potential pain experiences (Fitzgerald and Walker 2009). What this example highlights, however, is that evaluating pain processes is difficult when the patient cannot verbally describe their experiences to us.

And what about pain in animals? For many, attributing an awareness of pain and a capacity for that pain to lead to a negative affective state and potentially to suffering seems perfectly appropriate, but others are more skeptical and are unwilling to consider certain animals as having any awareness of their pain (Rose 2002; Fraser 2009). Although many forms of animal research legislation cover all vertebrates and sometimes include protection for the sophisticated cephalopod invertebrates, there are still debates about whether certain taxa such as fish are capable of experiencing pain (Key 2016; but see Braithwaite and Droege 2016).[1] So how do we distinguish between animals who feel pain and those who don't? To answer this question we need to consider the mechanisms that underlie pain processes and then investigate the capacity for these kinds of process to occur in different animals (Melzack and Wall 1988; Fitzgerald 2005; Braithwaite 2010).

If we reflect on why animals experience pain, we see that it is an adaptive process; animals who have the ability to respond to and withdraw from tissue-damaging events will be more likely to survive and protect themselves from injury in the immediate to short term (Kavaliers 1988, 1989). So, a nervous system that supports the sensory detection of tissue damage and helps the animal to react appropriately to noxious events would be highly adaptive (Bateson 1991; Broom 2001). It is not surprising, therefore, that almost all animals have mechanisms in place that allow them to detect damaging or noxious stimuli and help them move away or withdraw from this danger as quickly as possible. Such systems are called *nociceptive* systems.

After this initial nociceptive detection and response, however, some animals do more with this information. Some species become aware of the damage, and this awareness is even more beneficial, because it can help animals minimize further damage and promote swifter healing (Duncan 1996, 2006; Fraser 1999). Being aware of the pain might also improve

---

1. For recent examples, see the many comments (Balcombe 2016a; Brown 2016; Burghardt 2016; Mather 2016) in *Animal Sentience* that were written in response to the article published by Key (2016).

learning and memory associated with the event—the place or the circumstances that led to the damage—and a memory of such events might help an animal avoid something like that from happening again in the future (LaBuda and Fuchs 2000; Allen et al. 2005; Weary, Droege, and Braithwaite 2017). Thus, a felt, negative, hurting sensation might permit animals to respond in a more flexible manner than simply withdrawing and avoiding noxious stimuli (Droege and Braithwaite 2014). However, as I will argue in this chapter, this additional capacity to have an awareness of the hurt related to the noxious experience is not as widespread across the animal kingdom as nociception (Weary, Droege, and Braithwaite 2017).

Recognizing that pain is more than just a single process is helpful because it provides us with multiple capacities to look for in different animals (Fitzgerald 2005; Morrison, Perini, and Dunham 2013; Braithwaite 2014). But at the same time, this knowledge has opened up various pitfalls because there is a tendency to create lists of processes and capacities and then determine how many of these can be demonstrated to occur in certain animal species. While an interesting exercise in comparative biology, such an approach does not reveal much about felt pain unless processes on the list include awareness of the hurt, but such lists typically focus on processes associated with nociception and higher-order cognition (e.g., Elwood 2009; Sneddon et al. 2014). Higher-order cognition may be correlated with the capacity for awareness, but it is not the same thing, so using cognitive ability as a proxy for felt pain is misleading and should be avoided.

There is, however, another reason why recognizing that pain is more than one process is helpful; it contributes to the debate about the evolution of pain and whether it is appropriate to consider nociception (i.e., the detection of tissue damage) separately from the conscious detection of the physical hurt (Wall 2000; Fraser 2009; Dawkins 2015). The view that nociception and the sensory awareness of a painful experience can be dissociated is often eschewed by those working on human pain (Wall 2000; Key 2016), but being able to separate them is helpful if we want to explore the evolutionary history of nociception and felt pain. Thus, demonstrations of dissociation have been extremely valuable (Price 2000; Shriver 2009). Whether this dissociation is induced through the administration of opioids or it arises because of specific damage to key areas of the human cortex, the recognition that detection and the affective component can be separated is a significant observation for understanding the evolution of pain (Shriver 2014).

Typically in a chapter such as this, it would seem appropriate to begin with definitions, particularly for *pain*. I will define it eventually. But in order to define *pain*, we need to know the processes that are involved during a painful experience, so I will begin with a description and an overview of these. The importance of the distinction between nociception as opposed to when something causes an awareness of felt pain will be emphasized. After an initial description of pain-related processes, I will then turn to definitions. I will begin by exploring the definitions that are commonly used and consider how these have sometimes been misinterpreted, which has led to confusion. One of the problems we need to tackle here is that the topic of pain has invited research and scholarly contributions from a broad range of disciplines, and the language and understanding of people from these diverse backgrounds is not always consistent (Varner 1998; Allen 2004; Braithwaite 2010; Elwood 2012; Morrison, Perini, and Dunham 2013). To draw the chapter to a close, I will discuss how pain in animals has been studied and why these approaches tell us a lot about nociception but, so far, very little about felt pain.

Nociception and Nociceptive-Like Processes

The first part of the pain process, as we understand it in ourselves, is the sensory detection of a noxious stimulus that is relayed to the spinal cord and the brain. This is called *nociception*. We are not unique in having this capacity. Many animals have a whole array of apparatuses (i.e., specialized receptors and nerve fibers) dedicated to sensing when damage has occurred (Walters 2009; Dubin and Patapoutian 2010). Nociception, or nociceptive-like responses, appear to be initiated in response to something that is damaging, and these responses play a significant role in promoting animal survival. This is presumably why these kinds of process are found throughout much of the animal kingdom (Kavaliers 1988; Walters 2009). An animal that cannot detect or protect itself against something that is damaging is less likely to survive and reproduce and so would be at a selective disadvantage compared with animals who can detect and avoid injury (Bateson 1991; Braithwaite 2014).

In vertebrates, tissue damage is first detected by nociceptors. These are free nerve endings found in tissues such as the skin, oral and nasal membranes, skeletal muscle, and visceral organs (Walters 2009). The specialized receptors are triggered when extreme incidents arise, for instance, excessive temperatures, mechanical crushing, or noxious chemicals that can injure and damage the body. In vertebrates, there are a number of receptor types, and these respond to different forms of noxious stimulation (Dubin and Patapoutian 2010). Some receptors can detect more than one form of noxious stimulus; bimodal receptors such as mechanothermal nociceptors detect both mechanically damaging and excessive heat stimuli, or mechanochemical nociceptors respond to excessive mechanical and noxious chemical stimuli. There are also polymodal nociceptors that respond to noxious mechanical, heat, and chemical stimuli (Dubin and Patapoutian 2010). Combinations of these different receptors types are widely distributed throughout the skin. Nociceptors play a critical role in defense from damaging stimuli, and they are found in both invertebrates and vertebrates (Kavaliers 1988).

Research examining the effects of noxious mechanical and thermal stimulation in *Dropsophila* larvae has identified a particular form of invertebrate sensory neuron that has nociceptive properties—a multidentritic class IV neuron (Tracey et al. 2003; Hwang et al. 2007). Identification of this invertebrate larval nociceptive neuron has led to the investigation into the molecular mechanisms that control nociception in these flies (Zhong, Hwang, and Tracey 2010; Kim et al. 2012). Fly larvae who are unable to produce mRNA that encodes a protein required by a specific receptor ion channel family do not respond to a heated probe, a stimulus that normal, wild-type fly larvae react to with a strong stereotypical rolling escape response. The flies who lack the mRNA for this particular ion channel family were called *painless*. As I will explain below, this name is unfortunate, and a more accurate name for this mutant line would be *nociceptionless*. But the results of this research are, nevertheless, significant, because they are the first demonstration of a specific mechanism underlying nociception in an invertebrate. And although it had been predicted to be present, the actual nature of an insect's nociceptive system was not described until this *Drosophila* work was undertaken.

Within the vertebrates, specialized nociceptors have been extensively characterized in birds (Gentle et al. 2001) and mammals (Lynn 1994; Yeomans and Proudfit 1996), and more recently, their presence in bony fish was confirmed (Sneddon, Braithwaite, and Gentle 2003; Dunlop and Laming 2005; Norgreen et al. 2007; Roques et al. 2010). When noxious stimuli are detected by the nociceptors, information about the damage is transmitted within the nervous system through specialized types of nerve fibers. The International Association for the Study of Pain (IASP) has perhaps one of the most commonly used definitions for pain that I will cover further below, but I want to note here a crucial exclusion: "activity induced in the nociceptor and nociceptive pathways by a noxious stimulus is not pain, which is always a psychological state, even though we may well appreciate that pain most often has a proximate physical cause" (International Association for the Study of Pain 2012). Making this distinction and recognizing that there are two separate dimensions of the pain process are critically important (Price 2000; Allen et al. 2005; Shriver 2009).

In vertebrates, the processes of nociception and the psychological experience of the negative affect are highly connected with one modulating the other and vice versa. The tight linkage and interaction between these two has led researchers to claim that nociception and the affective component of felt pain should not be dissociated from each other (Wall 2000). Certainly a functioning vertebrate nervous system that is responding to a nociceptive signal involves complex interactions among afferent neurons, interneurons, different brain regions, and descending pathways, but demonstrating that nociception can occur independently of felt pain is a significant observation because it offers an insight into how nociceptive systems may have initially evolved and then become very widespread throughout the animal kingdom (Kavaliers 1988; Allen et al. 2005). The affective component of pain, by contrast, is likely to have evolved more recently and to have added a new dimension to how an animal responds to injury and tissue damage. Various suggestions have been made about the adaptive value of the affective component of pain (i.e., the hurt); however, the idea that it helps animals learn about what caused the noxious stimulus and how this can be avoided in the future is one of the more widely accepted explanations (LaBuda & Fuchs 2000; Allen et al. 2005). Nociception even restricted to the level of the spinal cord can influence learning (Allen, Grau, and Meagher 2009), and it seems likely that tying aspects of learning to an affective pain pathway could lead to more flexible forms of learning that would promote future avoidance of damage or injury (Droege and Braithwaite 2014).

Processes Involved in Feeling Pain

When a nociceptive experience is felt, it hurts. This subjective experience is a psychological process that causes an animal to experience a negative emotion (Panksepp 1998; Morris, Öhman, and Dolan 1998; Paul, Harding, and Mendl 2005). It is this affective component of the response to noxious stimulation that we refer to when we talk about the unpleasantness of pain (Fuchs 2000; Price 2000; Shriver 2014). Thus, the experience of pain requires that an animal is aware that it hurts, or to put it a different way, any animal that is capable of experiencing its pain must therefore be aware (Droege and Braithwaite 2014; Weary, Droege, and Braithwaite 2017).

In humans we know that the nociceptive signal moves though the nervous system and that the nociceptive information is then further relayed to a number of different brain areas where there appears to be cross talk between regions such as the prefrontal cortex, the anterior cingulate cortex (ACC), the insula, and somatosensory areas (Rainville et al. 1997; Ploner, Freund, and Schnitzler 1999; Romanelli, Esposito, and Adler 2004; Mano and Seymour 2015). And at some point, the cross talk leads to the conscious experience that something hurts (Wager et al. 2013).

Human subjective experience of pain can be influenced by attention, mood state, beliefs, and there is heritable genetic variation in sensitivity to pain that modifies how individuals respond differently to the same noxious stimulus. We have known for some time that the ACC and the insula are associated with the affective pain pathway and that the unpleasantness associated with pain is mediated by processing in these brain structures (Price 2000; Shriver 2014). Patients with damage to their ACC report that noxious stimulation is detected but the unpleasantness associated with the damage is no longer bothersome (Rainville et al. 1997). Such descriptions are very similar to the state that can be induced by the administration of morphine (Price 2000). That treatment with morphine more heavily influences the affective component of pain than the sensory detection of pain has been suggested to be a direct result of the fact that there are a higher number of opiate receptors in the affective pathway (Shriver 2014). The sensory pain processing system also has opiate receptors, and morphine can affect these, too, but the affective component seems to be under greater opiate influence.

A similar effect to treatment with morphine has been reported in patients who have lesions in the insula cortex, which produce a condition known as pain asymbolia, in which people report being aware of the pain, but they no longer show aversive reaction to the pain-inducing stimuli (Berthier, Starkstein, and Leiguards 1988; Shriver 2014). It is these kinds of observation that have revealed how the sensory and affective pain pathways can be dissociated. This has been an important advance in our understanding of pain processing, and this knowledge has the potential to help us develop more informed hypotheses regarding which animals will be capable of both pain pathways and which may be restricted to the sensory components only.

Definitions of Pain in Humans and Other Animals

Pain is still referred to as a "phenomenon" because there are parts of the pain process that continue to defy our understanding, and finding a definition that is universally agreed on has proved to be a challenge. Although different kinds of imaging studies are helping us to understand aspects associated with of the way the human brain processes pain (Wager et al. 2013), we are still some way from understanding similar processes within the brain of a nonhuman animal, and in being able to distinguish when an animal is actually experiencing feelings associated with pain (Allen et al. 2005; Braithwaite 2010).

The most widely quoted definition of pain comes from the IASP, which considers pain to be "an unpleasant sensory and emotional experience associated with actual or potential tissue damage, or described in terms of such damage" (Merskey 1964), and more recently, as discussed above, the IASP provided more specific details of what is or is not considered in cases of human pain (International Association for the Study of Pain 2012). For this definition to be useful with regard to pain in animals, we need to be able to determine whether the animal in question can detect unpleasant sensations and whether she has an awareness of unpleasant emotions (Weary, Droebe, and Braithwaite 2017). Investigating the ability to detect unpleasant sensation is fairly straightforward, but assessing a capacity for felt, valenced awareness of the hurt that tissue damage can induce is far more challenging. Thus, the human-centric definition of pain is a useful starting point, but it turns out that it has not been the best starting point for what we consider to be a definition for pain in animals.

In 1986 Zimmerman created a slightly different definition to describe animal pain. Although clearly based on the original IASP definition, Zimmerman's choice of words made a number of changes in describing pain in animals: "an aversive sensory experience caused by actual or potential tissue damage that elicits protective motor and vegetative reactions, results in learned avoidance and may modify species-specific behavior, including social behavior" (Zimmerman 1984). These changes to the IASP definition, while well intentioned, raise some problems. As discussed by Broom (2001), the change from "unpleasant" in humans to "aversive" for animals seems to suggest that animals are not able to experience something unpleasant but rather that something negative produces an aversive reaction. This is possibly true for all organisms—being able to recognize aversive stimuli is a basic need to help an organism survive; as discussed above, this is what we refer to as *nociception*. But using the term *aversive* here overlooks the growing evidence that indicates that there are several animal species who can experience emotions. So here, I would argue, the word *unpleasant* is both more accurate and appropriate for describing pain in these animals (Duncan 2006; Fraser 2009). In addition, Zimmerman removed the word *emotional*, presumably for similar reasons as the change to *aversive*. And finally, he also introduces behavioral responses, "protective motor and vegetative reactions," and recognition that the animal learns and changes his future behavior because of the noxious pain experience.

In a particularly clear review of nociception across the animal kingdom, Kavaliers (1988) points out that many researchers working on animal pain either consciously or unconsciously treat the processes of nociception and pain as the same thing, but as I have emphasized, they are not, and failing to make a distinction between them makes the task of studying pain in animals much harder. This is why I suggest that the *Drosophila* mutant called *painless* was poorly named and why *nociceptionless* would have been a more accurate description. I want to emphasize that investigating pain in animals needs us to distinguish between the sensory nociceptive detection of damage and the emotional, subjective component of the hurt.

Further confusion in the field seems to have arisen because, according to some who work on human pain, nociception (i.e., the sensory detection of injury) and the emotional reaction (i.e., the hurt) cannot be separated (Wall 2000). It is clear that in mammals, where we have the greatest understanding of the mechanisms involved, these processes are highly linked, and the complexity of the pain process arises by virtue of there being both feed-forward mechanisms that relay the signal to the brain and its cortex and also multiple feedback, modulatory processes that are essential to creating an appropriate pain response (Garry, Jones, and Fleetwood-Walker 2004). Despite the integrated nature of these processes, however, rare examples of patients with damage to the ACC or insula cortex challenge the idea that nociception cannot occur without an affective feeling of hurt and that these two processes are inextricably bound together (Shriver 2009). I suggest that this distinction and the dissociation is, in fact, essential if we are going to be able to recognize and define when and where in the animal kingdom something is more than aversive and where it becomes painful and hurts.

Some of the current debate regarding what we mean by animal pain occurs because we have relied too heavily on the definition of human pain. Doing so, however, sets up an unfair comparison. It suggests that an animal's pain will be like human pain in terms of how it comes about and how it is experienced, but it can never be (Allen et al. 2005). Without the same well-developed human brain, particularly the complex cortex, how can any animal process and experience the world, painful or not, in the same way? There are a number of authors who have put forward arguments that pain can only arise in animals who have the same kinds of neural apparatus as humans (e.g., Rose 2002; Rose et al. 2014; Key 2016). And they argue that animals who lack, for example, a well-developed cortex with discrete areas that process different kinds of information will not experience pain because their nervous systems are incapable of supporting conscious feeling. Along these lines, a heated discussion still continues about whether fish species have the capacity to experience pain (see chap. 8).

The problem with such arguments is that they use a top-down approach, setting ourselves as some kind of "gold standard" that other animals can't meet. From an evolutionary perspective, though, an approach that uses human pain as the reference point is back to front. If pain is the adaptive, protective mechanism that I describe above, it is unlikely that the hurt sensation associated with pain will have just suddenly appeared in humans. It is very likely that the feelings associated with pain became more refined in the human brain, but to suggest that the hurt sensation only appeared when the human brain evolved seems to overlook the adaptive value that feeling the hurt provides. Here, I suggest, it would be much more productive to investigate how animals with simpler nervous systems than our own perceive and experience the world and to devise ways to test whether these simpler organisms are capable of experiencing feelings, such as pain (Weary, Droege, and Braithwaite 2017). In other words, rather than starting at the top, which forces us to work back from the human to less sophisticated animal nervous systems, a more logical and informative way to tackle the problem is from the bottom up, that is, exploring the different brains and capacities of not only other vertebrate taxa but among invertebrates, too (Braithwaite 2010; Elwood 2012; Crook et al. 2014; Godfrey-Smith 2016).

The Problem with Lists

In fact, a number of researchers have already begun to document what happens when different taxa and species respond to noxious stimuli (Bateson 1991; Braithwaite 2010; Crook et al. 2011; Elwood 2012). While the approach has not focused yet on whether these animals are aware of their pain experiences, what this has done is provide us with lists of processes associated with nociception and events that follow from that. For example, a recent review suggested that to determine whether specific animal species experience pain, we should determine whether (i) they possess nociceptors, (ii) there are pathways from these nociceptors to a brain, (iii) they possess an opioid system, (iv) they show a reduction in aversive behavioral and physiological effects when an analgesic is administered, (v) they learn to avoid potentially damaging stimuli and this learning is rapid, (vi) their behavioral changes should be more than a reflex response, and (vii) their signs of "discomfort" should be accompanied by long-term changes in motivation (Sneddon et al. 2014). Comparisons across different species using lists of criteria such as this have revealed overlap in several nociceptive mechanisms, and this has allowed us to distinguish a series of stages that many animals go through when they respond to noxious events. For example, in vertebrates, when something noxious occurs, we recognize that specific nociceptors detect the damage. The nociceptive response also triggers cascades of chemical changes that alter physiological processes. Some components of the information are processed at the level of the spinal cord, while others are relayed to the brain, where specific areas process them and then send modulatory signals back down the spinal cord. Overall, there are behavioral changes, changes in attention, and typically decreased motivation to eat or perform normal activities.

Breaking the process of pain down into discrete steps provides clues to look for in assessing pain. Such lists can be useful in terms of identifying common mechanisms that occur across different animal groups, but it is crucial to remember that these lists are not a substitute for demonstrating that a species has a capacity to experience the affective component of pain, which is often missing from the lists (Allen 2004; Allen et al. 2005; Droege and Braithwaite 2014; Weary, Droege, and Braithwaite 2017). Yet this is the most important process, because it determines the capacity to feel the hurt. Unfortunately, cognitive ability and intelligence continue to crop up, promoting the idea that measures of certain kinds of cognitive ability can be a substitute for an affective component (Sneddon et al. 2014), but as I have argued above, they cannot.

In the last two decades, our understanding of pain processes, both in ourselves and in animals, has improved significantly. But there is still some way to go to understand pain in animals and in particular to be able to identify when an animal feels the hurt associated with pain. I suggest that to move the field of animal pain research forward we need to find a way to promote across-discipline exchange of knowledge and ideas and that such discussions would benefit from the development of an agreed, common terminology. In addition, we need to accept that for an organism to experience pain, we are admitting that the animal is capable of some level of consciousness (in order to experience the hurt). Thus, understanding and keeping abreast of advances being made in the field of animal consciousness will also be necessary. However, it is also likely that as we move forward, the field of pain research may make important contributions to research on consciousness. Looking back over the confusions and omissions that have been made in studies of animal pain in the recent past, we need to strive to think clearly about the distinctions between nociception and the affective component of pain. And using an evolutionary framework, it should be easier to understand when a mechanism that provides protection through nociception alone will be sufficient and when it is beneficial for an animal to truly experience pain and the hurt that that engenders.

Suggestions for Further Reading

Allen, Colin, P. N. Fuchs, A. Shriver, and H. Wilson. 2005. "Deciphering Animal Pain." In *Pain: New Essays on the Nature of Pain and the Methodology of Its Study*, edited by M. Aydede, 352–66. Cambridge, MA: MIT Press.

Braithwaite, Victoria A. 2010. *Do Fish Feel Pain?* Oxford: Oxford University Press.

Droege, Paula, and Victoria A. Braithwaite. 2014. "A Framework for Investigating Animal Consciousness." *Current Topics in Behavioral Neurosciences* 19: 79–98.

Paul, Elizabeth S., Emma J. Harding, and Michael Mendl. 2005. "Measuring Emotional Processes in Animals: The Utility of a Cognitive Approach." *Neuroscience and Biobehavioral Reviews* 29: 469–91.Shriver, Adam. 2014. "The Asymmetrical Contributions of Pleasure and Pain to Animal Welfare." *Cambridge Quarterly of Healthcare Ethics* 23: 152–62.

References

Allen, Colin. 2004. "Animal Pain." *Noûs* 38: 617–43.

Allen, Colin, P. N. Fuchs, A. Shriver, and H. Wilson. 2005. "Deciphering Animal Pain." In *Pain: New Essays on the Nature of Pain and the Methodology of Its Study*, edited by M. Aydede, 352–66. Cambridge, MA: MIT Press.

Allen, Colin, James. W. Grau, and Mary. W. Meagher. 2009. "The Lower Bounds of Cognition: What Do Spinal Cords Reveal?" In *The Oxford Handbook of Philosophy and Neuroscience*, edited by J. B. Bickle. 129–42. Oxford: Oxford University Press.

Barron, Andrew B., and Colin Klein. 2016. "What Insects Can Tell Us about the Origins of Consciousness." *Proceedings of the National Academy of Sciences* 113: 4900–4908.

Bartocci, M., L. L. Bergqvist, H. Lagercrantz, and K. J. Anand. 2006. Pain Activates Cortical Areas in the Preterm Newborn Brain. *Pain* 122: 109–17.

Bateson, Patrick. 1991 "Assessment of Pain in Animals." *Animal Behaviour* 42: 827–39.

Berthier, M., S. Starkstein, and R. Leiguards. 1988. "Asymbolia for Pain: A Sensory-Limbic Disconnection Syndrome." *Annals of Neurology* 24: 41–9.

Braithwaite, Victoria A. 2010. *Do Fish Feel Pain?* Oxford: Oxford University Press.

———. 2014. "Pain Perception." in *The Physiology of Fishes*, 4th ed., edited by David H. Evans, James B. Clairborne, and Suzanne Currie, 327–43. Boca Raton, FL: CRC.

Braithwaite, Victoria A., and Paula Droege. 2016. "Why Human Pain Can't Tell Us Whether Fish Feel Pain." *Animal Sentience.* http://animalstudiesrepository.org/cgi/viewcontent.cgi?article=1041&context=animsent.

Broom, Donald M. 2001. "Evolution of Pain." In *Pain: Its Nature and Management in Man and Animals*, edited by Lord Soulsby and D. Morton. 17–25. Royal Society of Medicine International Congress Symposium Series 246. London: Royal Society of Medicine Press.

Crook, Robyn J., K. Dickson, Roger T. Hanlon, and Edgar T. Walters. 2014. "Nociceptive Sensitization Reduces Predation Risk," *Current Biology,* 24: 1121–25.

Crook, Robyn J., Lewis, T., Hanlon, Roger T. and Walters, Edgar. T. 2011 "Peripheral Injury Induces Long-Term Sensitization of Defensive Responses to Visual and Tactile Stimuli in the Squid *Loligo pealeii*, Lesueur 1821." *Journal of Experimental Biology* 214: 3173–85.

Dawkins, Marian S. 2015 "Animal Welfare and the Paradox of Animal Consciousness." *Advances in the Study of Behavior* 47: 5–38.

Droege, Paula, and Victoria A. Braithwaite. 2014. "A Framework for Investigating Animal Consciousness." *Current Topics in Behavioral Neurosciences* 19: 79–98.

Dubin, A. E., and A. Patapoutian. 2010. "Nociceptors: The Sensors of the Pain Pathway." *Journal of Clinical Investigation* 120: 3760–72.

Duncan Ian, J. H. 1996. "Animal Welfare in Terms of Feelings." *Acta Agriculturae Scandinavica, Section A: Animal Science* 27: 29–35.

———. 2006. "The Changing Concept of Animal Sentience." *Applied Animal Behaviour Science* 100: 11–19.

Dunlop, Rebecca, and Laming, Peter. 2005. "Mechanoreceptive and Nociceptive Responses in the Central Nervous System of Goldfish (*Carassius auratus*) and Trout (*Oncorhynchus mykiss*)." *Journal of Pain* 6: 561–68.

Elwood, Robert W. 2009. "Pain and Suffering in Invertebrates?" *ILAR Journal* 52: 175–84.

———. 2012. "Evidence for Pain in Decapod Crustaceans." *Animal Welfare* 21 (S2): 23–27.

Fitzgerald, Maria. 2005. "The Development of Nociceptive Circuits." *Nature Reviews in Neuroscience* 6: 507–20.

Fitzgerald, Maria, and Suellen M. Walker. 2009. "Infant Pain Management: A Developmental Neurobiological Approach," *Nature Reviews Neurology* 5: 35–50.

Fraser, David. 1999. "Animal Ethics and Animal Welfare Science: Bridging the Two Cultures," *Applied Animal Behaviour Science* 65: 171–89.

———. 2009. "Animal Behaviour, Animal Welfare and the Study of Affect." *Applied Animal Behaviour Science* 118: 108–17.

Fuchs, Perry N. 2000. "Beyond Reflexive Measures to Examine Higher Order Pain Processing in Rats." *Pain Research Management* 5: 215–19.

Garry, Emer M., Emma Jones, and Susan M. Fleetwood-Walker. 2004. "Nociception in Vertebrates: Key Receptors Participating in Spinal Mechanisms of Chronic Pain in Animals." *Brain Research Reviews* 46: 216–24.

Gentle, Michael J. 2001. "Attentional Shifts Alter Pain Perception in the Chicken." *Animal Welfare* 10: S187–S194.

Godfrey-Smith, P. 2016. *Other Minds: The Octopus, the Sea, and the Deep Origins of Consciousness*. New York: Farrar, Straus and Giroux.

Hwang, R. Y., L. Zhong, Y. Xu, T. Johnson, F. Zhang, K. Deisseroth, K., and W. D. Tracey. 2007. "Nociceptive Neurons Protect *Drosophila* Larvae from Parasitoid Wasps." *Current Biology* 17: 2105–16.

International Association for the Study of Pain. 2012. "IASP Taxonomy." IASP. http://www .iasp-pain.org/Taxonomy.

Kavaliers, Martin. 1988. "Evolutionary and Comparative Aspects in Nociception." *Brain Research Bulletin* 21: 923–31.

———. 1989. "Evolutionary Aspects of the Neuromodulation of Nociceptive Behaviors." *American Zoologist* 29: 1345–53.

Key, Brian. 2016. "Why Fish Do Not Feel Pain. *Animal Sentience*. http://animalstudies repository.org/cgi/viewcontent.cgi?article=1011&context=animsent.

Kim, S. E., B. Coste, A. Chadha, B. Cook, and A. Patapoutian. 2012. "The Role of *Drosophila* Piezo in Mechanical Nociception," *Nature* 483: 209–12.

LaBuda, Christopher J., and Perry N. Fuchs. 2000. "A Behavioral Test Paradigm to Measure the Aversive Quality of Inflammatory and Neuropathic Pain in Rats." *Experimental Neurology* 163: 490–94.

Lynn, B. 1994. "The Fibre Composition of Cutaneous Nerves and the Classification and Response Properties of Cutaneous Afferents, with Particular Reference to Nociception." *Pain Review* 1: 172–83.

Mano, Hiroaki, and Ben Seymour. 2015. "Pain: A Distributed Brain Information Network?" *PLoS Biology* 13: e1002037.

Melzack, Robert, and Patrick Wall. 1988. *The Challenge of Pain*. London: Penguin Books.

Merskey, H. 1964. "An Investigation of Pain in Psychological Illness." DM thesis, University of Oxford.

Morris, J. S., A. Öhman, and R. J. Dolan. 1998. "Conscious and Unconscious Emotional Learning in the Human Amygdala." *Nature* 393: 467–70.

Morrison, India, Irene Perini, and James Dunham. 2013. "Facets and Mechanisms of Adaptive Pain Behavior: Predictive Regulation and Action." *Frontiers in Human Neuroscience*: doi:10.3389/fnhum.2013.00755.

Nordgreen, Janicke, Tor. E. Horsberg, B. Ranheim, and A. C. N. Chen. 2007. "Somatosensory

Evoked Potentials in the Telencephalon of Atlantic Salmon (*Salmo salar*) Following Galvanic Stimulation of the Tail." *Journal of Comparative Physiology A* 193: 1235–42.

Panksepp, Jaak. 1998. *Affective Neuroscience: The Foundations of Human and Animal Emotions.* New York: Oxford University Press.

Paul, Elizabeth S., Emma J. Harding, and Michael Mendl. 2005. "Measuring Emotional Processes in Animals: The Utility of a Cognitive Approach." *Neuroscience and Biobehavioral Reviews* 29: 469–91.

Ploner, M., H. J. Freund, and A. Schnitzler. 1999. "Pain Affect without Pain Sensation in a Patient with Postcentral Lesion." *Pain* 81: 211–14.

Price, Donald D. 2000. "Psychological and Neural Mechanisms of the Affective Dimension of Pain." *Science* 288: 1769–72.

Qiu, Jane. 2006. "Infant Pain: Does It Hurt?" *Nature* 444: 143–45.

Rainville, P., G. H. Duncan, D. D. Price, B. Carrier, M. C. Bushnell. 1997. "Pain Affect Encoded in Human Anterior Cingulate but Not Somatosensory Cortex." *Science* 277: 968–71.

Romanelli, P., V. Esposito, and J. Adler. 2004. "Ablative Procedures for Chronic Pain." *Neurosurgery Clinics of North America* 15: 335–42.

Roques, Jonathan A. C., W. Abbink, F. Geurds, Hans van de Vis, and Gert Flik. 2010. "Tailfin Clipping, a Painful Procedure: Studies on Nile Tilapia and Common Carp." *Physiology and Behavior* 101: 533–40.

Rose, James D. 2002. "The Neurobehavioral Nature of Fish and the Question of Awareness of Pain" *Reviews in Fisheries Science* 10: 1–38.

Rose, James D., Robert Arlinghaus, Steven J. Cooke, B. K. Diggles, W. Sawynok,E. Donald Stevens, and Clive D. L. Wynne. 2014. "Can Fish Really Feel Pain?" *Fish and Fisheries* 15(1): 97–133.

Shriver, Adam. 2009. "Knocking Out Pain in Livestock: Can Technology Succeed Where Morality Has Stalled?" *Neuroethics* 2: 115–24.

———. 2014. "The Asymmetrical Contributions of Pleasure and Pain to Animal Welfare." *Cambridge Quarterly of Healthcare Ethics* 23: 152–62.

Slater, R., A. Cantarella, S. Gallella, A. Worley, S. Boyd, J. Meek, and M. Fitzgerald. 2006. "Cortical Pain Responses in Human Infants." *Journal of Neuroscience* 26: 3662–66.

Sneddon, Lynne U., Victoria A. Braithwaite, and Michael J. Gentle. 2003. "Do Fishes Have Nociceptors? Evidence for the Evolution of a Vertebrate Sensory System. *Proceedings of the Royal Society B* 270: 1115–21.

Sneddon, Lynne U., Robert W. Elwood, Shelley Adamo, and Matthew C. Leach. 2014. "Defining and Assessing Pain in Animals." *Animal Behaviour* 97: 201–12.

Tracey, W. D., R. I. Wilson, G. Laurent, and S. Benzer. 2003. "Painless, a *Drosophila* gene essential for nociception." *Cell* 113: 261–73.

Varner, Gary. 1998. *In Nature's Interests? Interests, Animal Rights, and Environmental Ethics.* New York: Oxford University Press.

Wager, Tor D., L. Y. Atlas, M. Lindquist, M. Roy, C.-W. Woo, and E. Kross. 2013. "An fMRI-Based Neurologic Signature of Physical Pain." *New England Journal of Medicine* 368: 1388–97.

Wall, Patrick D. 1992. "Defining Pain in Animals." In *Animal Pain*, edited by C. E. Short and A. van Poznak, 63–79. New York: Churchill Livingstone.

———. 2000. *Pain: The Science of Suffering.* London: Weidenfeld and Nicolson.

Walters, Edgar T. 2009. "Chronic Pain, Memory, and Injury: Evolutionary Clues from Snail and Rat Nociceptors." *International Journal of Comparative Psychology* 22 (3): 127–40.

Weary, Daniel M., Paula Droege, and Victoria A. Braithwaite. 2017. Behavioural evidence of felt emotions: approaches, inferences and refinements. *Advances in the Study of Behaviour* 49:27–48.

Yeomans, D. C., and H. K. Proudfit. 1996. "Nociceptive Responses to High and Low Rates of Noxious Cutaneous Heating Are Mediated by Different Nociceptors in the Rat: Electrophysiological Evidence." *Pain* 68: 141–50.

Zhong, L., R. Y. Hwang, and W. D. Tracey. 2010. "Pickpocket Is a DEG/ENaC Protein Required for Mechanical Nociception in *Drosophila* larvae." *Current Biology* 20: 429–34.

Zimmerman, M. 1984. "Ethical Considerations in Relation to Pain in Animal Experimentation." In *Biomedical Research Involving Animals*, edited by Z. Bankowski and N. Howard-Jones, N. 132–39. Geneva: CIOMS.

18   PERSONHOOD

Colin Dayan

If animals did have voices, and they could speak with the tongues of angels—at the very least with the tongues of angels—they would be unable to save themselves from us. What good would language do? JOY WILLIAMS, *Ill Nature* (2001, 123)

There is perhaps no term more important in the legal attempt to give animals rights—and thereby spare them pain and abusive treatment—than *persons* or *personhood*. Animals are not humans. Especially since Descartes, this fact has made animals targets for extinction, torturous experimentation, and victims of sport—hunted, exhibited, and killed for the pleasure of humans. The Cartesian division between mind and body put nonhuman animals squarely on the side of bodies, emptied of consciousness, feeling, and awareness. Descartes argued that animals are machines, lacking both immortal souls and mental experiences. But he has not had the last word.

What is a person? How does it differ from people? In his *Essay on Human Understanding*, John Locke ([1700] 1975) offers a sharp distinction between *personhood* or *personal identity* and *human*. In the process, he blurs the divide between humans and animals, persons and things. There is, he admits, "something in us, that has a Power to think," but one can never be sure whether that "substance perpetually thinks, or no" (Locke [1700] 1975, 2.1.10, 109). Locke dreads the abuse of words and rejects confounding verbal games such as "*Humanity is Animality, or Rationality, or Whiteness*" ([1700] 1975, 3.8.1, 474). So he creates a template for understanding what it means to be human by setting up terms for comparison. But the task of definition remains in the end arbitrary. The assumed oppositions between animals and humans, brutes and men, lose their stability as he considers them.[1]

For Locke, then, the force of consciousness can adhere to things. Writing against the necessary physical continuity of substance—whether material or immaterial—he instead located personal identity in consciousness. He argues that to say there is a "living soul" in a corpse just because it has the "shape" of a man is as senseless "as [saying] that there is a rational soul in a changeling, because he has the outside of a rational creature, when his actions carry far less marks of reason with them, in the whole course of his life, than what are to be found in many a beast" ([1700] 1975, 4.4.14, 569–70). In this movement from corpse to changeling to beast, Locke destroys hierarchy, the easy verticality of a great chain of being.

Locke is best known for the passage in the *Essay* that characterizes "person" as a "forensic term, appropriating actions and their merit" and so belonging "only to intelligent agents, capable of a law, and happiness, and misery" ([1700] 1975, 2.27.26, 346). He also more broadly defines a person as "a thinking thing," which can have the same thoughts in different places, even hypothetically in different bodies, even in "the little Finger." But what makes a person—and Locke implies that a dog with awareness might also be a person—is responsibility: the capacity to appropriate past actions "to that present *self* by consciousness" ([1700] 1975, 2.27.26, 346).

For Locke, the word *human* pertained to the "frontispiece," the external frame or body. But "personal identity" or "personhood" was shared by human and nonhuman animals: "dogs or elephants" give "every demonstration of [thinking] imaginable, except only telling us that they do so" ([1700] 1975, 2.1.19).[2] He described the *self* as "that conscious thinking thing, whatever Substance made up of (whether Spiritual or Material, Simple or Compounded, it matters not), which is sensible, or conscious of Pleasure and Pain, capable of Happiness or Misery, and so is concerned for it *self* as far as that consciousness extends" ([1700] 1975, 2.27.17, 341). If you allow a piece of the body, say, the "little Finger," to depart from the body while carrying consciousness with it, "it is evident the little Finger would be the *Person*, the *same Person*; and *self* then would have nothing to do with the rest of the Body" ([1700] 1975, 2.27.17, 341).

1. The second book of Locke's *Essay*, "Identity and Diversity" (added in the second edition in 1694) caused numerous theological debates, most famously with Edward Stillingfleet, Bishop of Worcester.

2. For the best discussion of Locke's firm belief that "nonhuman animals have conscious experience," see Strawson (2011).

Locke's inquiries into personhood challenged the philosophy and science that sought to prove human superiority to animals as well as racial inferiority among humans. Given Locke's criterion of consciousness, the distinctions between internal and external remain unreliable, such as in the relation between mental capacity and skin color. Locke's demonstration that consciousness could as easily be in our little finger as in our mind presses us to test what we mean when we distinguish persons from things ([1700] 1975, 2.27.17, 341). He worked hard to suggest that dogs, parrots, and monkeys could be understood as "corporeal rational creatures" even if their physiologies were not those of men. If God can add thought to persons, then He can also add the thinking principle to matter.

Put simply, matter thinks in animals, and the differences between animal and human consciousness are only of degree. He thus called into question the boundaries between thing, animal, and human self. Even Jeremy Bentham, in a discussion of penal jurisprudence, considers that certain humans are "styled as persons." Such a formulation suggests that not all humans are persons and leaves open the possibility that nonhumans might similarly be "styled" as persons (Bentham [1789] 2007, 282).

Legal Persons

That personhood is not an innate feature of humans is clear in law, which, as I demonstrate in my 2011 book *The Law Is a White Dog: How Legal Rituals Make and Unmake Persons*, transforms, denies, and bestows personhood through a power both legal and magical. Whether made or unmade by court decisions, *personhood* becomes the term through which humans turn into slaves, ghosts turn into persons, or the biologically alive turn into the legally dead (Dayan 2011, 20). In particular, the legal racial slave, transformed into both self and property, haunts contemporary debates about personhood. The terminological history of slavery presses up against rituals of definition that depend on an unspoken, secret intimacy with the animals we order, the objects of our discipline and duplicity.

Throughout the Americas, under pressure of limitless punishment, the concept of personhood could be eliminated for the enslaved. In respect to civil rights and relations, slaves were not persons but things. Creatures of law, they were nonetheless dead to the law. The creation of a new, hybrid person in law, earmarked for domination, was a weighty matter: old legal rituals, once adapted to the new grammar of servility, were spectacular.

What is this species of embodied property? I will first generalize and then turn with greater specificity to animals, those objects with or without souls whose proximity to the *thing called slave* guarantees the conversion of the object of law into nonpersons. This negative person has no legal mind, and, unless apprehended as a criminal, is absolutely disabled from forms of responsibility. Animals summoned in varying rites of definition and similitude makes this juristic curiosity real and viable, give flesh to this artifice of law.

Thomas Cobb, in *Inquiry into the Law of Negro Slavery* (1858), turned to the horse and cow to find the right idiom of discipline for human chattel: "Like the horse, the cow, the domestication and subjection to service" of the slave "did not impair, but on the contrary improved his physical condition" (Cobb 1858, 5). He concludes, "Subjection was consistent with his natural development, and therefore not contrary to his nature." Though dogs, horses, cattle, undomesticated animals, *ferae naturae*, and slaves were recognized as something more than a package of goods, they were nothing more, in legal contemplation, than chattels. The analogy is especially striking in *Bailey v. Poindexter's Executor*, an 1858 case that grants value to that unique slave who performs a job well but also asks whether such performance can be considered a *civil* act.[3] The answer puts us in touch with a dog. Only if the service of "a well-trained and sagacious dog" in bringing his owner meat in a basket from a butcher could "in a legal sense" be understood as "the civil act of a dog."[4]

The incommensurability of persons and things was foundational to the institution of slavery. Rather than a "logical inconsistency," this paradox, Bryan Wagner (2009) argues, was "a precondition for the system's normal operation": "slavery's indignity is not about being turned from a person into a thing but rather about being in a position where it does not matter if you are a person or a thing" (74).

In the South the adaptation of Lockean notions of personal identity to slaves was inextricably bound up with the understanding of *person* as a forensic term and with the kind of legal incapacity and nonrecognition that signaled *negative personhood*. Thomas Morris (1996), in *Southern Slavery and the Law: 1619–1860*, identifies the essential legal fiction: "the slave was an object of property rights, he or she was a 'thing'" (57).[5] However,

3. Bailey v. Poindexter's Executor, 55 Va. 132 (1858).

4. Ibid., 20, 23.

5. In *Scenes of Subjection: Terror, Slavery, and Self-Making in Nineteenth-Century America*, Saidiya Hartman (1997) focuses on the nature of the "captive person in law," the mutilated subjecthood that "intensified the bonds of captivity and the deadening objectification of chattel status" (94).

what most occupied the thoughts of Virginia lawyers and judges arguing and hearing cases about personal rights on the eve of the Civil War was not any need to affirm the slave as property but rather to articulate the personhood of slaves in such a way that it was disfigured but not erased. Slave law depended on this juridical diminution. Proofs of animality and marks of reason entered the courtroom drama, as the double character of person and property was legally confirmed.

The masks of personhood remain the key not only to a specifically US legal history but also to the philosophical indeterminacy of the border between human and other animal species. Such masks come in many shapes and colors, but they are unconvincing. For instance, the claim that legal personhood is limited to those possessing moral agency only reinforces the ambiguity it attempts to eliminate. Such a presumption of moral claims to legitimacy distorts the constitutional definition of personhood and wreaks havoc on the genealogy of fetal personhood as well as advocacy on behalf of nonhuman animal rights. According to *Black's Law Dictionary*, the definition of *personhood* encompasses far more than moral agents: "In general usage, a human being (i.e. natural person), though by statute the term may include a firm, labor organizations, partnerships, associations, corporations, legal representatives, trustees, trustees in bankruptcy, or receivers." The law has already decided that moral action is not consequential to personhood.

Consider corporations. As Blackstone described them in his *Commentaries on the Laws of England*, corporations are "artificial persons" created by law "for the advantage of the public." In *Dartmouth College v. Woodward* (1819), US Chief Justice John Marshall explained the status of a corporation in the eyes of federal law:

> A corporation is an artificial being, invisible, intangible, and existing only in contemplation of law. Being the mere creature of law, it possesses only those properties which the charter of creation confers upon it, either expressly, or as incidental to its very existence. These are such as are supposed best calculated to effect the object for which it was created.[6]

If corporations can be persons and thus cannot be deprived of the power to acquire and utilize property without due process of law, why can't animals be endowed with such guarantees? If personhood is legally assigned

6. Dartmouth College v. Woodward, 17 U.S. 518 (1819).

rather than innate in a species, there is nothing to prevent animals obtaining it.

The legal person is sufficiently unreal to make claims on our habits of thought. But still more difficult to grasp is what happens when property itself—not the individual who owns or possesses it—becomes the defendant. If more or less human objects can be either property or persons in the eyes of the law, then they can equally be stripped of these attributes. We are obliged to consider the creation of a species of depersonalized persons, identified as offending, and hence ready for surrender to or extermination by the state. The legal demolition of personhood that began with slavery has been perfected through the logic of the courtroom and adjusted to apply to prisoners. The unspoken assumption remains: prisoners are not persons. Or at best they are a different kind of human so dehumanized that normal standards of decency do not apply.

Animal Persons

What are the conditions under which categories of identity are constructed and reconstructed? To answer such a question demands that we take up the challenge Eduardo Kohn presents in "How Dogs Dream: Amazonian Natures and the Politics of Transspecies Engagement." "All beings, and not just humans," he writes, "engage with the world and with each other as selves—that is, as beings that have a point of view" (Kohn 2007, 3; see also Kohn 2013). How do we delineate the limits of personhood when we know both that it is not equal to the human and that humans are not the only selves?

The definition of *person* matters now more than ever because proof of personhood has become the threshold between life and death for non-human animals, for if they possess personhood, they are also granted rights associated therewith. Spain and New Zealand have extended personhood rights to great apes. In 2015 the town council of tiny Trigueros del Valle, Spain, voted unanimously to define dogs and cats as "non-human residents," giving them rights similar to those of humans living there (Dawber 2015). Two years ago, the Indian government ruled that cetaceans, such as whales and dolphins, are "non-human persons" with their own specific rights (Ketler 2013).

It is dangerous, however, to consider nonhuman personhood only in the cases of animals who have sophisticated cognitive abilities—chimps, elephants, and dolphins, for instance. What is "morally relevant," legal scholar and activist Gary Francione says, is "sentience, or having subjective aware-ness. And most of the animals we routinely exploit every day—the cows, pigs, chickens, and fish—are sentient" (Francione 2013).

Animals experience pain and suffering. They remember. Their con-sciousness extends back over time. They also think, which has a great deal to do with sentience and emotions if not with reason or rationality. The stakes are clear. If sentience is the stuff of personhood, then animals are nonhuman persons with rights to be free from harm (see chap. 24). Given the appalling cruelty visited upon animals the world over, the granting of such rights seemingly could prevent a great deal of misery.

Steven Wise, of the Nonhuman Rights Project, has turned again and again to law in order to make a case for the personhood of chimpanzees. Arguing before Judge Barbara Jaffee of the New York Supreme Court on May 27, 2015, he contended that chimpanzees are so much like humans that they deserve the right to "bodily liberty" even if other rights are not meaningful to them. It was an unusual circumstance. Wise was responding to an earlier order from the judge asking to show cause and writ of habeas corpus on behalf of Hercules and Leo, two chimpanzees used for bio-medical experimentation at Stony Brook University on Long Island. Under New York law, only "legal persons" may have a writ of habeas corpus issued on their behalf. Wise was in effect arguing that the animals were persons for this one legal purpose if not necessarily for any other.

No concept has been more central to constitutional law than due pro-cess, exemplified by the writ of habeas corpus. Derived from Magna Carta, the writ guarantees that individuals cannot be imprisoned or restrained in their liberty without due process of law. "No free man," says Magna Carta, "shall be taken, imprisoned, disseised, outlawed, banished, or in any way destroyed, nor will we proceed against or prosecute him, except by the lawful judgment of his peers and by the law of the land." The extreme con-ditions suffered by Hercules and Leo, according to Wise, violate the prohi-bition against "degrading, punitive, and unconstitutional" treatment. On behalf of the university, Christopher Coulson, an assistant state attorney general, responded, "There's simply no precedent anywhere of an animal getting the same rights as a human" (Wise 2015). But, Wise argued, "These animals are . . . autonomous, self-determining beings" (McKinley 2015).

Though the court's initial judgment seemed favorable to the view that Hercules and Leo have the right to "bodily liberty," with the judge ordering Stony Brook to appear in court to explain the animals' detainment, the court, wary of setting precedent, ultimately denied the petition. Hercules and Leo were returned to New Iberia Research Center (NIRC), that leased them to Stony Brook. In March 2018, NIRC released Hercules and Leo, and will eventually release the other two hundred chimpanzees, to Project Chimps sanctuary in Blue Ridge, Georgia.[7]

Yet what if the court had the courage to recognize Hercules and Leo as bearers of actionable rights? The granting of rights to be free from captivity appears to ensure the best outcome for animals because courts would be required to respect their autonomy and free them from detainment. But there is a downside as well: the granting of rights maintains the hierarchy of human largesse and animal dependency. Giving animals what we think they need or deserve in terms of human conceptions of right and wrong, capacity or incapacity, maintains their subordination.

The terminology of human rights—or rights writ large—is not natural. It has a history both paradoxical and vexing. In *The Origins of Totalitarianism*, Hannah Arendt (2004) observes that not all persons have rights, despite repeated claims to universality. Further, beneficent efforts to recognize rights closely parallel those that direct sentimentality toward animals. Arendt's interpretation is bracing and instructive:

> All societies formed for the protection of the Rights of Man, all attempts to arrive at a new bill of human rights were sponsored by marginal figures—by a few international jurists without political experience or professional philanthropists supported by the uncertain sentiments of professional idealists. The groups they formed, the declarations they issued, showed an uncanny similarity in language and composition to that of societies for the prevention of cruelty to animals. (Arendt 2004, 292)

In *Animals, Property, and the Law*, Francione warns about the calculated risks of rights talk. The word *rights* undergoes changes in meaning and reach depending on whom or what it applies to. The malleability of legal language as applied to slavery, imprisonment, and allowable suffering in animal experimentation proves his point: what is considered "'humane' treatment" or "'unnecessary' suffering" may, Francione (1995) explains, "differ considerably from the ordinary-language interpretation of those terms" (4–5).

Gradations of personhood are articulated in law, whether stigmatized

7. http://www.sciencemag/org/news/2016/05/world-s-largest-chimpanzee-research-facility -release-its-chimps.

bodies or devalued minds. As a locus of embodied history, law becomes crucial to understanding what it meant when slaves, formerly property, were freed into another kind of status that simply exchanged one kind of bondage for another. Even after emancipation, to the extent that former slaves were allowed personalities before the law, they were regarded chiefly—almost solely—as potential criminals. The Thirteenth Amendment to the Constitution abolished slavery "except as a punishment for crime where of the party shall have been duly convicted." With that exemption, the legal rhetoric of protection and allowable injury continued—not just in the treatment of prisoners, or "slaves of the state," but also in the treatment of other mammals. It is in the treatment of animals and the projection of "animal welfare" that we see how easily status can not only sustain prejudicial harm but lead to the possibility of further abuse, defined as "minimal," for example, in the Animal Welfare Act of 1966. To apply personhood to animals is not invariably to secure them from harm but only to subject them to law.

In "Should Animals Have Rights?" Lori Gruen argues, "the rights approach . . . tends to reduce our relationships to those in which we value similarities and overlook important differences that may help us to rethink who is valuable and why" (Gruen 2014). Animal rights talk gives animals what it is we think we get as bearers of rights and obligations in standard liberal and moral terms. But can we push beyond rituals of law and the residual humanism that lingers in considerations of the rights of personhood? What, beyond personhood, might make something an object of moral concern?

Beyond Personhood

Mary Midgely asks this question in another way: how, she wonders, can we bring "some creatures nearer to the degree of consideration which is due to humans." Her qualification is more pressing still, going beyond legal personhood and toward an ethical approach to animal and human interaction: "What elements in 'persons' are central in entitling them to moral consideration"? What counts, she explains, is "sensibility, social and emotional complexity of the kind which is expressed by the forming of deep, subtle, and lasting relationships" (Midgley 2005, 132–43).

We thus turn to a remote and uncertain reservoir of connection on which all creatures might draw but from which most humans have learned to cut themselves off. Instead of opposing humans to animals, we need to question the boundaries of humanity, or, more precisely, the making and management of human boundary objects. In *With Dogs at the Edge of Life*, I ask, "What does conscience look like at the boundaries of humanity, at the edge of a cherished humanism?" (Dayan 2016, 10). I resist the terms of our epistemological debates and cast doubt on the robustness and transportability of the ontological partitions they presuppose: body and mind, animality and humanity, nature and culture—and most of all, persons and things. Empathetic entanglement, upsetting these divisions, may be the means out of rationality and the human assumptions that have bedeviled other species and our environment.

Empathy's renewed intimacy of contact promises to lead us out of thought and into feeling—sensation (of the senses) but not sentiment—that extends the reach of the political (see chap. 9). It also promises a possible escape from the obsession with personhood and its rights, to be replaced with something more like respect for what is not reduced to human terms. As Gruen argues in *Entangled Empathy: An Alternative Ethic for Our Relationships with Animals*, the goal is "caring perception"—a demanding reciprocity, a being together in pain that can be healed if shared. In becoming acquainted with what lies outside the self, we enter into another kind of knowing. "The wellbeing of another grabs the empathizer's attention; then the empathizer reflectively imagines himself in the position of the other. . . . This sort of empathy doesn't separate emotion and cognition and will tend to lead to action because what draws our attention in the first place is another's experiential wellbeing" (Gruen 2015, 51).

In responding to the state of injury, the pain and violence of this world, we need to step back and ask how we can know feeling that is not tied to our assumptions. Indeed, to risk losing ourselves in what is beyond our ken is to experience what it might mean to *feel sufficiently*. This means that "personhood" and its opposite, "depersonalization" or "depersonification," do not supply the terminological framework we need in the effort to include all sentient beings in a new ethics of replenishment and redemption from harm and injury.

Let me take the drama of vodou, and its *lwa*—its spirits or gods—as exemplary. The *lwa* can only manifest in the corporeal envelope: in lineaments both human and nonhuman, spirits experience life and unfold their potential. The epistemology of vodou therefore offers a context for reconfiguring our understanding of the supernatural—not impalpable or ideal but rather all too natural, or natural to the nth degree. Vodou's understanding of sentient life takes us beyond personhood by exploding terms such as *humanism*, as it bridges the gaps between body and mind, dead and living, human and nonhuman.

Perhaps animality is what we should be thinking about rather than claims for humanity. Animals live on the track between the mental and the physical and sometimes seem to tease out a near-mystical disintegration of the bounds between them. Such knowing has everything to do with perception, attentiveness that unleashes another kind of intelligibility beyond the world of the human.

The question we face is what such a struggle beyond the limit and reach of personhood and rights will look like. Faced with compelling questions— why shouldn't all animals have equal rights, including a right to life?—it is difficult to describe a new program outside the human-centered enterprise of law and personalist conceptions of morality. What is the particular terrain for human cruelty, and who gets to command its shifts in terms of species and race? Can we locate in granular and theoretical registers the often invisible nexus of animality and human marginalization?

In *Creaturely Poetics*, Anat Pick eases some of this difficulty. She thinks with the vulnerability of bodies, not the force of thought (see chap. 28). In other words, she endeavors to understand the "creaturely" as a register of being distinct from "persons": "Contact with the flesh and blood vulnerability of beings—whether human or not" projects an "antiphilosophical" reading, beyond an "anthropocentric perspective." This is "the move *from rights to lives*" (Pick 2011, 3, 5, 11, emphasis in original). Beyond personhood we might sketch a landscape of resistance that skirts transcendence and goes beyond human-centered ideas of personal identity. The world of the vulnerable and the violated prompts us both conceptually and practically to appreciate the creatureliness of *all* things. Quite simply, we have to recognize how *uncreaturely* we are for opposing humans to animals.

Suggestions for Further Reading

Claxton, Guy. 2015. *Intelligence in the Flesh: Why Your Mind Needs Your Body Much More Than It Thinks*. New Haven, CT: Yale University Press.

Naimou, Angela. 2015. *Salvage Work: U.S. and Caribbean Literatures amid the Debris of Legal Personhood*. New York: Fordham University Press.

Perry, John, ed. 1975. *Personal Identity*. Berkeley: University of California Press.

Radin, Margaret Jane. 1993. *Reinterpreting Property*. Chicago: University of Chicago Press.

Rorty, Amelie Oksenberg, ed. 1976. *The Identities of Persons*. Berkeley: University of California Press.

Stone, Christopher D. 1996. *Should Trees Have Standing? And Other Essays on Law, Morals, and the Environment*. Dobbs Ferry, NY: Oceana.

Strawson, Galen. 2011. *Locke on Personal Identity: Consciousness and Concernment*. Rev. ed. Princeton, NJ: Princeton University Press,

Tamen, Miguel. 2001. *Friends of Interpretable Objects*. Cambridge, MA: Harvard University Press.

Wise, Steven M. 2002. *Drawing the Line: Science and the Case for Animal Rights*. Cambridge, MA: Perseus Books.

References

Arendt, Hannah. (1951) 2004. *The Origins of Totalitarianism*. San Diego, CA: Harcourt Brace Jovanovich. Reprint, New York: Schocken Books.

Bentham, Jeremy. (1789) 2007. *An Introduction to the Principles of Morals and Legislation*. London: T. Payne. Reprint, New York: Dover.

Cobb, Thomas Read Rootes. 1858. *Inquiry into the Law of Negro Slavery*. Philadelphia: T. and J. W. Johnson; Savannah: W. T. Williams.

Dawber, Alistair. 2015. "Human Rights for Cats and Dogs: Spanish Town Council Votes Overwhelmingly in Favour of Defining Pets as 'Non-human Residents.'" *Independent* (London), July 22.

Dayan, Colin. 2011. *The Law Is a White Dog: How Legal Rituals Make and Unmake Persons*. Princeton, NJ: Princeton University Press.

———. 2016. *With Dogs at the Edge of Life*. New York: Columbia University Press.

Francione, Gary. 1995. *Animals, Property, and the Law*. Philadelphia: Temple University Press.

———. 2013. Comment on Andrew C. Revkin, "A Closer Look at 'Nonhuman Personhood' and Animal Welfare." *NYTimes.com*, July 28. http://dotearth.blogs.nytimes.com/2013/07/28/a-closer-look-at-nonhuman-personhood-and-animal-welfare.

Grimm, David. 2016. "World's largest chimpanzee research facility to release its chimps," *Science*, May 4. http://www.sciencemag.org/news/2016/05/world-s-largest-chimpanzee-research-facility-release-its-chimps.

Gruen, Lori. 2014. "Should Animals Have Rights?" *Dodo*, January 20, https://www.thedodo.com/community/lorigruen/should-animals-have-rights-396291626.html.

———. 2015. *Entangled Empathy: An Alternative Ethic for Our Relationships with Animals*. New York: Lantern Books.

Hartman, Saidiya. 1997. *Scenes of Subjection: Terror, Slavery, and Self-Making in Nineteenth-Century America*. New York: Oxford University Press.

Ketler, Alanna. 2013. "India Declares Dolphins and Whales as 'Non-human Persons,' Dolphin Shows Banned," *Collective Evolution*, September 17. http://www.collective-evolution.com/2013/09/17/india-declares-dolphins-whales-as-non-human-persons.Kohn, Eduardo. 2007. "How Dogs Dream: Amazonian Matures and the Politics of Transspecies Engagement." *American Ethnologist* 34 (1): 3–24.

———. 2013. *How Forests Think: Toward an Anthropology beyond the Human*. Berkeley: University of California Press.

Locke, John. (1700) 1975. *An Essay on Human Understanding*. 4th ed. Edited by Peter H. Nidditch. London: Awnsham and John Churchill. Reprint, New York: Oxford University Press.

McKinley, James C., Jr. 2015. "Arguing in Court Whether 2 Chimps Have the Right to 'Bodily Liberty,'" *New York Times*, May 27. https://www.nytimes.com/2015/05/28/nyregion/arguing-in-court-whether-2-chimps-have-the-right-to-bodily-liberty.html.

Midgley, Mary. 2005. "Is a Dolphin a Person?" In *The Essential Mary Midgley Reader*, edited by David Midgley, 132–43. New York: Routledge.

Morris, Thomas D. 1996. *Southern Slavery and the Law, 1619–1860*. Chapel Hill, NC: University of North Carolina Press.

Naimou, Angela. 2015. *Salvage Work: U.S. and Caribbean Literatures amid the Debris of Legal Personhood*. New York: Fordham University Press.

Perry, John, ed. 1975. *Personal Identity*. Berkeley: University of California Press.

Pick, Anat. 2011. *Creaturely Poetics: Animality and Vulnerability in Literature and Film*. New York: Columbia University Press.

Radin, Margaret Jane. 1993. *Reinterpreting Property*. Chicago: University of Chicago Press.

Rorty, Amelie Oksenberg, ed. 1976. *The Identities of Persons*. Berkeley: University of California Press.

Stone, Christopher D. 1996. *Should Trees Have Standing? And Other Essays on Law, Morals, and the Environment*. Dobbs Ferry, NY: Oceana.

Strawson, Galen. 2011. *Locke on Personal Identity: Consciousness and Concernment*. Rev. ed. Princeton, NJ: Princeton University Press.

Tamen, Miguel. 2001. *Friends of Interpretable Objects*. Cambridge, MA: Harvard University Press.

Wagner, Bryan. 2009. *Disturbing the Peace: Black Culture and the Police Power after Slavery*. Cambridge, MA: Harvard University Press.

Williams, Joy. 2001. *Ill Nature: Rants and Reflections on Humanity and Other Animals*. New York: Lyons Press.

Wise, Steven M. 2002. *Drawing the Line: Science and the Case for Animal Rights*. Cambridge, MA: Perseus Books.

———. 2015. "That's One Small Step for a Judge, One Giant Leap for the Nonhuman Rights Project." *Nonhuman Rights Blog*. http://www.nonhumanrightsproject.org/2015/08/04/thats-one-small-step-for-a-judge-one-giant-leap-for-the-nonhuman-rights-project.

19 POSTCOLONIAL

Maneesha Deckha

A "postcolonial" analysis typically highlights how narratives emanating from institutional, epistemic, and geopolitical sites read as "Western" create problematic cultural and racial representations that operate as "truths" about non-Western peoples (Bhambra 2014; Darian-Smith 1996; Gandhi 1998). How, then, does the postcolonial orientation figure into Animal Studies? Three related postcolonial dimensions to Animal Studies may be identified by (1) demonstrating how representations of race and culture are deeply mediated by constructs of animality and species, (2) undergirding but also responding to criticism that the vegan praxis many Animal Studies scholars promote reflects white and Western privilege, and (3) motivating calls for greater presence of non-Western epistemologies in maturing the field. Each of these postcolonial conversations/debates in Animal Studies will be discussed below.

Imbrication of Race with Species in Civilizational Discourses on the Human

A first, and perhaps less contentious, manner in which Western/ non-Western power relations register in Animal Studies is through arguments attesting to the co-constitution of coloniality, race, and racism with animality, species, and anthropocentrism (Anderson 2000; Salih 2007; Wolfe 2012). Postcolonial scholars have explained how ideas about humans, animals, humanness, and animality shaped the racist belief in Western civilizational superiority and white normativity that propelled imperialist missions and the creation and practice of colonial governance. In particular, racist narratives about colonized peoples as "wild," closer to nature, and subhuman—bolstered by scientific assertions about the bestial biological and physiognomic resemblances between nonwhites and animals—enabled European metropoles to be constituted as civilized and modern and whiteness to become the exemplar of humanity (Condis 2015; Haraway 1989; Kim 2015; Lugones 2010; Peterson 2013). Associations of wildness and animality shaped a range of difference-based discourses relating to gender, class, and ability as well (Anderson 2000). In essence, animality was a metric of race, gender, class, and civilizational status (and vice versa), and thus, as Animal Studies scholars observe, a critical humanizing agent (Kim 2015). Conversely, scholars have also noted how racialized civilizational thinking sustained the mythic narratives of human exceptionalism and universality (Corbey 2005).

Recognizing that colonial modernist logic pivoted on a constellation of discriminatory, mutually constituting discourses, Animal Studies scholars have also discussed how the colonial racialized associations regarding nature, the wild, and animals authorized European colonial governance over and violence against colonized and racialized humans as well as animals (Anderson 2000). As a particularly prominent example, colonial regimes frequently passed laws in their colonies against animal cruelty. Such laws purporting to cultivate compassion for (certain) animals facilitated colonial civilizing missions—they were presented as vehicles to inculcate "humane" sentiments among the colonized against "unnecessary suffering" toward animals (Deckha 2013). These anticruelty laws formed part of "humane" legislative packages directed at cultivating overall humanity in the colonies, a modernist effort that Samera Esmeir (2012) has identified as creating "juridical humanity" for colonial subjects whose natural human status was disavowed. Such laws were indelibly biased because they were crafted according to the colonizers' perceptions of what constituted cruelty toward animals. Even where *cruelty* and *unnecessary suffering* went undefined in the legislation, these laws never targeted violent animal-based industries or uses that were important to and normalized in European metropoles (such as slaughterhouse practices, hunting, or vivisection). *Cruelty*, instead, was a label applied to "non-Western" animal-based practices (Esmeir 2012).

Animal Studies scholars also point to the prominence of species demarcations for colonial mind-sets and material practices in the present day. Influentially, Glen Elder, Jennifer Wolch, and Jody Emel (1998a, 1998b) have argued that our differentiated ethical responses to animal practices today, despite similarity in the violence and terror animals endure, continue to establish racial and cultural hierarchies. They acknowledge that the present-day counterpart to the colonial racist thinking outlined above now foregoes (at least overtly in public discourse) mentioning that humans can be measured and ranked on physiological characteristics and divided into races/species accordingly. Public discourse, however, still relies on animal bodies as points of racial demarcation in terms of measuring (human) cultures or safeguarding whiteness and civilization. How racially minoritized cultures use animals, for example, continues to come under dominant legal scrutiny through anticruelty legislation as evidence of cultural and racial backwardness and inferiority despite the gross exploitation that dominant cultural practices—unmarked as "cultural" but rather normalized as mainstream—entail (Deckha 2013; Kim 2015; Wadiwel 2015).

Guided by Elder, Wolch, and Emel's innovative intervention into their home discipline of geography with this insight, scholars have discussed the differentiated responses to animals in urban environments. They have tracked anthropocentric—but also colonial and racialized—sensibilities among city dwellers about which animals legitimately belong in urban spaces, revealing how the discourses used invoke colonial-inflected ideologies about "native, "nonnative," and "invasive" species (Deckha and Pritchard 2016; Jerolmack 2008; Palmer 2003). Of course, many urban residents live in cities with zoos—an exemplar of a racialized, anthropocentric, colonial institution. Scholars have demonstrated how the originating impetus of zoos was to bring back "exotic" animals from the colonies to European metropoles. Zoos still traffic today in exotica, creating species-based "conservation" spectacles pitched at "teaching" Westerners about the implicitly racialized non-Western world of "wildness" and "nature" (Milstein 2009).

Feminist animal care scholars and ecofeminists who write about animal oppression as a branch of Animal Studies have also contributed to showcasing the relationship between colonialism and anthropocentrism by emphasizing the Westernness of dominating epistemologies and ontologies that subordinate animals. Scholarship here demonstrates how the human-nonhuman divide is entwined and correlated with long-standing, interrelated Cartesian hierarchical dualisms of man versus woman, culture versus nature, reason versus emotion, mind versus body, and West versus non-West in Western epistemologies (Adams 1991; Adams and Donovan 1995; Adams and Gruen 2014; Gaard and Gruen 1993; Kheel 1985; Plumwood 1993). While gender has received primary attention in much of this literature as an analytical lens (Deckha 2012), some ecofeminist scholarship has centered postcolonial concerns in their feminist analyses to work toward nonviolent animal human relations (Gaard 2013, 2001, 1993; Gaard and Gruen 1993; Plumwood 2004).

Through these and other realities of colonial discourse, most postcolonial scholars are keenly aware of the species subtext of colonial racial coding and have accepted that the concepts of race, culture, gender, and species in the eighteenth and nineteenth centuries were deeply intertwined and generative of each other (Lugones 2010; Peterson 2013; Hund, Mills, and Sebastini 2015). Yet despite the centrality of the tropes of the nonhuman and animality to colonial narratives, most postcolonial scholars still do not include animals in their accounts of colonized subjects (Narayanan 2017).

Veganism as Cultural Imperialism?

This disconnect may have something to do with the critique of cultural imperialism directed at Animal Studies when it is associated with the mainstream animal rights movement. The majority of those identified as adherents to the animal rights movement in the United States and Canada identify and are perceived to be white and middle class. This demographic factor, along with campaigns that controversially compare the treatment of animals today to human slavery and genocide and do not acknowledge the capitalist exploitation inherent in processed vegan products, have stimulated the critique that the movement fosters postracialism and white privilege (Harper 2010; Wren 2016).[1] Against this backdrop, scholars within Animal Studies and those external to it have raised concerns about the hegemony of Western values in relation to the call for a vegan praxis, arguing that an ethic that requires all humans to refrain from animal consumption espouses Western values that originate from an elitist and culturally imperialist worldview. In particular, critics have highlighted two examples of marginalized racialized groups whose needs are erased by the promotion of veganism: (1) poor urban communities in the United States, where plant-based diets may be inaccessible and too expensive because of the phenomenon of food deserts (Harper 2010), and (2) indigenous societies that have historically relied on sustenance killing of animals and seek

1. It is important to note that the overwhelming majority of animal advocates are also female. See Emily Gaarder, *Women and the Animal Rights Movement* (New Brunswick, NJ: Rutgers University Press, 2011). How this demographic factor and the systemic and institutional sexism that attends its affects the "white" portrayal and representation of animal advocacy in settler states like Canada and the United States is undertheorized.

to recuperate and uphold these practices today in the name of cultural tradition and self-determination (Nadasdy 2016; Wenzel 1991; see chap. 27). As Will Kymlicka and Sue Donaldson (2014) have argued, such critiques have dissuaded critical scholars on the left from endorsing and promoting animal justice claims out of the fear of being seen as racist or culturally imperialist and "performing whiteness" (122).

Distortions and Ironies

The critique of calls for universal veganism as culturally imperialist, racist, or otherwise elitist strikes many critical animal scholars as distorted, misplaced, and even ironic. Several factors account for this reaction. First, scholars note the richness of traditional plant-based diets in non-Western cultures as opposed to Western ones as well as resistance to anthropocentrism in non-Western cultural and religious traditions (Adams and Donovan 1995, 227; Twine 2014, 199). For example, the growth of vegetarianism in the West is credited to Westerners' exposure during British imperialism to ancient philosophies of nonviolence to animals in India (Stuart 2006, xxi). While not claiming that these countries are havens for animals, scholars point to the long-standing traditions in many non-Western societies that do not view animals as mere objects (Aristarkhova 2012; Narayanan 2016).

They also note that animals in the global South fare worse today than their counterparts in previous times because of the importation of Western industrial farming methods and other dualistic thinking that has commodified animals and altered traditionally plant-based diets worldwide (Bailey 2007; Kymlicka and Donaldson 2014; Twine 2014). To now see veganism as a Western phenomenon would appear to deny trajectories of historical geopolitical influence of non-Western philosophies promoting peaceful human-animal relations (Stuart 2006). As Richard Twine (2014) has commented, "critical scrutiny toward a goal of universal veganism, though vitally important, takes on a tragicomic aura when one actually considers current trends around food practices and universalizing tendencies . . . [namely] the present-day universalization of Western food practices which include, of course, increasing global trajectories of meat and dairy consumption" (193). Twine observes that the assumption that vegetarianism and veganism are Western-born phenomena can itself be viewed as an example of imperial epistemology—erasing as it does the histories and contributions of non-Western societies and ignoring the history of animal agricultural intensification and the cultural and economic export of those practices as part of globalization. Instead of fixating on the question of the ethnocentric impulses of vegan advocacy, Twine (2014) asks, "What if more critical discourse was directed at this significant cross-cultural change as being enmeshed within ethnocentrism and colonialism?" (205).

A further reason that critical animal scholars challenge the association of veganism with cultural imperialism stems from the intersectional ethic against all forms of exploitation that historically catalyzed vegan advocacy in the West (Cole 2014). Branding the movement as culturally imperialist, racist, or elitist today, when many adherents still understand veganism to be an all-encompassing antiexploitation ethic, is said to misunderstand its inclusive political origins. To the extent that vegan advocacy in the West is unreflexive about whiteness and white privilege (Harper 2010) is, of course, a serious deficiency that needs to be reversed (Kim 2015; Twine 2014). But if vegan advocacy can be said to suffer from a lack of reflexivity regarding white normativity, it is not dissimilar to other social movements in the West. Consider, for example the postcolonial criticisms of Western feminist (Mohanty 1988; Grewal and Kaplan 1994) and queer (Ahmed 2006; Puar 2007) theory and advocacy. Postcolonial criticisms of feminist and queer movements, however, have not dismissed these movements and their causes as imperialist, racist, or elitist despite the white normativity in their dominant iterations.

How can we explain what appears to be a different response to veganism and concern for animals? Cathryn Bailey offers one suggestion. She notes that because dietary practices are so constitutive of identity, they elicit heightened "visceral moral and emotional impact" when challenged (Bailey 2007, 53). The "regulation" of nonflesh eaters through the one-sided critique of white and male privilege that is meant to dismiss plant-based diets that Bailey (2007, 44) discusses may also be an instance of what Claire Jean Kim has identified as "multiculturalism go(ing) imperial," that is, a situation where advocates for racial and cultural equality adopt imperial attitudes toward the suffering of animals (Kim 2007) and disavow the legitimacy of condemning animal suffering (Kim 2015). Racialized vegan scholars, critical of American vegan advocacy that promotes white ideals, have also written about the need to frame veganism instead as a decolonizing bodily practice in the context of a nutrition landscape that encourages racialized and marginalized humans to consume highly processed and nutritionally poor foods despite alarming health inequities (Harper 2013).

A More Nuanced Debate: Contextual Moral Veg(etari)anism

A more legitimate query within the postcolonial debate about the potentially culturally oppressive ethic of exalting a vegan diet for all humans wonders whether universal prescriptions in general are inherently culturally imperialist. The objection to veganism where it encodes a universal ethic that all humans abstain from eating animals is an issue that various ecofeminists writing about animals have carefully addressed (Adams and Donovan 1995; Curtin 1991; Gaard 2001; Gruen 1993; Kheel 2004; Plumwood 2004). As leading ecofeminists who write on animals have recently defined it, "ecofeminism addresses the various ways that sexism, heteronormativity, racism, colonialism, and ableism are informed by and support speciesism and how analyzing the ways these forces intersect can produce less violent, more just practices" (Adams and Gruen 2014, 1). Committed to internal scrutiny of its core tenets from this intersectional perspective, ecofeminists have conscientiously engaged with the critique that a universal vegan ethic is a culturally imperialist position. As Twine's discussion of these authors and their varying approaches reveals, many eschew universalism as a masculinist theoretical ambition in general and instead have called for a "contextual moral vegetarianism" whereby animal consumption is placed in full social context in order to assess ethically any disputed eating practice (Twine 2014, 194). For most, this typically means that animal consumption would be ethical in emergency contexts where human life is threatened and survival depends on consumption (Twine 2014).

There is less consensus among those who subscribe to "contextual moral vegetarianism" about whether indigenous practices to hunt and eat animals today qualify as such a need. Claims due to geographic location that foreclose a plant-based diet fall under the exception of emergency/necessity (Curtin 1991). Where subsistence-hunting claims are justified on the ground of maintaining cultural tradition, however, nonindigenous feminist responses, trying to hold together respect for animals and respect for indigenous communities, have been mixed. Some ecofeminists maintain that a vegan ethic is ultimately incompatible with indigenous and other ecological perspectives that emphasize the interconnectedness of all life forms including an inevitability of predation and prey identities (Plumwood 2004). Others claims that a vegan ethic is compatible with postcolonial and indigenous sensibilities. Nonindigenous feminists have asserted that reliance on tradition as a ground for killing animals must be carefully scrutinized, noting that cultures are not static and that traditions are frequently violent and problematic (Kemmerer 2004). Some have concluded that, ideally, such scrutiny should occur within indigenous communities, taking care to ensure that dissenting voices participate (Deckha 2007; Gaard 2001). Others have concluded that external critique is legitimate when properly contextualized and informed of the multiple power dynamics underpinning a given debate and phenomenon (Kim 2015).

The insights of indigenous vegan and animal advocates on this debate also emphasize that cultures are open to change and that veganism can be an expression and not a sublimation of indigenous cultural values regarding respect for animals. Margaret Robinson (2014), for example, has described her veganism as an indigenous person as an obvious choice to convey respect for animals while living in an urban environment. She also contests the fundamental premise animating the critique of cultural imperialism aimed at veganism, namely, the characterization of veganism as a white practice. She points out that this characterization problematically disputes the authenticity of her indigenous cultural identity given that she is vegan. Constance MacIntosh (2015), however, suggests that animal scholars who advocate veganism violate indigenous communities' rights to self-determination as well as misunderstand indigenous cosmologies that view the killing of an animal as nonviolent because it is done with the animal's consent. Will Kymlicka and Sue Donaldson seek to guide our appraisal of animal agency when hunted by humans in a way that takes both animal and indigenous claims such as MacIntosh's seriously. Addressing the argument that animals killed for sustenance by indigenous peoples consent to their deaths, Kymlicka and Donaldson (2014) suggest that the human community needs some way to assess the validity of such claims rather than just presume consent. Craig Womack (2013), a Native American scholar, would reject MacIntosh's "consent" argument altogether. Womack argues that respect and hunting are incompatible concepts, stating that nonviolence toward animals should trump cultural claims to traditional hunting where it is not necessary to kill animals to survive physically. He contests the view that animals consent to their deaths by asking those who believe this to consider how animals feel when they are hunted and before they are killed as well as to acknowledge that animals, like humans, are part of kin networks that they do not likely wish to leave. Kim (2015) further counsels that indigenous peoples who believe in animal consent to legitimate hunting practices must be mindful that their worldviews are still human interpretations and thus anthropocentric. This openness must be part of the mutual avowal between animal advocates and racialized and indigenous communities in general, that is, a willingness to acknowledge the legitimacy of the others' views about injustice while still subjecting them to critique.

Postcolonial Turn for Developing Postanthropocentric Relations

The cultural imperialism debate surrounding veganism is unlikely to settle anytime soon. A position all critical scholars in the debate would likely adopt, however, is to strive to counter the epistemic violence against animals that Western animal ontologies create (Wadiwel 2015). Several scholars have urged Western scholars and audiences to learn from models of less violent human-animal relations in non-Western societies. In conceptually moving past critiques emphasizing animal difference and alterity (though they are important as counters to arguments that would reward animals with moral status only when they are sufficiently similar to humans), Matthew Calarco (2015) suggests that we need to build scholarship that envisions what a postanthropocentric ontology would look like in practice. In discussing the possibilities for this line of inquiry, Calarco encourages his readers to look to societies less influenced by Western dualistic epistemological traditions, such as postcolonial and indigenous ones. Feminist philosophers seeking to chart new interspecies ethics have also suggested that those of us immersed in Western epistemologies can find resources for imagining relations with animals anew by looking to non-Western cultures in the global South as well as indigenous traditions worldwide (Gaard 2001; Gruen 2011; Willett 2014).

This academic call for pluralism, then, approaches the relationship between cultural diversity—and, specifically, non-Western/racialized cultures—and animals from a position of epistemological humility and postcolonial resistance. That is, looking to these cultures for better, harmonious, and nonviolent ways of living with animals corrodes the longstanding colonial narrative discussed above that Western cultures are superior when it comes to animal treatment. Indeed, and paradoxically, it is this imperialist framing of animal cruelty that enables the misperception that veganism is a praxis tied to white and Western privilege. A broader postcolonial remit for our understanding of Animal Studies places into sharp relief the deficiencies in Western ontologies about animals and animality. From this perspective, it is peculiar to think of veganism as a white or Western "thing." Scholarship directed at Western audiences that actually does this transnational and transcultural work of learning from human cultural traditions typically cast as Other within Western-based epistemologies is still comparatively nascent in the field of Animal Studies (Aristarkhova 2012; Belcourt 2015; Corbey and Lanjouw 2013; Dalal and Taylor 2014; Kemmerer 2012). One can only hope that this subfield will continue to mature.

If postcolonial critiques expose the partial nature of Western projects to define the human and make visible the Western subject parading as a universal figure of humanity, then the importance of animality to postcolonial analyses is evident. From historical and contemporary examples of analyses attesting to the relevance of race, culture, and colonialism to thinking about animals and species, postcolonial concerns centrally animate human-animal relations and, increasingly, scholarship in Animal Studies. Animal studies emphasizes the deeply co-constitutive nature of "truths" about animals, animality, species, and species difference with racial and cultural ideologies usually in an effort to demonstrate shared roots of injustice and the need for other critical theories to abandon their anthropocentric premises. At the same time, the hallmark postcolonial concern about "cultural imperialism" has worked to impede academic coalition building between other critical schools of thought and Animal Studies. This phenomenon is most pointed in the argument that veganism is a white, Western, and overall privileged dietary practice. Animal studies scholars have noted the ironies in this characterization of a plant-based diet given histories and trends of global dietary practices but have also provided nuanced reflection of the claim. While the "cultural imperialism" debate will no doubt continue, the call for Animal Studies to turn to non-Western cultures to learn from less anthropocentric models of human-animal relations can be viewed as a clear commitment within Animal Studies to further cultivate a decolonizing postanthropocentric ethic.

Suggestions for Further Reading

Anderson, Kay. 2000. "The Beast Within: Race, Humanity, and Animality." *Environment and Planning D: Society and Space* 18 (3): 301–20.

Bailey, Kathryn. 2007. "We Are What We Eat: Feminist Vegetarianism and the Reproduction of Racial Identity." *Hypatia* 22 (2): 39–59.

Deckha, Maneesha. 2013. "Welfarist and Imperial: The Contributions of Anti-Cruelty Legislation to Civilizational Discourse." *American Quarterly* 65:3 515–48.

Elder, Glen, Jody Emel, and Jennifer Wolch. 1998. "Race, Place, and the Bounds of Humanity." *Society and Animals* 6 (2): 183–202.

Harper, Amie Breeze. 2010. "Race as a 'Feeble Matter' in Veganism: Interrogating Whiteness, Geopolitical Privilege, and Consumption Philosophy of 'Cruelty-Free' Products." *Journal for Critical Animal Studies* 8 (3): 5–27.

Kim, Claire Jean. 2015. *Dangerous Crossings: Race, Species, and Nature in a Multicultural Age*. Cambridge: Cambridge University Press.

Kymlicka, Will, and Sue Donaldson. 2014. "Animal Rights, Multiculturalism, and the Left." *Journal of Social Philosophy* 45 (1): 116–35.

Twine, Richard. 2014. "Ecofeminism and Veganism: Revisiting the Question of Universalism." In *Ecofeminism: Feminist Intersections with Other Animals and the Earth*, edited by Carol J. Adams and Lori Gruen, 191–208. New York: Bloomsbury.

References

Adams, Carol J. 1991. *The Sexual Politics of Meat: A Feminist-Vegetarian Critical Theory*. New York: Continuum.

Adams, Carol J., and Josephine Donovan, eds. 1995. *Animals and Women: Feminist Theoretical Explorations*. Durham, NC: Duke University Press.

Adams, Carol J., and Lori Gruen. 2014. "Introduction." In *Ecofeminism: Feminist Intersections with Other Animals and the Earth*, edited by Carol J Adams and Lori Gruen, 1–6. New York: Bloomsbury.

Ahmed, Sara. 2006. *Queer Phenomenology: Orientations, Objects, Others*. Durham, NC: Duke University Press.

Anderson, Kay. 2000. "The Beast Within: Race, Humanity, and Animality." *Environment and Planning D: Society and Space* 18 (3): 301–20.

Aristarkhova, Irina. 2012. "Thou Shall Not Harm All Living Beings: Feminism, Jainism, and Animals." *Hypatia* 27: 636–50.

Bailey, Kathryn. 2007. "We Are What We Eat: Feminist Vegetarianism and the Reproduction of Racial Identity." *Hypatia* 22 (2): 39–59.

Belcourt, Billy-Ray. 2015. "Animal Bodies, Colonial Subjects: (Re)Locating Animality in Decolonial Thought." *Societies* 5: 1–11.

Bhambra, Gurminder K. 2014. "Postcolonial and Decolonial Dialogues." *Postcolonial Studies* 17 (2): 115–21.

Calarco, Matthew. 2015. "Animal Studies." 2015. *The Year's Work in Critical and Cultural Theory* 23 (1): 20–39.

Cole, Matthew. 2014. "'The Greatest Cause on Earth': The Historical Formation of Veganism as An Ethical Practice." In *The Rise of Critical Animal Studies*, edited by Nik Taylor and Richard Twine, 203–24. New York: Routledge.

Condis, Megan. 2015. "She Was a Beautiful Girl and All of the Animals Loved Her: Race, the Disney Princesses, and Their Animal Friends." *Gender Forum* 55. http://gender forum.org/wp-content/uploads/2017/01/201509CompleteIssueAnimals.pdf.

Corbey, Raymond. 2005. *The Metaphysics of Apes: Negotiating the Animal-Human Boundary*. New York: Cambridge University Press.

Corbey, Raymond, and Annette Lanjouw. 2013. *The Politics of Species: Reshaping Our Relationships with Other Animals*. New York: Cambridge University Press.

Curtin, Deane. 1991. "Toward an Ecological Ethic of Care." *Hypatia* 6 (1): 60–74.

Dalal, Neil, and Chloe Taylor. 2014. *Asian Perspectives on Animal Ethics: Rethinking the Nonhuman*. Abingdon, Oxon: Routledge.

Darian-Smith, Eve. 1996. "Postcolonialism: A Brief Introduction." *Social and Legal Studies* 5 (3): 291–99.

Deckha, Maneesha. 2007. "Animal Justice, Cultural Justice: A Posthumanist Response to Cultural Rights in Animals." *Journal of Animal Law and Ethics* 2: 189–230.

———. 2012. "Toward a Postcolonial Posthumanist Feminist Theory: Centralizing Race and Culture in Feminist Work on Nonhuman Animals." *Hypatia* 27 (3): 527–45.

———. 2013. "Welfarist and Imperial: The Contributions of Anti-Cruelty Legislation to Civilizational Discourse." *American Quarterly* 65 (3): 515–48.

Deckha, Maneesha, and Erin Pritchard. 2016. "Recasting Our 'Wild' Neighbours: Toward an Egalitarian Legal Approach to Urban Wildlife Conflicts." *UBC Law Review* 49 (1): 161–202.

Elder, Glen, Jennifer Wolch, and Jody Emel. 1998a. "Race, Place, and the Bounds of Humanity." *Society and Animals* 6 (2): 183–202.

———. 1998b. "Le pratique sauvage: Race, Place, and the Human-Animal Divide." In *Animal Geographies: Place, Politics, and Identity in the Nature-Culture Borderlands*, edited by Jennifer R Wolch and Jody Emel, 72–90. London: Verso.

Esmeir, Samera. 2012. *Juridical Humanity: A Colonial History*. Stanford, CA: Stanford University Press.

Gaard, Greta. 1993. "Ecofeminism and Native American Cultures: Pushing the Limits of Cultural Imperialism." In *Ecofeminism: women, animals, nature*, edited by Greta Gaard, 295–314. Philadelphia: Temple University Press.

———. 2001. "Tools for a Cross-Cultural Feminist Ethics: Exploring Ethical Contexts and Contents in the Makah Whale Hunt." *Hypatia* 16 (1): 1–26.

———. 2013. "Toward a Feminist Postcolonial Milk Studies." *American Quarterly* 65 (3): 595–618.

Gaard, Greta, and Lori Gruen. 1993. "Ecofeminism: Toward Global Justice and Planetary Health." *Society and Nature* 4: 1–35.

Gandhi, Leela. 1998. *Postcolonial Theory: A Critical Introduction*. New York: Columbia University Press.

Grewal, Inderpal, and Caren Kaplan, eds. 1994. *Scattered Hegemonies: Postmodernity and Transnational Feminist Practices*. Minneapolis: University of Minnesota Press.

Gruen, Lori. 1993. "Dismantling oppression: An Analysis of the Connection between Women and Animals. In *Ecofeminism: Women, Animals, Nature*, edited by Greta Gaard, 60–90. Philadelphia: Temple University Press.

Gruen, Lori. 2011. *Ethics and Animals: An Introduction*. New York: Cambridge University Press, 2011.

Haraway, Donna. 1989. *Primate Visions: Gender, Race, and Nature in the World of Modern Science*. New York: Routledge.

Harper, Amie Breeze. 2010. "Race as a 'Feeble Matter' in Veganism: Interrogating Whiteness, Geopolitical Privilege, and Consumption Philosophy of 'Cruelty-Free' Products." *Journal for Critical Animal Studies* 8 (3): 5–27.

———. 2013. "Doing Veganism Differently: Racialized Trauma and the Personal Journey towards Vegan Healing." In *Doing Nutrition Differently*, edited by Jessica Hayes-Conroy and Alison Hayes-Conroy, 133–50. London: Ashgate.

Hund, Wulf D., Charles W. Mills, and Silvia Sebastini, eds. 2015. *Simianization: Apes, Gender, Class, and Race*. Zurich: Lit.

Jerolmack, Colin. 2008. "How Pigeons Became Rats: The Cultural-Spatial Logic of Problem Animals." *Social Problems* 55 (1): 72–94.

Kemmerer, Lisa. 2004. "Hunting Tradition: Treaties, Law, and Subsistence Killing." *Animal Liberation Philosophy and Policy Journal* 2 (2): 1–20.

Kemmerer, Lisa. 2012. *Animals and World Religions*. Oxford: Oxford University Press.

Kheel, Marti. 1985. "The Liberation of Nature: A Circular Affair." *Environmental Ethics* 2: 135–49.

———. 2004. "Vegetarianism and Ecofeminism: Toppling Patriarchy with a Fork." In *Food for Thought: The Debate over Eating Meat*, edited by Steve F. Sapontzis, 327–41. Amherst, NY: Prometheus Books.

Kim, Claire Jean. 2007. "Multiculturalism Goes Imperial." *Du Bois Review* 4(1): 233–249.

———. 2015. *Dangerous Crossings: Race, Species, and Nature in a Multicultural Age*. Cambridge: Cambridge University Press.

Kymlicka, Will, and Sue Donaldson. 2014. "Animal Rights, Multiculturalism, and the Left." *Journal of Social Philosophy* 45 (1): 116–35.

———. 2015. "Animal Rights and Aboriginal Rights." In *Canadian Perspectives on Animals and the Law*, edited by Peter Sankoff, Vaughan Black, and Katie Sykes, 159–86. Toronto: Irwin Law.

Lugones, Maria. 2010. "Toward a Decolonial Feminism." *Hypatia* 25 (4): 742–59.

MacIntosh, Constance. 2015. "Indigenous Rights and Relations with Animals: Seeing beyond Canadian Law." In *Canadian Perspectives on Animals and the Law*, edited by Peter Sankoff, Vaughan Black, and Katie Sykes, 187–208. Toronto: Irwin Law.

Milstein, Tema. 2009. "'Somethin' Tells Me It's All Happening at the Zoo': Discourse, Power and Conservationism." *Environmental Communication: A Journal of Nature and Culture* 3 (1): 25–48.

Mohanty, Chandra Talpade. 1988. "Under Western Eyes: Feminist Scholarship and Colonial Discourses." *Feminist Review* 30 (1): 61–88.

Nadasdy, Paul. 2016. "First Nations, Citizenship and Animals; or, Why Northern Indigenous People Might Not Want to Live in Zoopolis." *Canadian Journal of Political Science* 49 (1): 1–20.

Narayanan, Yamini. 2016. "Where Are the Animals in Sustainable Development? Religion and the Case for Ethical Stewardship in Animal Husbandry." *Sustainable Development* 24: 172–180.

———. 2017. "Street Dogs at the Intersection of Colonialism and Informality: 'Subaltern Animism' As a Posthuman Critique of Indian Cities." *Environment and Planning D: Society and Space* 35 (3): 475–94.

Palmer, Clare. 2003. "Colonization, Urbanization, and Animals." *Philosophy and Geography* 6 (1): 47–58.

Peterson, Christopher. 2013. *Bestial Traces: Race, Sexuality, Animality*. New York: Fordham University Press.

Plumwood, Val. 1993. *Feminism and the Mastery of Nature*. London: Routledge.

———. 2004. "Gender, Eco-feminism and the Environment." In *Controversies in Environmental Sociology*, edited by Rob White, 43–61. Cambridge: Cambridge University Press.

Puar, Jasbir. 2007. *Terrorist Assemblages: Homonationalism in Queer Times*. Durham, NC: Duke University Press.

Robinson, Margaret. 2014. "Animal Personhood in Mi'kmaq Perspective." *Societies* 4: 672–88.

Salih, Sara. 2007. "Filling Up the Space between Mankind and Ape: Racism, Speciesism and the Androphilic Ape." *Ariel* 38: 95–111.

Stuart, Tristam. 2006. *The Bloodless Revolution: A Cultural History of Vegetarianism from 1600 to Modern Times*. London: W. W. Norton.

Twine, Richard. 2014. "Ecofeminism and Veganism: Revisiting the Question of Universalism." In *Ecofeminism: Feminist Intersections with Other Animals and the Earth*, edited by Carol J. Adams and Lori Gruen, 191–208. New York: Bloomsbury.

Wadiwel, Dinesh. 2015. *The War against Animals*. Leiden: Brill.

Wenzel, George. 1991. *Animal Rights, Human Rights: Ecology, Economy and Ideology in the Canadian Arctic*. Toronto: University of Toronto Press.

Willett, Cynthia. 2014. *Interspecies Ethics*. New York: Columbia University Press.

Wolfe, Cary. 2012. *Before the Law: Humans and Other Animals in a Biopolitical Frame*. Chicago: University of Chicago Press.

Womack, Craig. 2013. "There Is No Respectful Way to Kill an Animal." *Studies in American Indian Literatures* 25 (4): 11–27

Wrenn, Corey Lee. 2016. "An Analysis of Diversity in Nonhuman Animal Rights Media." *Journal of Agricultural and Environmental Ethics* 29 (2): 143–65.

20 RATIONALITY

Christine M. Korsgaard

According to a traditional philosophical view dating back to Aristotle and shared by Immanuel Kant and many others, what makes human beings different from the other animals is that human beings are "rational" or have "reason." But consider, for example, Sultan, one of the chimpanzees Wolfgang Köhler used in his famous experiments on animal problem solving. Without previous training, Sultan stacked up crates to retrieve bananas that Köhler had hung from the ceiling and put together two sticks to form an implement long enough to reach bananas outside his cage. So Sultan thought about how to achieve his end and did what was necessary to achieve it. Was Sultan rational?

Both colloquially and in academia, the terms *reason* and *rational* are used in many different ways. *Reason* may be used to denote sanity, as when people say that a mentally deranged person has "lost his reason." *Rational* may be used to mean orderly and intelligible, or in accordance with causal laws, as when people say they believe that the world is a "rational place." *Reasoning* may be taken to describe the activity of working out what to do or believe by thinking, in which case "reason" is simply the capacity to do that. Charles Darwin (1981) uses the terms *reasoning* and *reason* in this way in *The Descent of Man* when he says

> Few persons any longer dispute that animals possess some power of reasoning. Animals may constantly be seen to pause, deliberate, and resolve. And it is a significant fact that the more the habits of any particular animal are studied by a naturalist, the more he attributes to reason and the less to unlearnt instincts. (Darwin 1981, 46)

In this passage, *reason* is opposed to *instinct*. Actions from instinct are unlearned, automatic behavioral responses. When the terms are used in this way, acting from reason is acting intentionally and intelligently. Although some people have believed that all animal behavior is governed by instinct in Darwin's sense, the modern study of animal minds supports Darwin's nonchalant dismissal of that view. Sultan and his companions—as well as many other primates, corvids, and cetaceans— have demonstrated impressive problem-solving abilities that could not possibly be instinctive. And in many animals, behaviors that are probably instinctive are refined by learning: animals as they grow up get better at doing the things they instinctively do.

Social Scientific Uses of "Reason" and "Rational"

Some uses of *reason* and *rational* are more specific than the ones described above. In the social sciences, acting rationally usually means acting prudently (doing what is in your own best interests)[1]—or acting with instrumental rationality (doing what will get you whatever ends you wish to achieve, whether they are in your best interests or not). When we say that someone acts rationally in these senses, we usually also mean that she is subjectively guided by the standards of prudence or instrumental rationality. That is, we mean that she is motivated to do what she *believes* is in her own best interests or will get her whatever ends she wishes to achieve, and she is motivated *by* that belief. This subjective dimension to the idea of being rational is worth emphasizing. If we take rationality to be a purely objective standard (one that does not refer at all to the subjective states of the agent), and we take survival and reproduction to be the ends of life, most animal behavior will be rational in the sense that it tends to achieve these ends. But, of course, this need not mean that the animals themselves grasp or employ the standards of prudence or instrumental rationality or even that they are deliberately aiming at these ends. Evolution itself produces behavior that efficiently promotes these ends.

In general, it is important to distinguish between the evolutionary function of an animal's behavior and what, from the animal's own point of view,

---

1. Although being "rational" in the social scientific sense is acting in your own best interests, being "reasonable" in ordinary parlance sometimes means almost the opposite: being unwilling to impose undue burdens on others for your own sake.

she is trying to do, if indeed there is anything that she is trying to do. In the case of automatic instinctive behavior, especially in animals not otherwise cognitively sophisticated, perhaps there is not. But when we consider more intelligent animals, there are real questions about how their own motives and intentions are related to the evolutionary function of their behavior, and it is easy to blur the line between these two things. For example, it is often suggested that the evolutionary function of altruistic behavior is to increase an animal's own reproductive fitness in some indirect way. But this does not mean that an animal who helps another does so with the intention of increasing her own reproductive fitness or that there is something secretly self-interested about her conduct.

But as cases like that of Sultan show, not all instrumentally effective animal behavior is instinctive and automatic. Intelligent animals do figure out how to do things that instinct does not teach them, and there is no reason to doubt that from their own point of view that is exactly what they are trying to do: figure out how to achieve their ends. But philosophers in the Kantian tradition would argue that even when an animal like Sultan does engage in intelligent instrumental thinking—that is, when he figures out how to achieve a desired end—that does not necessarily show that his motivation is "rational." These philosophers argue that rational motivation involves the awareness that the consideration on which you act *is a reason* for acting that way. When we say that an agent has a reason for what she does, we imply that there is a standard of evaluation for her action—the action is in some way "reasonable" or "rational"—and that the agent has gone some way toward meeting that standard.[2] So knowing that you yourself are acting for a reason means knowing that an evaluative standard applies to your conduct, that there is a way you should act or ought to act or that it is good or correct to act, and being motivated in part by that awareness.[3] An animal like Sultan who has figured out that taking a certain means will get him an end that he wants might be moved by that conclusion to take the means without any thought about the normative correctness of being motivated in that way. We will come back to this conception of rationality later on.

2. "Gone some way toward" because you might have *a* reason for what you do while still doing something that is not what the balance of reasons favors.

3. For a discussion of this view of rationality and the view of reasons that it requires, see Korsgaard (2008).

Philosophical Conceptions of Reason and Rationality

In the philosophical tradition, reason is often taken to refer to the active as opposed to the passive or receptive aspects of the mind. "Reason" in this sense is contrasted with perception, sensation, and emotion, which are thought of as forms of passivity or at least as involving passivity. The contrast is not unproblematic, for it seems clear that the kind of receptivity or responsiveness involved in sensation, perception, and emotion cannot be understood as wholly passive. The perceived world does not simply enter the mind as through an open door. In sensing and responding to the world, our minds interact with it, and the activity of our senses themselves makes a contribution to the character of the world as we perceive it. All of this is undoubtedly true of the minds of the other animals as well. In fact, if, as people usually suppose, nonhuman animals are in some sense "less intelligent" than human beings, it is all the more important that their perceptions should present their environment to them in ways that are already practically interpreted for them. An animal who cannot figure out what to do through reasoning must perceive the world in ways that guide his behavior without further thought—he must perceive the world around him as consisting, say, of predators who are to be avoided and prey who are to be eaten and potential mates who are to be courted and offspring who are to be cared for and so on. His perceptual faculties must do the work of thinking for him; in that sense, they shape the inputs of perception. But that is not something that the animal himself, considered as a conscious subject, deliberately does.

The mental activity associated with reason is activity that *is* attributable to the reasoner himself. Reasoning is self-conscious, self-directing activity through which we deliberately give shape to the inputs of receptivity. This happens both in the case of theoretical reasoning, when we are constructing a scientific account of the world, and in the case of practical reasoning, where its characteristic manifestation is choice based on deliberation. An animal who is rational in this sense exerts a kind of deliberate control over her own mental processes, at least those that issue in belief and action. It is the possession of reason in this sense—the exercise of deliberate control over our own mental lives—that it is most plausible to suppose is a distinctively human attribute.

Reason has also traditionally been identified with the ability to grasp, employ, or simply conform to certain principles, which are usually conceived as *a priori*. These include the principles of logic, the principles that guide the construction of scientific theories, mathematical principles, and the principles of practical reason, including the principles of prudence and instrumental rationality mentioned in connection with the social scientific view described above. According to some philosophers, moral principles are also rational principles. In Kant's theory, for example, we evaluate our reasons by applying the categorical imperative—by asking ourselves whether the principles in which those reasons are embodied could possibly serve as universal laws. This conception of rationality is not at odds with the conception of reason as the active dimension of the mind: the principles of rationality may be taken to describe the activity in which the reasoner consciously engages when she considers whether her reasons are good ones. Even when we do not deploy these principles consciously, however, they are the principles in accordance with which we reason, the principles that describe how we think out what to do or believe. A person is called reasonable or rational when his beliefs and actions conform to the dictates of those principles or when he is subjectively guided by them.

Reason is also identified with the capacity that enables us to identify "reasons," the particular considerations that count in favor of belief or action. Reasons, in this sense, may be the considerations picked out by rational principles or directly grasped by "reason" in the sense of the active capacity of the mind. Ordinarily, we take reasons both to justify and to explain the actions for which they are done, or at least to make them intelligible. When we understand what an agent's reasons are, we see how the situation looked to him and why it motivated him to act as he did. In this sense of *reason* it is plain that the other animals, like human beings, sometimes act for reasons and sometimes do not. Suppose that an animal is banging himself against the bars of his cage, and we ask why he is doing that. If the answer is that he is trying to escape from the cage by breaking or bending the bars, we can see what his reason is in the sense that we have some grasp of how the situation looks to him and what he intends to do about it. On the other hand, if the answer is that being caged has made the animal mentally ill, his behavior has a cause, but it is not something done for a reason. In this sense of reason, it seems plain that Sultan's reason for putting the two sticks together was that it enabled him to reach the bananas outside of his cage.

Reason as a Distinctively Human Attribute

No doubt some who have believed that only human beings are "rational" or have "reason" have thought that all of animal behavior is automatic, unthinking, and instinctive. But that is not the only form the view can take. Philosophers who think that only human beings are "rational" or have "reason" may believe something along the lines that many animals may have the capacity to think about their situations and to be motivated to act intelligently in those situations to get what they want and avoid what they don't. But intelligence is not the same as reason. Only human beings have the capacity to think about their motivations themselves—about the potential reasons for their actions—and to ask whether those potential reasons meet certain normative or evaluative standards. Human beings have a distinct form of self-consciousness that enables them to be aware of the motivations or potential reasons on which they are tempted to act, to evaluate those potential reasons, and to be moved to act accordingly. A parallel point may be made about the considerations on the basis of which we are tempted to believe. The standards of evaluation for our potential reasons are given by the principles of reason or rationality.

According to these views, nonhuman animals may have reasons for what they believe and do in the sense that they believe and act on the basis of the *contents* of their perceptions, desires, fears, and instincts. But they do not think about these reasons themselves or the fact that they have them, and they do not ask themselves whether their reasons are good ones or not. Human beings, however, do, and this gives our beliefs and actions a normative character that the beliefs and actions of the other animals lack. If you are rational, when you believe something you also think that there is a sense in which you *should* believe it; when you do something, you think there is some sense (not necessarily a moral one) in which it is the appropriate or correct thing to do.

The theory that rationality in this more demanding sense marks a distinctive difference between human beings and the other animals will be most tempting if we can appeal to it to explain other things that appear to be distinctive about human life. For example, we might suppose that the fact that human beings think about the quality of the reasons for our beliefs is part of what enabled us to develop the scientific method. Many people believe that morality itself is a distinctive characteristic of human beings. If morality is a manifestation of a form of rationality that is unique to human beings, that would explain why that is so.

Some would argue that the idea that *any* capacity unique to human beings is at odds with the theory of evolution. Darwin himself, who believed that morality is a distinctively human attribute, explained it as a result of the interaction between two capacities we share with some other animals: social instincts and advanced cognitive faculties.[4] Evolutionary thought need not be taken to imply that all differences between human beings and the other animals must be matters of degree. Some distinctively human attributes might emerge from unique combinations of evolved powers. The apparently immense difference in "intelligence" between human beings and the other animals might itself be better explained by the interaction of intelligence with rationality in this more demanding sense than as a simple quantitative leap. I have already suggested that rationality is tied to a special form of self-consciousness, one that makes us aware of and capable of evaluating our reasons themselves. If this is so, perhaps rationality itself can be explained as a result of the interaction between having an advanced "theory of mind"—the ability to attribute mental states to ourselves and others—and having other advanced cognitive capacities. These powers combined might enable us to think about what our own point of view and the points of view of others have in common and to realize that those common elements set evaluative standards for our actions and beliefs.

Rationality and the Value of Human Life

Many people hold that rational beings or rational life has a greater value than nonrational beings or nonrational life. They think that this value makes rational beings either uniquely morally important or more morally important than the other animals. This view is naturally associated with the idea that it is rationality that makes us moral beings, since that is an attribute that might appear to give rational beings a special value. But it is unclear exactly how the argument is supposed to go. Even if the capacity for moral *goodness* is a form of superiority, the fact that human beings alone are moral animals, if it is a fact, also makes us uniquely capable of

---

4. Darwin (1981) says, "I fully subscribe to the judgment of those writers who maintain that of all the differences between man and the lower animals, the moral sense or conscience is by far the most important" (70) and that "we have no reason to suppose that any of the lower animals has this capacity" (88–90). But he proposes that "any animal whatever, endowed with well-marked social instincts, would inevitably acquire a moral sense or conscience, as soon as its intellectual powers had become as well-developed, or nearly as well-developed, as in man" (71–72).

moral evil. Perhaps the thought that rationality conveys a unique or superior value originates with the idea that rationality is one of the attributes in virtue of which human beings can be said to be made "in the image of God."[5] Without putting it on this theological footing, Immanuel Kant (1997) held that rational beings as "persons" are "ends in themselves," while nonrational animals are "things" that may legitimately be used as mere means to the ends of rational beings (37). Many writers in the philosophical literature assume that what Kant meant is simply that rationality has a kind of intrinsic value that in turn confers value on those who have it. This does not make it clear why only rational beings should be the objects of moral consideration, however, and philosophers in the utilitarian tradition protest that sentience—the capacity for suffering—is sufficient to make a creature the object of moral concern.

A more promising argument for the conclusion that only rational beings are worthy of moral consideration, and one closer to the one Kant actually had in mind, holds that morality is a system of *reciprocal* rights and obligations. You cannot have the rights unless you are also capable of meeting the obligations. Kant (1997) argues that to have a duty to another is to be constrained by that other's will (192). As members of a "Kingdom of Ends," a community of all rational beings, we all legislate moral laws for ourselves and each other and in this way exercise moral constraint over each other (Kant 1997, 42–43). But the other animals, who are incapable of willing moral laws, are no part of this system. Aristotle seems to have this kind of argument in mind as well. In the *Nicomachean Ethics* he argues at one point that neither justice nor friendship can exist between a (free) human being and a horse or an ox, or a slave qua slave, but that a free human being can be friends or enter into relations of justice with the slave qua human. He explains, "for there seems to be some justice between any man and any other who can share in a system of law or be a party to an agreement" (Aristotle 1984, 8.11.1161b1–1161b8). Both philosophers conclude that nonhuman animals have only an instrumental value, but that conclusion is supposed to follow from the other animals' inability to participate in the moral system rather than to be the reason for it. Whether this is a good reason for excluding nonrational or nonmoral creatures from moral consideration is of course a matter for further argument.

Another possible source of the idea that rational life is more morally important than nonrational life is an argument that goes roughly like

5. For a protest against this line of thinking, see Coetzee (1999, 22–23).

this. (1) Only rational animals are self-conscious. (2) Therefore only rational animals can think about themselves and their temporally extended lives. (3) Therefore only rational animals really have genuinely temporally extended identities. (4) Not much except local pleasures and pains can be important to you if you lack a temporally extended identity. And therefore (5) not much can be very important to a nonrational animal. According to this argument, nonrational life is less morally important because the lives of nonrational animals are less important to those animals themselves.[6] All of the steps in this argument can be challenged, of course, but there is also an element of truth in some of them. Rather than taking the argument to ground a blanket distinction between rational and nonrational life, we might take it to explain why some things, such as the continuance of life, might legitimately be taken to matter more to animals who are cognitively and emotionally sophisticated in certain ways than they do to others.[7]

Normative and Descriptive Conceptions of Rationality, Rational Defects, and Marginal Cases

Those who believe that rationality is *not* what makes human beings worthy of moral consideration often point out that some of the human beings to whom we extend moral consideration are "not rational." We extend moral consideration to infants, the comatose, the demented, and those with severe mental illnesses or cognitive defects. This is sometimes taken to show that it is actually "speciesism"—a prejudice in favor of our own species—and not a view about the special value of rational life that motivates people to withhold moral consideration from nonhuman animals. This argument, sometimes called the argument from marginal cases, raises questions about what we mean when we say of a being that he or she "is rational."

A first point to notice is that the terms *rational, reasonable, reason,* and their cognates may be used in either what we may call a descriptive or a normative way. When we say that someone "is rational" in the descriptive sense, we mean that there is some consideration that she believes to be relevant and on the basis of which she acts or believes, that she knows what

6. Coetzee (1999, 64) puts this view into the mouth of a fictional philosopher, Thomas O'Hearne, and challenges it through the reflections of his central character, Elizabeth Costello.

7. For a more detailed treatment of this kind of argument, see McMahan (2002, chap. 3).

that consideration is, and that she would offer it in answer if you asked her why she believes or does what she does. Being rational in this sense contrasts with being deranged, or out of control, or engaging in some form of expressive action—slapping someone in rage, screaming in terror—or acting in an automatic, unreflective way—ducking, wincing, or doing something entirely out of habit. In these kinds of cases we can explain what the person does, but the explanation does not take the form of citing a consideration on the basis of which she made a decision or formed a belief. When we say that someone "is rational" in the normative sense, we mean that she is rational in the descriptive sense *and* that the consideration of the basis of which she acted or formed a belief is a *good* reason, measured by the evaluative standards by which rational beings judge the quality of our reasons. When we use the terms normatively, we are prepared to call a consideration "a reason" only if it is a good reason.

Suppose, for example, that Edna tells us she plans to vote for a certain presidential candidate whom we know to be racially prejudiced, unintelligent, ignorant of foreign policy, on the take from lobbies, etc., and we ask her why. She says he is the nominee of the XYZ party, and she always votes with the XYZ party. Edna is rational in the descriptive sense: her action is based on a consideration that she takes to be relevant, that the candidate has been nominated by her party. We may pose an objection to her plan either by saying "that's a bad reason to vote for him" or "that's no reason to vote for him." If we say it's "a bad reason," we are using the term *reason* in the descriptive sense: a consideration on the basis of which someone acts. If we say it's "no reason," we are using the term *reason* in a normative sense, essentially to mean a good reason. Someone can reason badly and still be "a rational being" in the descriptive sense.

There is one kind of case that cuts across the distinction in an interesting way: cases in which someone believes or acts on the basis of a consideration but one whose efficacy for her she is not aware of and whose relevance she might consciously deny. We think of Freudian slips, prejudices, implicit biases, and various forms of self-deception as working in this way. Perhaps Edna is in fact drawn to vote for the XYZ candidate because he shares a prejudice of hers or because the opposing party's candidate is a member of a racial minority that Edna unconsciously views as inferior. A critic of Edna might say that that is "the real reason" she is voting for the XYZ candidate. The consideration is in a sense operating as a cause rather than as a reason, because Edna is not aware of its influence, but it makes her conduct intelligible in something like the way a reason does: it shows us how the situation looks through her eyes and how her action is a response to that.

We generally assume that being rational in the descriptive sense puts someone in the way of being rational in the normative sense; if you can reason, you can figure out or at least be taught how to reason well. But the categories of persons singled out in the argument from marginal cases, for various reasons, are not capable of reasoning well and may not be capable of reasoning at all. Are they therefore "not rational beings" even in the descriptive sense?

The trouble with drawing this conclusion is that rationality is not merely a property, like having blue eyes. It is a way of functioning—a way of determining what to believe and to do. More generally, an organism is not merely a collection of properties but a functional unity whose various parts and capacities work together to produce a certain form of life. A rational being who lacks some of the properties that together make rational functioning possible is not nonrational but rather defectively rational and therefore unable to function well. For example, a small child or someone with severe cognitive disabilities may be aware of the desires that tempt her to act and capable of refraining from acting on them if she knows that it would be wrong to do so but unable to evaluate them correctly.[8] An addict, on the other hand, may be able to evaluate her desires correctly but be unable to refrain from acting on them. A comatose person does not have a different way of functioning but rather is unable to function at all. It is not as if you could simply subtract "rationality" from a human being and you would be left with something that functions like a nonhuman animal. What we owe to defectively rational persons is not to treat them however it is right to treat nonhuman animals but rather to treat them with duties of care that are responsive to the defects in their capacity to function rationally.

The point of advancing these considerations is not necessarily to object to the intended conclusion of the argument from marginal cases—that sentience is sufficient to make a creature the object of moral concern. Even if we agree with that conclusion, we may think that the argument from marginal cases makes morally significant errors about how the objects of moral concern should be identified. For example, categorizing "infants" as a type of creature conflates the subject of a temporal phase of a life with

8. For a more sophisticated view along these lines, of the rationality of children, see Schapiro (1999).

the subject of a life as a whole, which arguably is the proper object of moral concern.[9] The subject of a life is not a mere collection of the subjects of various temporal phases of the life but something that is connected in a certain way throughout the life. The way it is right to treat infants of every species is responsive not just to the properties they exhibit now but to the fact that they are going to grow up in a certain way. In a similar way, whether a being is rational does not depend on the properties he exhibits now. A rational being is one "designed" to function in a certain way, and a creature so designed but with defects does not therefore become a different kind of creature. The proper moral response to such a creature reflects those facts.

The Moral Implications of Rationality

As we have seen, *rationality* refers to a range of properties that characterize the mental lives of the creatures deemed *rational*. It can mean simply sane and well oriented toward the world, and it can mean capable of acting intentionally and intelligently. In these senses, nonhuman animals like Sultan can be "rational." Philosophers who suppose rationality is a unique human attribute or one that human beings have to a uniquely high degree usually mean something more specific. Rationality in this sense involves an awareness of the considerations that tempt us to believe and act in certain ways—of our potential reasons—and the ability to evaluate those reasons in accordance with a priori principles that determine whether they are good and sufficient reasons or not.

Some people believe that rationality in this sense makes human beings uniquely worthy of moral consideration, or more morally important than the other animals. This may be because they believe that only human beings are moral animals and that morality itself is a system of reciprocity in which nonrational creatures cannot participate. Or it may be because they believe that the opposite of a rational mental life is a mental life without much temporal extension and that therefore nothing but local experiences can be very important to nonhuman animals themselves. These views can certainly be challenged, both empirically and philosophically. But even if rationality is a distinctively human attribute, it may be the source not of a unique moral value but rather of a different kind of special moral standing. If human beings alone are rational in the sense that

9. I borrow the phrase "subject of a life" from Tom Regan (1983), who identifies such subjects as the objects of moral concern.

includes the capacity for the moral evaluation of our reasons, then we can ask ourselves whether we really have good and sufficient reasons for doing experiments on Sultan, or treating him as bush meat, or destroying the habitat on which he and his fellows depend. The possession of rationality will imply that human beings, alone among the animals, have moral obligations to the other animals with whom they share the world.[10]

Suggestions for Further Reading

Coetzee, J. M. 1999. *The Lives of Animals*. Edited by Amy Gutmann. Princeton, NJ: Princeton University Press.

Darwin, Charles. 1981. *The Descent of Man and Selection in Relation to Sex*. Princeton, NJ: Princeton University Press.

Kant, Immanuel. 1997. *Groundwork of the Metaphysics of Morals*. Translated by Mary Gregor. Cambridge: Cambridge University Press.

Korsgaard, Christine M. 2008. "Acting for a Reason." In *The Constitution of Agency*, 207–30. Oxford: Oxford University Press.

———. 2018. *Fellow Creatures: Our Obligations to the Other Animals*. Oxford: Oxford University Press.

References

Aristotle. 1984. *The Nicomachean Ethics*. Translated by W. D. Ross. Revised by J. O. Urmson. Vol. 2 of *The Complete Works of Aristotle*, edited by Jonathan Barnes. Princeton, NJ: Princeton University Press.

Coetzee, J. M. 1999. *The Lives of Animals*. Edited by Amy Gutmann. Princeton, NJ: Princeton University Press.

Darwin, Charles. 1981. *The Descent of Man and Selection in Relation to Sex*. Princeton, NJ: Princeton University Press.

Kant, Immanuel. 1997. *Groundwork of the Metaphysics of Morals*. Translated by Mary Gregor. Cambridge: Cambridge University Press.

Korsgaard, Christine M. 2008. "Acting for a Reason" In *The Constitution of Agency*, 207–30. Oxford: Oxford University Press.

———. 2018. *Fellow Creatures: Our Obligations to the Other Animals*. Oxford: Oxford University Press.

McMahan, Jeff. 2002. *The Ethics of Killing: Problems at the Margin of Life*. Oxford: Oxford University Press.

Regan, Tom. 1983. *The Case for Animal Rights*. Berkeley: University of California Press.

Schapiro, Tamar. 1999. "What Is a Child?" *Ethics* 109 (4): 715–38.

10. A detailed explication of this position is available in Korsgaard's *Fellow Creatures: The Moral and Legal Standing of Animals* (forthcoming).

21 REPRESENTATION

Robert R. McKay

I recently had a series of conversations with librarians at the university at which I work about whether the library's online catalog system could allow users to find easily all of its holdings in the field of Animal Studies. It became clear fairly quickly that it could not. The various data that the cataloging system draws on to classify academic books and journals (title, subject keywords, author, date, and so on), which is to say the elements of those works that are *representative* of them in the logic of the cataloging system—are insufficient for such a task. This is a problem, primarily, of isolating what exactly "Animal Studies" is. In this case, the library's catalog cannot specifically *represent* something called "Animal Studies" in the form of a list of holdings except by finding this term in the titles and related data about the works. And even this method only works up to a point: books on critical Animal Studies in the humanities are mixed up with published reports of scientific experiments that use animals, also confusingly referred to as "Animal Studies"; and of these there are a great many. Even the term *Animal Studies*, it would seem, does not adequately *represent* Animal Studies.

We might also well imagine that this academic field reaches beyond the narrow usage of the specific term that names it, and so we want to find it in the full range of published writing about animals that is held by a university library. Here it is evidently not possible for the catalog to discriminate between (to select only the first four returns when searching for "animals" in it) *Animals*, a Swiss journal that appeared in 2010, which publishes articles "relevant to any field of study that involves animals, including zoology, ethnozoology, animal science, animal ethics and animal welfare"; *Animals*, a zoological textbook by Robert McNeill Alexander from 1990, which details the "structure, physiology and ways of life of the major groups of animals"; "Animals," a quietly accusatory poem by Matthew Sweeney published in 2000 which is, on the surface at least, about what can happen when human priorities are disturbed by animal presences and the kinds of intolerance, forbearance, and violence this annoyance might provoke; and *Society and Animals*, a journal first published in 1993 that originally focused on "social scientific studies of the human experience of other animals" but has expanded to include humanities scholarship and which has a good claim to be the most important early vehicle for work in Animal Studies. The relative diversity here in terms of intellectual or academic focus might seem to indicate the challenge for any catalog of works to *represent* Animal Studies: a conventional work of zoology; a literary work which may or may not be interpreted as speaking for (or *representing*) the interests of nonhuman animals at all; one journal based in animal science that reaches outward to consider attendant social and philosophical issues; and another that works somewhat in the other direction.

As will be suggested if you cast your eye over the highlighted words in the preceding paragraphs, the challenge of distinguishing Animal Studies can be understood in terms of a series of quandaries related to the issue of representation. Indeed, we might draw on the range of meanings of the term to ask how a particular example of academic inquiry represents animals. How does it *portray* them? How does it look for, *characterize*, or evaluate their lives and their interests? Does it *speak for* them (i.e., for animals), and if so, how? Are animals studied in their own right or do they *stand for* something else (i.e., human interests or concerns)?

It is particularly noteworthy that running through these kinds of questions is a tension that exists between two key meanings, *portrayal* and *advocacy*, that are vexed within the concept of representation. Understanding this tension is crucial for making sense of contemporary thought about animals, and so I want to take some time to explore it. It is important because of the position that animals are often allotted in social frameworks of knowledge: in fundamental ways they are regarded as like humans in some ways *but definitely not the same*—belief in an abiding and radical difference (a belief known as human exceptionalism) conditions any notional similarity. In this state of affairs, animals (or certainly many, if not most, animals, unlike machines or stones) are easily assumed to experience their own lives and to have interests, and these are surely understandable, or at least recognizable in some way. And although assuredly unlike human lives and interests, animal lives and interests are not so radically other as to be unimaginable in these terms. And so, they can indeed be more or less accurately portrayed, *but certainly not by animals*. Animals cannot speak for themselves: so goes the refrain; someone (human) must speak for them. It is the pressure created by the sense that animals clearly do have lives and interests but that these cannot be portrayed by them within the circuit of discourse that is presumed to be a human preserve, that instigates the key dilemma of representation for Animal Studies. Just how fully can human portrayals of animals achieve the ecstatic feat of complete accuracy, figuring forth animals' lives and interests in just the way that they themselves would, if they could? This would be a portrayal that could thereby authentically and justifiably stand in for an animal's conceivable but impossible self-portrayal. As such, it could be said fully to speak on behalf of animals, to *represent* them.

Re-presenting Representation

Representation, I hope it is becoming clear, is a term that has a rich array of meanings at some variance with one another. It can also be used loosely in ways that obscure this fact or confuse the issue with significant implications for what is actually being said. So, having introduced how some of the different senses of representation might affect the thought and argument about animals that we find in Animal Studies and in order to develop this same point, it will be helpful to step back and map the concept out more definitively.

In his brilliant and wonderfully clear conceptual précis of the term *representative*, the cultural critic Raymond Williams (1983) claims that "the group of words in which represent is central is very complex, and has long been so" (267). This idea of complexity is precisely right, because the different senses of representation shade into and are imbricated with one another. But we can preliminarily distinguish two fundamental senses of the verb to represent: *to make present again* (a literal derivation from the word's Latin root) and *to stand in place of*.

To explain the first of these two senses, we can say that the appearance of a ghost would be a *representation* of the living being. The term has also conventionally been used in referring to dramatic performances as "theatrical representations" that make present before an audience an already existing story. The sense is of replication or reappearance across time or space. But an especially important usage of *representation* in this sense within the field of Animal Studies appears in the context of the study of cognition, which speaks of "mental representations" of nonmental or external things. The notion of mental representation is crucial in the fields of cognitive science and the philosophy of mind, and there is much debate therein among scholars who seek to analyse animal consciousness about the type, extent, or very existence of mental representations in animals.[1] Mental representations, to be sure, are different from nonmental objects in themselves and so might also be said to stand in place of them; and yet (especially in what are called "strong representationalist" theories of the mind) such representations are the only way in which the world is perceived, and so it makes sense to approach this usage in terms of making present again. For the purposes of my broader discussion here, in any case, what is remarkable in this first sense of the term is the confidence it expresses in a full or at least potentially full correspondence between the representation and the element represented. As we will see, this confidence, or rather the absence of it, comes to haunt the notion of the representation of animals.

The second broad sense of *representation* is diametrically opposed to the first; rather than meaning that the same thing is present again, it refers to something standing in place of, or standing for, something else. The key shift in sense, then, is that here there are a range of possible relations of correspondence between the representation (often called the "subject of representation") and the thing represented (often called the "object of representation"). The notion of "similar but not the same," which I suggested pertains importantly in human-animal relations, returns here at the heart of the logic of representation. Perhaps the easiest to grasp of such relations of correspondence are forms of symbolism, like metaphor or iconography, in which one thing (called the vehicle) represents another (called the tenor). An example is when, drawing on biblical sources such as Revelation 5:5, imagery of the lion represents Christ in church decoration or indeed in C. S. Lewis's *Chronicles of Narnia* books. Now, particularly when compared to the first sense of representation, there is a temptation to say that by the very fact of representing Christ, such lions are not *really* lions at all; however, this would misunderstand the very purpose and meaningfulness

1. See Allen and Trestman (2016) and Allen and Bekoff (1997). Tom Tyler (Tyler 2011, 109–25) offers a careful explication of the underlying logic of mental representation in the context of animals but focused on its elaboration by the philosopher Immanuel Kant.

of representation in this second sense. Better to say that ecclesiastical lion statues or Lewis's Aslan are both lion and Christ.[2]

The first of these two fundamental senses of *representation* is submerged with the important effect of giving rise to the major tension I outlined earlier within the two strands of meaning in the contemporary usage of the term. These, to cite the Oxford English Dictionary definition, are (1) "the action of standing for, or in the place of, a person, group or thing," and (2) "senses relating to depiction or portrayal." They can be helpfully understood as follows. First, we have representations that "speak for" their object in the sense of advocating for it or relaying its interest. The classic examples here are parliamentary or legal representation, in which political or legal representatives speak for their constituencies or clients. Second, there are representations that "speak of" their object in the sense of figuring or portraying an image of it. Representation as portrayal takes myriad forms and appears in every imaginable cultural form—linguistic, pictorial, poetic, filmic, dramatic, narrative, journalistic, photographic, sculptural. Within Animal Studies over the past twenty to thirty years, different domains of inquiry have placed significantly different emphases on these two different ways of thinking about representation. Work emergent from philosophy, social, legal, and political studies, for example, may tend to develop understanding of the former, while work in disciplines analyzing history and the production and reception of cultural texts tends to focus on the latter.[3]

My claim here, however, is that taking care to understand the relations between the different senses of *representation* is particularly important for Animal Studies. Because a fundamental tension runs through the meaning of the concept, stemming from the etymology of representation as making present again and relating to the relative plenitude of presence of the thing represented in any representation, scholars of human-animal relations thus might find themselves asking just how total is, or can ever

2. Animal symbolism has traditionally been a rich area of study in the humanist history of aesthetics, but the significant change wrought by Animal Studies approaches in this arena has been to take seriously the interactions between symbolic representations and the material realities of animals. The classic study is Berger (1980), but a useful introduction to debates is Simons (2001). The semiotics of animal symbolism is a huge field of inquiry, and the best critical analysis of how issues of representation weigh on the understanding of animals, especially in the context of the vexed question of anthropomorphism, is Tyler (2011, 9–75).

3. An important example of the philosophy of political and legal representation of animals is Hadley 2015. An exemplary analysis that links visual representation and historical understanding of animals is Donald 2007.

be, the act of representing an animal or animals? This tension thus ineluc-
tably leads to a range of evaluative considerations. If one-to-one correspondence is impossible between the object of representation (the animal
or animals) and the human representative (in the sense of speaking for)
or portraying representation (in the sense of speaking of), then what
kinds of correspondence are desirable? What manner of representation
re-presents well, or badly? Are representations that approach or pretend
to achieve complete correspondence ideal or merely duplicitous? And
on what basis would attempting to represent animals in such ways be
valid? (See Suen 2016.)

Representing Animals

An especially significant site in Animal Studies for critical brow furrow-
ing about the limits of representation is the issue of anthropomorphism,
the representation of animals in ways that ascribe to them features that
are seen to be human preserves. Arguments that attempt to rescue this
concept from the taboo status it has in scientific discourse about animals
because it is seen to compromise objectivity, such as ethologist Franz de
Waal's (1999) important concept of "animalcentric anthropomorphism,"
(255–80) can be understood as attempts to move beyond the hope for total
or perfect representation of animal being. By abjuring the commitment
to an accuracy that is imagined achievable only by scientific discourse and
method, they seek a looser kind of correspondence between the portrayal
and what it represents that does a concomitantly fuller job of speaking
politically for the animal.[4]

Worry about the politics of representation took systematic root in academic thought in the 1980s, especially in the critical humanities and social
sciences, under the influence of the critique of the concept in Michel Foucault's seminal work *The Order of Things*. What was termed then "the crisis
of representation" focused on pessimistically analysing how the act of representation (here primarily in the sense of portrayal) is always a dispensa-
tion of power (see Hall 1997). This is an important intellectual context for
the rethinking of human-animal relations that occurs in Animal Studies.

4. See Daston and Mittman (2006). Similar arguments play out in the context of the cultural
representation of animals; see the notion of "strong anthropomorphism" in Simons (2001,
116–39).

On the other hand, it is again fair to say that the disposition toward representation differs in different domains of academic inquiry. In general terms, questioning how representation (as speaking for) works has long been part of mainstream political theory in the context of working analytically toward the ideal form of human political representation. However, largely in response to the need to recognize the presence of nonhuman interests in the political arena, some theorists have drawn on the kind of tension I am attending to here to become comparably suspicious of and pessimistic about the effectiveness of representation as a fundament of the democratic process. Let me explain, then, how this tension regarding the more or less complete presence of the represented in the act of representation can be found in the two key senses of representation as standing in place of. I will focus on three particularly significant contexts for representing animals: legal representation, political representation, and aesthetic portrayals.

In the case of legal representation, the very existence of a law profession is evidence that it is likely preferable to have one's interests represented by an expert in the law than to represent oneself (a notion that itself quietly reveals a distinction between one's actual interests and what can be portrayed of those interests in court). An abiding concern in any just legal process is thus the extent to which the best interests of the client are adequately represented (i.e., made present again in the proceedings) by a lawyer in her "representations" to the court. Indeed, evidence of a lack of adequate representation offers some of the most significant grounds for appeal in many legal systems. This reveals how they presume the possibility of full enough representation and indeed require it. To turn specifically to the question of the legal representation of animals—a significant scholarly development of the last thirty years, which has run in parallel with the academic development of Animal Studies—the faith that legal systems place in the principle of advocacy and the necessity of adequate representation is something of a double-edged sword. On the one hand, the expectation that someone does not speak but should be represented in court means that animal plaintiffs are in a structurally similar position to human ones; animals' claims can in principle be heard like humans'; animals are not so evidently excluded by the notion of a full and rational voice traditionally granted to humans alone. On the other, faith in the possibility of adequate representation ensures that the legal process relies in the last instance on the capacity to speak for one's own interests, a legal agency that certainly is conventionally and dogmatically reserved to the human (or rather, to healthy adult ones).[5]

In the case of parliamentary democracy, a significant theoretical distinction is made between two key kinds of political representation. There is first the notion of the political representative as *delegate* (i.e., someone who is mandated by electors principally to make their views present in parliament). Then there is, by contrast, the notion of the representative as *trustee* (i.e., someone who is appointed to speak for constituents by exercising his or her own judgement about their best interests without being bound by their say-so). Mainstream political theory has conventionally regarded it as impossible for animals to be represented by delegates on the humanist presumption that animals simply cannot express their interests.[6] A more progressive strain in political thought argues from the trustee model for proxy representation of animals by human advocates in the political process.[7] The significant difference between these ways of thinking about political representation can be well understood as a difference in how substantively present constituents are in the parliamentary setting. When compared with the delegate model, the trustee model can be regarded as a form of political representation (speaking for) that tends toward representation as portrayal. Here, the trustee proffers a portrayal, in his or her own terms, of the best interests of the represented. Recently, however, political theorists influenced by Foucauldian critique, have argued that the ideal of complete representation implied here is impossible (for humans just as much as for animals). We should understand that acts of political representation simply cannot make constituencies present again in the political arena. Instead, such acts portray those constituencies (e.g., animals) and their interests in various incomplete ways and can even be said potentially to shape and constitute those interests in the very act of claiming to speak for them. As the major theorist of the concept of representation in this vein, Michael Saward (2010), puts it:

5. The twists and turns of this position are exemplified well in the documentary film *Unlocking the Cage*, directed by D. A. Pennebaker and Chris Hegedus, which follows the work of Steven Wise and the Nonhuman Rights Project in representing the captive apes Tommy, Hercules, and Leo in US courts. On the legal representation of animals see Sunstein (2004), Wise (2000), and Favre (2008).

6. The classic study is Pitkin (1967). A recent introduction to the optimistic political theory of representation is Brito Vieira and Runciman (2008).

7. An early example is Nearing (1965); see also Dobson (1996).

"Animals can be engaged with, looked for, traced, understood, and appreciated in new ways by humans opening up themselves to new ways of 'reading' and 'writing' them. . . . But to do this is to tap into new ideas of what it means to represent, and to make representations, in the senses of both what it can involve and who can do it (115)."[8] Such a view firmly resituates political representation (speaking for) on the field of aesthetic portrayal (speaking of).

And it is easy enough to see that representation as portrayal has a very complicated relationship to whatever is notionally represented. In this understanding of representation, there is a wide range of possible forms of correspondence between the representation and its object, which is also to say that there are a range of possible relationships between the object of representation and whoever or whatever is doing the representing. It is very important to understand the sheer complexity of this relationship with respect to the issue of how fully present the represented thing is in the portrayal. Another way of making this point is captured by the ordinary usage of the term *realistic*. If we think, for instance, in terms of the visual portrayal of animals, we can certainly distinguish along a spectrum of how "realistic" the representation is, from a taxidermized animal; a photograph of the same animal; a conventionally realist painting of that animal; a cartoon, sketch, or woodcut of that animal; and an abstract painting that evokes something about that animal (or another one like it). Even here, though, there would be disagreement about whether a taxidermized animal is most "realistic"—that is, a representation in which the thing represented is more seemingly present—than a photograph, and this *despite the fact* that many parts of the actual body of the living animal represented are materially present. Suffice to say, the act of portraying animals (which is of course as old as aesthetics itself) is impossible to reconcile with any notion of "restoring" the real animal to full visibility through representation; this is a fact that will continue to frustrate any critic whose ideal aesthetic portrayal of animals defeats anthropocentric ideology by virtue of being true to animals themselves.[9]

8. I am grateful to Lucy Parry for introducing me to Saward through her unpublished doctoral work on the political representation of animals.

9. A very important version of this argument is found in the notion of restoring the absent referent; see Adams (2000). A development of the point I am making here appears in Steve Baker's (Baker 2013, 227–39) response of the critique of representation in the work of Cary Wolfe.

Contradictory Representations

So, there is an internally contradictory set of expectations about the representation of animals embedded in Animal Studies. In a profound and extensive but not necessarily explicit way, the study of animals and production of knowledge about them is always implacably bound up with matters of how animals are portrayed and the issue of speaking for them. I would now like to look back to two of the earliest significant adventures in concerted academic inquiry about animals, which with hindsight can be seen to begin the shaping of a field before "Animal Studies" became its recognized name; both of them took the concept of representation as their focus.

In 1999, the Center for Twentieth-Century Studies at the University of Wisconsin–Milwaukee issued a call for papers for a conference titled "Representing Animals at the End of the Century."[10] Proposing to "focus on the ways in which animals have been imagined in cultures over the course of the last century," the call went on to frame this more specifically. "By tracing the development of how animals have been represented in different contexts, in different practices, and by different disciplines, the conference will explore the connections between our understandings of animals and the historical and cultural conditions in which those understandings are formed." From this interest in the human imagination and in representation as portrayal, the organizers go on to specify that discussions will "move through discussions of the material presence and functional role of animals . . . to broader conversations about how contemporary ways of imagining the significance of animals are shaping our fundamental cultural expectations of animals, of ourselves, and of our environments" (Rothfels and Isenberg 1999, 121–22). We see here, at a foundational moment in the development of Animal Studies, that under the aegis of understanding the ways in which animals are represented, inquiry about animals is expected to articulate between animals' place in human cultural imaginaries in the form of portrayals of them, animals' actual material lives across time and space, changes in how the value of animals is understood (here, representation as advocacy for animals comes in), and the impact of these changes on social attitudes about humans, animals, and the environment.

The "Representing Animals" conference was followed in 2001 by a special issue of the journal *Society and Animals* on the topic of "The Representation

---

10. The conference was organized by Nigel Rothfels and Andrew Isenberg; a selection of revised papers from it was published in Rothfels 2002.

of Animals," edited by the cultural theorist and art historian Steve Baker, for which the call for papers was circulated at Milwaukee. This document states,

> Human understanding of other animals is shaped, to a considerable extent, by representations of animals rather than by direct experience of them. Art, film, literature, the mass media, the language of scientific studies and the structure of museum and zoo displays employ many different forms of representation. Is it possible in any of these areas to claim to represent aniamls [sic] objectively and accurately, or do representations inevitably reflect and perpetuate human preconceptions?

The terms of the question as posed here—in which objectivity is aligned with accuracy and opposed to anthropocentric preconceptions—frame the issue of representation as about the very possibility of complete correspondence between portrayal and animal reality. (The misspelling of the word *animals* in the call for papers seems ironically to hint at an answer that it is a mistake to harbor any hopes of such accuracy.) This framing has an implicitly moral dimension that resides in a supposition that objective accuracy is better than preconceptions; this is picked up when the call for papers notes that more understanding is needed of how cultural conditions shape representations of animals and asks whether such representations "necessarily entail ethical judgements, and can they ever effectively challenge the values of the society in which they are used?" It is important to mark the movement or slippage from worry about whether or not representations are accurate to whether or not (even though inaccurate) representations can be useful in terms of changing values. The moral issues embedded here can helpfully be thought of as emerging from the tension that is implicit in the concept of representation: a crucial element in understanding portrayals of animals is the extent to which such representations speak for (that is to say, represent) animals themselves.

Of course the call for papers is a manner of textual representation in itself; it is a form with some conventional ways of handling ideas, such as tactically posing issues in starker terms than it is assumed will be addressed in more nuanced form by the authors who respond to the call.[11] That is to say, both animals themselves and the question of their representation are

11. Discussing representation easily becomes convoluted and self-referential for this reason: it is not possible to discuss representations except by way of further representations. For an explanation and response to this prevalent way of thinking about representation in critical theory, see Prendergast (2000, 1–16).

represented in a particular way here for the very reason that this document is an academic call for papers. Another form of representation could portray the situation entirely otherwise. As Baker (1993) puts it in his major work on the topic, "any understanding of the animal, and of what the animal means to us, will be informed by and is inseparable from our knowledge of its cultural representation" (4).

But as Baker makes very clear in the rest of his work, this is certainly not to say (as is often and too hastily assumed) that it is impossible to "truly" speak for animals because of the layers of cultural representation that stand in the way of accurate knowledge of them. Such an either/or view misunderstands and imagines to be straightforwardly resolvable what is in fact the irresolvable tension between portrayal and advocacy that constitutes the notion of representation itself. It would be better to say, with Baker (1993, 5), that "any effective cultural strategies on behalf of animals"—and in the broadest sense this is one way of describing what Animal Studies hopes to develop—"must be based on an understanding of contemporary cultural practice," that is, on representations. And yet, an impulse to hold such representations to account in terms of their relative interest in actual animals' real ways of life, their interests, and so on, properly persists, and this must surely be based on the developing knowledge about animals that academic inquiry delivers by offering new portrayals of them. Such is the dynamic of representation in Animal Studies.

Suggestions for Further Reading

Baker, Steve. 1993. *Picturing the Beast: Animals, Identity and Representation*. Manchester: Manchester University Press.
Rothfels, Nigel, ed. 2002. *Representing Animals*. Bloomington: Indiana University Press.
Saward, Michael. 2010. *The Representative Claim*. Oxford: Oxford University Press.
Suen, Alison. 2016. *The Speaking Animal: Ethics, Language and the Human-Animal Divide*. London: Rowman and Littlefield.
Tyler, Tom. 2011. *Ciferae: A Bestiary in Five Fingers*. Minneapolis: Minnesota University Press.

References

Adams, Carol. 2000. *The Sexual Politics of Meat*. London: Continuum.
Allen, Colin, and Marc Bekoff, eds. 1997. *Species of Mind: The Philosophy and Biology of Cognitive Ethology*. Cambridge, MA: MIT Press.
Allen, Colin, and Michael Trestman. 2016. "Animal Consciousness." In *The Stanford*

Encyclopedia of Philosophy, edited by Edward N. Zalta. Plato.stanford.edu, October 24. https://plato.stanford.edu/archives/win2016/entries/consciousness-animal/.

Baker, Steve. 1993. *Picturing the Beast: Animals, Identity and Representation*. Manchester: Manchester University Press.

———. 2013. *Artist/Animal*. Minneapolis: University of Minnesota Press.

Berger, John. 1980. "Why Look at Animals?" In *About Looking*, 1–26. New York: Pantheon.

Brito Vieira, Mónica, and David Runciman. 2008. *Representation*. Cambridge: Polity.

Daston, Lorrain, and Gregg Mitman, eds. 2006. *Thinking with Animals: New Perspectives on Anthropomorphism*. New York: Columbia University Press.

de Waal, Franz. 1999. "Anthropomorphism and Anthropodenial: Consistency in Our Thinking about Humans and Other Animals." *Philosophical Topics* 27: 255–80.

Dobson, Andrew. 1996. "Representative Democracy and the Environment." In *Democracy and the Environment*, edited by W. Lafferty and J. Meadowcroft, 124–39. Cheltenham, VT: Elgar.

Donald, Diana. *Picturing Animals in Britain*. New Haven, CT: Yale University Press.

Favre, David S. 2008. *Animal Law: Welfare, Interests, Rights*. New York: Aspen.

Hadley, John. 2015. *Animal Property Rights*. London: Lexington.

Hall, Stuart, ed. 1997. *Representation: Cultural Representations and Signifying Practices*. London: Sage.

Nearing, Scott. 1965. *The Conscience of a Radical*. Harborside, ME: Social Science Institute.

Pitkin, Hanna. 1967. *The Concept of Representation*. Berkeley: University of California Press.

Prendergast, Christopher. 2000. *The Triangle of Representation*. New York: Columbia University Press.

Rothfels, Nigel, ed. 2002. *Representing Animals*. Bloomington: Indiana University Press.

Rothfels, Nigel, and Andrew Isenberg. 1999. "Call for Papers: 'Representing Animals at the End of the Century.'" *Anthrozoös* 12 (2): 121–22.

Saward, Michael. 2010. *The Representative Claim*. Oxford: Oxford University Press.

Simons, John. 2001. *Animal Rights and the Politics of Literary Representation*. Houndmills: Palgrave.

Suen, Alison. 2016. *The Speaking Animal: Ethics, Language and the Human-Animal Divide*. London: Rowman and Littlefield.

Sunstein, Cass R. 2004. "Can Animals Sue?" In *Animal Rights: Current Debates and New Directions*, edited by Cass R. Sunstein and Martha Nussbaum, 251–62. Oxford: Oxford University Press.

Tyler, Tom. 2011. *Ciferae: A Bestiary in Five Fingers*. Minneapolis: University of Minnesota Press.

Unlocking the Cage. 2016. Directed by D. A. Pennebaker and Chris Hegedus. BBC Films.

Williams, Raymond. 1983. *Keywords: A Vocabulary of Culture and Society*. Oxford: Oxford University Press.

Wise, Steven. 2000. *Rattling the Cage: Towards Legal Rights for Animals*. New York: Basic Books.

22 RIGHTS

Will Kymlicka and Sue Donaldson

The term *animal rights* is used in academic and public debate in contradictory ways. In some contexts, it encompasses any commitment to improving the conditions or treatment of nonhuman animals, such as increasing the size of cages for chickens raised and killed for food or enriching the cages of laboratory animals raised and killed for medical research. Many people support animal rights in this very broad sense: public opinion polls show widespread support for measures to enhance animal welfare.[1] Defined this way, *animal rights* is used as an umbrella term that covers any and all forms of advocacy for better treatment of animals.

However, in other contexts, *animal rights* is used in a much narrower way to refer to the idea that animals are the bearers of inviolable moral rights that prohibit them being harmed or sacrificed for the benefit of humans. For defenders of animal rights in this strict sense, animals have a right not to be experimented on even if humans would benefit from the knowledge gained by such experimentation. Animal rights, so understood, is not about improving the welfare of animals in the process of raising and killing them; it is about recognizing that animals have their own lives to lead, and humans have no right to use them as resources or as means to enhance human welfare.

In this chapter, we will use the term *animal rights* (hereafter AR) in this narrower sense of rights that cannot be violated for the benefit of others. Defined this way, AR is a radical position, endorsed by only a small segment

1. Even producers engaged in factory farming trumpet their support for animal rights when announcing welfare reforms in their industry. Francione and Charlton (2015) quote veal producers who claim to be firm believers in "animal rights."

of the public,[2] and is deeply controversial inside and outside the animal advocacy movement. Outside the movement, concerted efforts have been made by corporate and state interests to label AR an "extremist" position, to delegitimize it, and to discourage other social justice movements from aligning with AR groups.[3] But even within the movement, debates rage about whether the appeal to rights is helpful, politically or philosophically. For some, an explicit endorsement of AR represents the "moral baseline" for any ethically acceptable approach to human-animal relations: anything short of AR entails complicity in the oppression of animals (Francione 2012). But for other animal advocates, a dogmatic insistence on AR is both a strategic failure and a philosophical mistake, occluding other better ways of formulating our ethical responsibilities.

We begin by briefly reviewing why the language of rights has become so prominent in the past fifty years in what is often called "the rights revolution." We will then outline the arguments for extending this rights revolution to animals before considering a series of both pragmatic and principled objections to the reliance on rights.

The Rise of Rights

While the idea of rights can be traced back to ancient Rome, the "rights revolution" in the West dates to the 1960s. The 1960s was a decade of intellectual ferment after a period of relative conservatism during the height of the Cold War in the 1950s. In many Western countries, the premises of liberal democracy were being challenged. For many American progressives, for example, the status of African Americans became a pivotal test for assessing the adequacy of liberal democracy. Whether liberal democracy was worthy of people's allegiance—whether it had the resources to provide a decent life for its citizens—would be tested by its capacity to respond to the claims of African Americans.

Faced with this challenge, it became apparent that traditional forms of liberal-democratic political philosophy were woefully inadequate. For most of the nineteenth and twentieth centuries, liberal democracy had been

2. Few surveys ask specifically about the idea of inviolable rights, but insofar as a belief in AR entails that we cannot kill animals for food or clothing, the number of ethical vegans provides a rough proxy for the number of people who endorse AR. In many Western societies, this is perhaps 1–2 percent of the population (or 5–10 percent if we count vegetarians).

3. On how the targeting of AR groups as "extremists" and "eco-terrorists" has disrupted possible solidarities, see Dauvergne and Lebaron (2014).

defended on essentially utilitarian grounds, most famously by Bentham and Mill. Utilitarianism seeks "the greatest good for the greatest number," and the first waves of liberal reform seemed to instantiate this goal insofar as they challenged the privileges of small elites in the interests of disadvantaged majorities. Extending political rights to the working class, say, or to women, or to overseas colonies, advanced the interests of the "greatest number" at the expense of a privileged minority and hence fit well with utilitarian forms of argument.

But in the case of the African American civil rights struggle, a minority was making claims against the wishes of the majority of society. If one simply counted up people's preferences, the majority's preference in many American states was to maintain racial segregation. In this context, desegregation seemed to require an antiutilitarian, countermajoritarian moral framework—one that could defend the claims of small and stigmatized minorities even when this contradicted the preferences of the majority. So too with gay rights, indigenous rights, or the rights of people with disabilities—each of these required defending the claims of small and stigmatized minorities in opposition to the deeply rooted traditions of the community and the strong preferences of majority heterosexual, European-origin, or able-bodied citizens.

Various utilitarians tried valiantly to show that overriding the preferences of the majority would nonetheless lead in the long run to the greatest happiness of the greatest number of people (e.g., because discrimination leads to economic inefficiencies, or creates a feeling of insecurity among other groups). But even where utilitarians were able to invent such reasons, they got the right answer for the wrong reasons. For surely the segregation of African Americans was inherently morally wrong, since it failed to respect their humanity and common citizenship, regardless of its long-term impact on overall preference satisfaction. The civil rights struggle cried out for a countermajoritarian moral argument.

To make sense of this counterutilitarian intuition, liberal political theorists turned to the idea of inviolable rights: the idea that an individual's most basic interests cannot be sacrificed for the greater good of others. In Dworkin's famous phrase, rights are "trumps" that cannot be violated no matter how much others would benefit from their violation (Dworkin 1984). For example, a person cannot be killed in order to harvest her body parts even if dozens of other humans might benefit from her organs, bone marrow, or stem cells. Nor can she be made a subject of nonconsensual medical experimentation, no matter how much the knowledge gained from experimenting on her would help others. Nor can she be stripped of her citizenship or subjected to expulsion or segregation no matter how much others stigmatize her race, religion, ability, or sexual orientation. Inviolable rights in this sense are a protective circle drawn around an individual, ensuring that she is not sacrificed for the good of others.

A similar argument underpins Rawls's *A Theory of Justice* (1971). He, too, argued that utilitarianism was unable to account for the wrongness of sacrificing individuals for the good of others. An adequate defense of liberal democracy, he believed, required a more "Kantian" conception of respect for individuals, which emphasizes that we are "self-originating sources of valid moral claims," and as such should not be treated simply as a means for the good of society.[4]

These rights-based theories quickly became hegemonic, displacing earlier attempts to ground liberal democracy in utilitarianism or other theories that left minorities vulnerable to the contingencies of majority sentiments or practices (such as pragmatism or conventionalism). And this philosophical shift reflected and supported a dramatic strengthening of rights protections in practice, as ideas of inviolable rights have become the basis of much of our medical ethics, of domestic bills of rights, and of international human rights law.[5]

From Human Rights to Animal Rights

While the idea of inviolability is now widely accepted in relation to human beings, few people have accepted that animals, too, might possess inviolable rights. Even those who acknowledge that animals matter morally and that they deserve to be treated more humanely often think that when push comes to shove, animals can be violated—endlessly sacrificed—for the greater good of others. Whereas killing one human to harvest organs to save five other humans is unacceptable, to kill one baboon to save five humans is permissible and perhaps even morally required. As McMahan puts it, animals are "freely violable in the service of the greater good," whereas human persons are "fully inviolable" (McMahan 2002, 265). Nozick famously summarized this view under the label "utilitarianism for animals, Kantianism for people" (Nozick 1974, 39).

---

4. For a more extended account of this post-1960s shift from utilitarian to rights-based theories in political philosophy, see Kymlicka (2002, chap. 2).

5. On this "rights revolution," see Walker (1998).

Defenders of AR, however, insist that the arguments for inviolability do not stop at the boundary of the human species. Sentient animals need rights for the same reason humans do: like humans, they have their own subjective experience of the world, have their own lives to lead, and as such should be seen as "self-originating sources of valid moral claims" who are not resources for others. If it is wrong to kill a human for her organs, even if we can save five people by doing so, so, too, is it wrong to kill a pig for his organs.[6] For AR theorists, it is simply discriminatory or "speciesist" to ascribe inviolability to humans but violability to animals. In this sense, as Cavalieri puts it, we should "take the human out of human rights" (Cavalieri 2001), or as Cochrane puts it, we should move "from human rights to sentient rights" (Cochrane 2013).

It is important to note that while Peter Singer's *Animal Liberation* is widely seen as a founding text of the AR field, he did not in fact argue for inviolable rights, and indeed the AR position emerged as a critical response to Singer's work. Singer is a utilitarian, and so he does not believe in inviolable rights for either humans or animals. He decried the tendency to invoke double standards when weighing and aggregating human and animal interests, but his response to this bias was to insist on a more rigorous and evenhanded application of utilitarian reasoning to both humans and animals. His arguments for the improved treatment of animals are, therefore, based on empirical claims that most of the harms we inflict on animals do not in fact serve the overall good rather than on the rights-based claim that it would be wrong to harm animals even when it would serve the greater good.

From an AR perspective, this utilitarian approach to animals is inadequate for the same reason it has been found wanting in relation to humans. It may get the right answer in cases where huge numbers of animals are harmed for trivial human benefits—say, killing seals to make fur coats. But it runs into trouble when the numbers are reversed, and the few are

6. Inviolability is not unconditional. There are circumstances, in both the human and animal cases, where inviolable rights can be overridden. One example concerns self-defense, where we recognize the right of individuals to protect themselves from grievous assault by injuring or even killing their attacker. Another example is the temporary forcible confinement of an individual with a deadly contagion who poses an immediate threat to others and refuses to undertake voluntary quarantine. In such cases, the inviolable rights of individuals can be overridden in extremis, when they pose an immediate threat to the basic inviolable rights of others (or, in some cases, when they pose such a threat to themselves). Inviolable rights are "trumps" against being sacrificed for the greater good of others but are not a license to harm others.

being sacrificed for the greater good of the many, as perhaps with some forms of animal experimentation. Singer himself bites this bullet: he agrees that lethal experiments on animals may be justified on utilitarian grounds. And having opened the door to violating animals for the greater good of others, he has trouble ruling out other ways of harming animals. If millions of humans enjoy seeing a gorilla at a zoo or a whale in an aquarium, utilitarian reasoning seems to endorse capturing a few of these animals, separating them from their families, and confining them for life in cages or pools for the pleasure of others. Even utilitarian opposition to raising and killing animals for food rests on uncertain speculations about its aggregate long-term effects.

The contemporary AR movement arose precisely as a response to these deficiencies in Singer's utilitarian animal ethics,[7] just as the human rights revolution arose as a response to the failings of utilitarian human ethics. Utilitarianism simply fails to respect individuals as self-originating sources of moral claims who cannot be sacrificed for the greater good of others. Regan's *The Case for Animal Rights* (1983) is widely cited as the first systematic modern statement of a distinctly rights-based approach to animals, but several other AR theorists have since elaborated related positions, including Cavalieri (2001), Francione (2000, 2008), Steiner (2008), Rowlands (2009) and our own work (Donaldson and Kymlicka 2011).

The AR position has been controversial both inside and outside the animal advocacy movement. Defenders of the status quo have tried to reaffirm the existing species hierarchy in relation to inviolable rights, disputing the AR claim that it is arbitrary to view humans as inviolable and animals as violable. In the rest of this entry, however, we will focus on critical responses from *within* the animal advocacy movement, questioning whether the language of rights is the only or best framework for theorizing ethical relations with animals. While AR theorists view themselves as making a frontal attack on species hierarchy, not everyone in the movement views this as beneficial for animals, either strategically or philosophically.

Strategic Objections

Let us start with the strategic concern. For many activists, even if inviolable rights are a philosophically defensible and attractive long-term goal, they are

---

7. For critiques of Singer's utilitarianism from an AR perspective, see Regan (1983), Francione (2000), and Davies (2016).

utopian at the moment, and it may be ineffective, even counterproductive, to frame animal advocacy in terms of rights.[8] We need to start with more modest welfarist reforms, which accept that humans are entitled to harm animals for our purposes but that more stringently regulate this harm.

AR theorists disagree among themselves about such strategic choices (Francione and Garner 2010). Many AR theorists would accept that such compromises are inevitable and that reformist measures can not only reduce discrete harms but also raise broader public awareness of and empathy for animals and thereby help build support for eventual rights-based claims. Other AR theorists, however, argue that all such reformist measures presuppose and further entrench the idea that humans are entitled to use animals for our benefit. Unless and until we get animals out of the category of property and into the category of rights-bearing subjects, any welfarist reforms to existing practices will be superficial and offset by all of the new ways that science and industry discover to instrumentalize animals.

In our view, this strategic debate about how or when reformist measures can advance an AR future rests on a number of empirical claims for which we have limited evidence. Do reformist measures serve to stabilize human supremacist assumptions, preempting more radical critiques of animal use? Or do these reforms gradually operate to weaken supremacist assumptions, subjecting our taken for granted assumptions about animal use to open debate and critical reflection? We have little reliable evidence to assess these competing claims about the strategic effects of welfarist reforms.

In addition to these strategic concerns, many animal ethicists raise more principled objections to rights talk, pointing out some potential blind spots of formulating human ethical obligations to animals in terms of rights. We will discuss four such concerns: (i) that a rights framework relies on animals being like humans and thereby is implicitly anthropocentric, (ii) that it defines ethical relations with animals in terms of negative duties of noninterference rather than positive relational duties, (iii) that rights conceive of relations with others as self-interested and antagonistic, not caring or loving, and (iv) that rights are empty and formalistic and do not address issues of power.[9] In our view, each of these objections raises important concerns about specific versions of AR theory but not against the fundamental idea of inviolable rights, and indeed each in its own way helps to clarify why we need such rights.

8. See Peters (2016) for the obstacles to incorporating a rights perspective into animal law.

9. This is a partial sample of such critiques; see Donovan and Adams (2007) and Suen (2015) for a broader list.

Rights as Anthropocentric

According to many critics, the basic logic of AR theory takes human beings as the moral standard and evaluates animals' claims based on whether they are sufficiently "like us" to warrant inviolable rights. AR theory looks for evidence that animals are sufficiently similar to humans to be treated as rights-bearing subjects rather than respecting animals in their uniqueness and difference from us (Gruen 2014). By taking humans as the gold standard for the entitlement to rights and then looking to see whether animals are sufficiently similar, AR may be able to extend rights to some animals, but it implicitly reinforces the idea that "man is the measure of all things" (see chap. 3). It thereby reinforces anthropocentrism and erases animal difference in the very act of extending rights to animals.

In response, we think it is important to distinguish the general moral logic of AR theory from certain specific legal strategies used to advance AR. The two most high-profile campaigns aiming to secure inviolable rights for animals are the Great Apes Project (GAP) and the Nonhuman Rights Project (NhRP), both of which involve a strategic decision to focus on primates as the starting point for AR advocacy. While the ultimate aim is to secure inviolable rights for all sentient animals, this is seen as utopian in the foreseeable future. It may be possible, however, to create a precedent by extending rights to one or two species in the hope and expectation that this will make it easier to extend rights to all sentient animals down the road. The strategic question then becomes which species is most likely to set a successful precedent, and in answering that strategic question, both GAP and NhRP have selected primates in large part because of their similarities to humans. In their advocacy, both GAP and NhRP highlight the evolutionary, genetic, physiological, cognitive, emotional, and social continuities between humans and primates in the hope that this will demonstrate the moral arbitrariness of drawing personhood at the line of *Homo sapiens* and so will lead courts or legislators to extend inviolable rights to primates. The strategy is to first get primates into the category of rights-bearing subjects, by emphasizing how like us they are, in the hope that this will set a precedent for all sentient animals down the road.[10]

This strategy of pursuing inviolable rights by singling out and prioritizing species that are most "like us" is controversial among AR theorists,

10. For descriptions and defenses of this strategy, see Cavalieri (2015, on GAP) and Prosin and Wise (2016, on NhRP).

precisely because it deviates from the general moral logic of AR. The basic logic of AR is to insist that inviolable rights are owed to all who need them—that is, to all sentient beings who subjectively experience their lives as going better or worse. And the moral purpose of rights is not to protect what matters to humans but rather to protect what matters most to the animals themselves: the function of rights is to protect their unique subjective experience. Viewed this way, the GAP/NhRP strategy of prioritizing animals who are most like us is a high-risk strategy: it may create a useful precedent that leads to a broader debate on the rights of all animals or it may unintentionally reinforce the idea that the value of animals is measured by their proximity to humans.[11]

A related question concerns the link between rights and "personhood" (see chap. 18). From a legal point of view, rights are owed to "persons." For example, many constitutions around the world state that "all persons" have a right to life or a right to bodily integrity. If we want to ensure that animals also have these rights, then one obvious strategy is to argue that sentient animals should also be seen as "persons." And this indeed is what many AR theorists have argued. Francione, for example, accepts the idea that personhood grounds rights but redefines personhood so that all sentient beings qualify as "persons." (Indeed, Francione's 2008 book is titled *Animals as Persons*). This is captured in the familiar AR slogan "from property to personhood." To achieve justice for animals, we need to move them from the property box to the personhood box.

Others, however, argue that the term *personhood* is too tied up with human supremacist ideologies (and indeed gender and racial supremacist ideologies) to be usable for animal justice. Historically, the idea of personhood has been used in the law and in philosophy to designate an exalted class of humans who have "higher" mental capacities or "refined" moral sensibilities.[12] Given this legacy, so long as we stick with the term

11. This is similar to the strategic dilemma facing welfarist reforms, and here, too, we lack robust evidence to evaluate the likely effects of the different strategic choices.

12. And it is precisely because they possess such higher capacities that "persons" have been seen as deserving of inviolable rights: inviolable rights (on this traditional view) are used to mark our special respect for the distinctive worth of those beings who have these sophisticated capacities. From an AR perspective, this fundamentally misconstrues the moral purpose of inviolable rights, which are owed to all who *need* them, not just to those who "merit" them based on some criteria of moral autonomy or moral agency. To limit inviolable rights to those who pass some threshold of moral agency would in fact radically undermine the existing practice of human rights, including the rights of children and people with cognitive disabilities.

personhood, animals (and nonneurotypical humans) will always appear as somehow deficient. Instead, theorists should develop some new category that can ground claims to inviolability without the anthropocentric, assimilationist, and ableist connotations of personhood. Proposals include subjecthood, selfhood, creatureliness, or beingness.[13]

In short, while AR theorists are united in arguing that sentience grounds inviolable rights for animals, they disagree strategically about whether the solution is to include sentient animals within the legal category of "persons" or to argue that personhood is not a requirement for inviolable rights.

Rights as Negative Duties

A second concern about AR is that it conceives our ethical obligations to animals solely in terms of negative duties not to harm or kill rather than in terms of positive duties to enter into relationships of care and respect. AR theorists condemn the exploitative use of animals, but they offer no positive story about what sorts of relations with animals are ethically desirable and indeed seem to implicitly assume that we should not have such interspecies entanglements.

In our view, this is indeed a valid critique of one influential strand of AR theory, sometimes called the "abolitionist" or "extinctionist" approach to AR, associated with the work of Gary Francione (2000) (see chap. 1). He assumes that any exercise of power by humans over animals will be exploitative, and so the only way to ensure respect for AR is to leave wild animals alone and to render domesticated animals extinct. He offers us no picture of how humans and animals could live and work together in a way that respects animals' rights. In Acampora's words, this strand of AR seems to envisage a world of "species apartheid" (Acampora 2005): humans living among humans, animals living among animals, and never the twain shall meet.

But nothing in the general logic of rights pushes us in this direction. Rights do not generally preclude relationships, as is clear in the intrahuman case. A child has a fundamental inviolable right not to be experimented on for the good of others, but this negative right coexists with a raft of positive relational obligations, including duties of care (on the part of parents, teachers, medical professionals) and positive membership rights as a

13. See, for example, Dayan (2011) and Pick (2012).

citizen (e.g., to education, to political representation). Far from precluding such ethically significant relational obligations, inviolable rights form a secure framework within which layers of such relations can be built.

And so, too, we would argue, in the case of animals. The inviolable rights of animals not to be killed, caged, or experimented on provide a framework within which humans and animals can then enter into a range of ethically significant interspecies relationships. In our own work, we have explored different models of co-citizenship, denizenship, and sovereignty with different animals, each with its own set of positive relational duties (Donaldson and Kymlicka 2011). But none of these ethical relations are possible unless and until humans renounce the right to sacrifice animals for our good.

Put another way, any plausible AR theory needs to supplement inviolable rights with a richer story about positive relational obligations. If inviolable rights are seen as the sum total of our ethical obligations to animals, this quickly leads to the species apartheid view characteristic of the abolitionist/extinctionist position. However, the same argument holds in reverse: any plausible relational ethic needs to be supplemented with inviolable rights. If relational obligations are seen as the sum total of our ethical responsibilities, without any inviolable rights, we quickly end up with justifications for the continued exploitation of animals. For example, both Haraway's *When Species Meet* (2007) and Rudy's *Loving Animals* (2011) disavow inviolable rights in the name of a relational ethic and predictably end up justifying the killing of animals for human benefit.[14]

Rights as Antagonistic

A related concern is that rights presuppose and perpetuate relations of indifference or even antagonism between individuals. Rights presuppose that we need to protect ourselves from the hostile intentions of others and work by expecting individuals to "stand up" for their rights in adversarial legal processes. They may protect individuals from certain harms, but only by reproducing a general ethos of individualism and distrust of others.

This is a long-standing objection to the rights revolution generally, not just in the animal case. Many indigenous scholars, for example, have

---

14. For critiques of how the language of relational ethics operates to obscure the instrumentalization of animals in Rudy and Haraway, see Stanescu (2013) and Weisberg (2009), respectively.

complained that the Anglo-American model of rights is antithetical to indigenous traditions that aim to promote solidarity and reconciliation. In Porter's words, resolving rights disputes "is a process of structured aggression in which the parties, assisted by lawyers, engage in a self-interested pursuit of justice" (Porter 1997, 263)

This is both unattractive in itself and moreover unlikely to be effective for animals, given that animals are unable to "stand up" for themselves in court and are likely to lose whenever social relations are constructed in antagonistic rather than solidaristic ways. A similar concern has been raised about the rise of rights talk in relation to children. Insofar as rights claiming is seen as a process of "structured aggression," vulnerable children who are dependent on the goodwill of others will be unlikely to stand up for their rights: they cannot afford to have their relations with caregivers reframed in antagonistic terms (Wall 2011).

This suggests at a minimum that we need to think about nonadversarial mechanisms for upholding rights. And of course there are many such mechanisms, including ombudspersons, ethics committees, advocates, and trustees tasked with the mandate of monitoring and securing the rights of vulnerable populations, including those unable to stand up for their own rights. But some commentators draw a more radical conclusion. Insofar as animals currently suffer from the hostile intent of others, the solution is not to grant animals rights that they (or their advocates) can invoke to ward off this hostile intent but rather to directly tackle the hostile intent and to convert hostility into empathy. Diamond advances a version of this argument, suggesting that the question of whether animals have rights is "a totally wrong way of beginning the discussion" (Diamond 2004), and that the real question is why we view animals as "food" in the first place. The task isn't to give animals rights against being eaten as food; the task is to stop humans from seeing animals as food.

Diamond is surely right that the goal is to live in a world where people wouldn't even think of eating animals and hence where there is no need or occasion for animals (or their advocates) to assert their right not to be eaten (or experimented on, or held in captivity). But history suggests that invoking rights can help us get there. Today, we hope that no one would think of enslaving a minority racial group, or experimenting on people with cognitive disabilities. Most people simply do not view other humans as potential slaves or potential (nonclinical) research subjects. But the fact that these are increasingly unthinkable is not evidence that rights are the "totally wrong way of beginning the discussion." Although minorities and people with disabilities were never seen as "food," many people historically denied that they were worthy of inviolable rights, and it was a struggle to assert their right not be sacrificed for the greater good of others—a struggle grounded precisely in an appeal to "rights." Today, that commitment has become internalized and habituated, and that surely is the goal and culmination of all rights struggles—for rights to set habituated ethical boundaries on our relationships with one another. We want to live in a society where everyone is always already seen as "fully inviolable" and no one is seen as "freely violable in the service of the greater good" (McMahan 2002, 265). There may be little need for anyone to "stand up" for their rights in such a world, but such a world is rights based, and we can only achieve it if we articulate an ethics of inviolability.

Rights as Empty

One final objection is that rights too often are "paper guarantees" disconnected from real-world political struggles and the exercise of power. As Honig notes, some liberals are prone simply to wait for the unfolding of a "chrono-logic" of the liberal "rights machine," as though the sheer reasonableness of ideas of animal rights has political agency in the absence of political struggles for power. Ideas of a logical unfolding of rights, too, often displace the realities of political struggle:

> Looking backward, we can say with satisfaction that the chrono-logic of rights required and therefore delivered the eventual inclusion of women, Africans, and native peoples into the schedule of formal rights. But what actually did the work? The impulsion of rights, their chrono-logic, or the political actors who won the battles they were variously motivated to fight and whose contingent victories were later credited not to the actors but to the independent trajectory of rights as such? (Honig 2009, 67)

Rights are inert on their own. They only become meaningful when the victims of injustice engage in ongoing struggle, resistance, and claims making. And if so, this suggests that animal rights are doomed to futility. Even in the case of human rights, commentators argue that rights are often ineffective. In Brown's words, "while [rights] formally mark personhood, they cannot confer it; while they promise protection from humiliating exposure, they do not deliver it" (Brown 1995, 127). Or as West puts it, rights formally repudiate entrenched social injustices without providing the tools needed to actually challenge these hierarchies, and as a result, they operate in practice to obscure injustice (West 2011). Only when rights are backed by politically organized contestation and struggle by the oppressed can they tackle injustice. And since animals cannot advocate or resist, at least not in an organized way, the prospect that animal rights will secure justice is minimal.

This is indeed a salutary reminder that achieving legal rights on paper is just one stage, not the end, of the political struggle. But it is not clear that a successful politics of rights requires that it is the victims of injustice themselves who organize and contest. Anderson (2011) argues that child labor reform in Britain during the Industrial Revolution is an important counterexample, and one that is particularly relevant for the AR movement. Children themselves played little or no role in the movement to ban child labor, but a powerful coalition emerged to defend their interests, both in the initial formulation of children's rights and in the subsequent struggle to ensure their effective implementation. Anderson suggests that this provides a model for a successful politics of AR.

Moreover, it is a mistake to assume too quickly that animals cannot participate in the process of claims making. We have elsewhere argued that animals can indeed be political actors, although it requires creating new kinds of political spaces in which animals' preferences are solicited and collective decisions are made responsive to those preferences.[15]

To sum up, the term AR is used loosely in Animal Studies, from the vaguest commitment to improving the welfare of animals to a very specific "abolitionist" program to replace existing entanglements between humans and animals with hands-off duties of noninterference. We have suggested that AR is most usefully understood as a commitment to inviolability, which sets limits on the extent to which individuals can be sacrificed for the greater good of others. This is the core meaning of the rights revolution of the late twentieth century, and it is a fundamental question for the twenty-first century whether this concept can or should guide human relations with animals. We have noted various concerns that have been raised about "rights talk" in relation to animals but have argued that

15. Farmed animal sanctuaries provide an interesting site for such experiments in animal citizenship (Donaldson and Kymlicka 2015). Not all AR theorists agree that a commitment to AR entails a commitment to animal agency. Cochrane (2012), for example, defends what he calls "animal rights without liberation," arguing that animals have basic needs and objective interests that require inviolable rights but that can be identified without enabling animals to control basic decisions about their lives. We elsewhere argue that rights should not be defined solely in

rights, understood in terms of inviolability, are an essential requirement of justice. They are not, however, sufficient for justice or for flourishing and indeed are best understood as providing a moral framework within which a diversity of relationships and entanglements occur and develop. Too often, rights and relationships have been seen as conflicting approaches: the abolitionist rights approach denies the need for ongoing relationships, and influential relational approaches have denied that relations need to be constrained within a framework of rights protection. The task, then, is to creatively integrate rights with relationships, inviolability with entanglement.

Suggestions for Further Reading

Cavalieri, Paola. 2001. *The Animal Question: Why Nonhuman Animals Deserve Human Rights*. Oxford: Oxford University Press.

Donaldson, Sue, and Will Kymlicka. 2011. *Zoopolis: A Political Theory of Animal Rights*. Oxford University Press.

Edmundson, William. 2015. "Do Animals Need Rights?" *Journal of Political Philosophy* 23 (3): 345–60.

Francione, Gary. 2000. *Introduction to Animal Rights*. Philadelphia: Temple University Press.

Gruen, Lori. 2014. *Entangled Empathy: An Alternative Ethic for Our Relationships with Animals*. New York: Lantern Books.

Regan, Tom. 1983. *The Case for Animal Rights*. Berkeley: University of California Press.

Sunstein, Cass, and Martha Nussbaum. 2004. *Animal Rights: Current Debates and New Directions*. Oxford University Press.

References

Acampora, Ralph. 2005. "*Oikos* and *Domus*: On Constructive Co-habitation with Other Creatures." *Philosophy and Geography* 7: 219–35.

Anderson, Jerry. 2011. "Protection for the Powerless: Political Economy History Lessons for the Animal Welfare Movement." *Stanford Journal of Animal Law and Policy* 4: 1–63.

Brown, Wendy. 1995. *States of Injury*. Princeton, NJ: Princeton University Press.

Cavalieri, Paola. 2001. *The Animal Question: Why Nonhuman Animals Deserve Human Rights*. Oxford: Oxford University Press.

relation to objective welfare standards see chap. 29), which are invariably too narrow (food, shelter, veterinary care, but not freedom, control, pleasure, intellectual challenge), and that we need to recognize that animals are often in the best position to understand their interests, so we need either to get out of the way or to create conditions in which they can take meaningful action.

———. 2015. "The Meaning of the Great Ape Project." *Politics and Animals* 1 (1): 16–34.

Cochrane, Alasdair 2012. *Animal rights without liberation*. Columbia University Press.

———. 2013. "From Human Rights to Sentient Rights." *Critical Review of International Social and Political Philosophy* 16 (5): 655–75.

Dauvergne, Peter, and Genevieve Lebaron. 2014. *Protest Inc: The Corporatization of Activism*. London: Polity.

Davies, Ben. 2016. "Utilitarianism and Animal Cruelty: Further Doubts." *De Ethica* 3 (3): 5–19.

Dayan, Colin. 2011. *The Law Is a White Dog: How Legal Rituals Make and Unmake Persons*. Princeton, NJ: Princeton University Press.

Diamond, Cora. 2004. "Eating Meat and Eating People." In *Animal Rights: Current Debates and New Directions*, edited by Cass Sunstein and Martha Nussbaum, 93–107. Oxford: Oxford University Press.

Donaldson, Sue, and Will Kymlicka. 2011. *Zoopolis: A Political Theory of Animal Rights*. Oxford University Press.

———. 2015. "Farmed Animal Sanctuaries: The Heart of the Movement?" *Politics and Animals* 1: 50–74.

Donovan, Josephine, and Carol Adams, eds. 2007. *The Feminist Care Tradition in Animal Ethics*. New York: Colombia University Press.

Dworkin, Ronald. 1984. "Rights as Trumps." In *Theories of Rights*, edited by Jeremy Waldron, 153–67. Oxford: Oxford University Press.

Francione, Gary. 2000. *Introduction to Animal Rights*. Philadelphia: Temple University Press.

———. 2008. *Animals as Persons*. New York: Columbia University Press.

———. 2012. "The Paradigm Shift Requires Clarity about the Moral Baseline: Veganism." *Animal Rights: The Abolitionist Approach*. http://www.abolitionistapproach.com /the-paradigm-shift-requires-clarity-about-the-moral-baseline-veganism.

Francione, Gary, and Anna Charlton. 2015. "Animal Rights." In *Oxford Handbook of Animal Studies*, edited by Linda Kalof, 25–42. New York: Oxford University Press.

Francione, Gary, and Robert Garner. 2010. *The Animal Rights Debate: Abolition or Regulation?* New York: Columbia University Press.

Gruen, Lori. 2014. "Should Animals Have Rights?" *Dodo*, January 20. https://www.the dodo.com/community/LoriGruen/should-animals-have-rights-396291626.html.

Haraway, Donna. 2007. *When Species Meet*. Minneapolis: University of Minnesota Press.

Honig, Bonnie. 2009. *Emergency Politics*. Princeton, NJ: Princeton University Press.

Kymlicka, Will. 2002. *Contemporary Political Philosophy*. Oxford: Oxford University Press.

McMahan, Jeff. 2002. *The Ethics of Killing: Problems at the Margins of Life*. Oxford: Oxford University Press.

Nozick, Robert. 1974. *Anarchy, State and Utopia*. New York: Basic Books.

Peters, Anne. 2016. "Global Animal Law: What It Is and Why We Need It." *Transnational Environmental Law* 5 (1): 9–23.

Pick, Anat. 2012. "Turning to Animals between Love and Law." *New Formations* 76 (1): 68–85.

Porter, Robert. 1997. "Strengthening Tribal Sovereignty through Peacemaking: How the Anglo-American Legal System Destroys Indigenous Societies." *Columbia Human Rights Law Review* 28: 235–305.

Prosin, Natalie, and Steven Wise. 2016. "The Nonhuman Rights Project: Coming to a Country Near You." *Global Journal of Animal Law* 2. https://ojs.abo.fi/ojs/index.php/gjal/article/view/1380.

Rawls, John. 1971 *A Theory of Justice.* Oxford: Oxford University Press.

Regan, Tom. 1983. *The Case for Animal Rights.* Berkeley: University of California Press.

Rowlands, Mark. 2009. *Animal Rights: Moral Theory and Practice.* London: Palgrave Macmillan.

Rudy, Kathy. 2011. *Loving Animals: Toward a New Animal Advocacy.* Minneapolis: University of Minnesota Press.

Singer, Peter. 2009. *Animal Liberation: The Definitive Classic of the Animal Movement.* 2nd ed. New York: Harper Collins.

Stanescu, Vasile. 2013. "Why "Loving" Animals Is Not Enough: A Response to Kathy Rudy." *Journal of American Culture* 36 (2): 100–110.

Steiner, Gary. 2008. *Animals and the Moral Community.* New York: Columbia University Press.

Suen, Alison. 2015. *The Speaking Animal.* New York: Rowman and Littlefield.

Walker, Samuel. 1998. *The Rights Revolution.* New York: Oxford University Press.

Wall, John. 2011. "Can Democracy Represent Children? Toward a Politics of Difference." *Childhood* 19 (1): 86–100.

Weisberg, Zipporah. 2009. "The Broken Promises of Monsters: Haraway, Animals and the Humanist Legacy." *Journal for Critical Animal Studies* 7 (2): 22–62.

West, Robin. 2011. "Tragic Rights: The Rights Critique in the Age of Obama." *William and Mary Law Review* 53: 713–46.

23 SANCTUARY

Timothy Pachirat

The Sanctuary Here,
I have nothing to be afraid of.
I climb up this hill.
I get down in this valley.
I watch my face floating in the water of this river.
Shady trees make me sleep.
Birds wake me up.
Beasts give me way and return my greetings.
ASHUTOSH DUBEY (2005)

Sanctuary: (n.) a nation or area near or contiguous to the combat area
that, by tacit agreement between the warring powers, is exempt from
attack and therefore serves as a refuge for staging, logistics, or other
activities of the combatant powers. *Oxford Essential Dictionary of the U.S.
Military* (2001)

Domestication means domination: the two words have the same root
sense of mastery over another being—of bringing it [*sic*] into one's
house or domain. YI-FU TUAN (1984)

Most etymologies of sanctuary trace its derivation from the Latin *sanctuarium*, for holy place, and emphasize its close connection to *sacer*, for sacred.[1]
Sacred, holy places, spatially demarcated by boundaries that render inoper-
able the profane powers of worldly law. Sacred, holy places, to which those
persecuted by that law may flee and say, with poet Ashutosh Dubey, "I have
nothing to be afraid of." Sacred, holy places, specific and material in their

---

1. See, for example, entries for sanctuary in *The Oxford English Dictionary*, *The Oxford Dictionary of Phrase and Fable*, *The Oxford Dictionary of World History*, and *The Oxford Dictionary of
World Religions*, among others.

demarcations. Orestes is safe from the Furies so long as he touches the *omphalos*, a sacred stone at Delphi.[2]

But how are we to think sanctuary in the political context of a brutal, unsparing human violence (Wadiwel 2015) and exercise of human tyranny (Donaldson and Kymlicka 2014) against the other sentient creatures of the earth, sea, and sky? How are we to reconcile the sanctity of the *sanctuarium*, the Edenic register of the poet's "I have nothing to be afraid of," with the strategic instrumentality of the *Oxford Essential Dictionary of the U.S. Military*, where sanctuary serves not as utopian (no-place) refuge but as specific staging grounds for resistance?

At a planetary scale, the Anthropocene—rather dryly defined by Google as "the current geological age in which humans are the dominant influence on climate and the environment"—provides a decent magnetic north for navigating this tension between peace and antagonism. Usefully unresolved debates about whether to fix the beginning of the Anthropocene to the rise of the Industrial Revolution two hundred and some years ago or much further back to the Neolithic Revolution ten thousand years ago (Broswimmer 2002; Wilson 2002) are nonetheless rooted in a shared consensus that as the world's human population detonated from four million ten thousand years ago to 7.4 *billion* today, the consequences for the atmosphere, oceans, plants, and nonhuman animals constitute nothing short of an ongoing and accelerating ecocide, the sixth—and perhaps most consequential—mass extinction event in the earth's 5.54 billion year existence (Kolbert 2014; Dawson 2016; Leakey and Lewin 1995).[3]

We, the *Homo esophagus colossus* (Broswimmer 2002) exercise violence against other animals along two distinct yet synergistic fronts: an extensification of existential threats to wild animals and an intensification of domination over domesticated animals. Each front shapes particular possibilities and perils for reconciling the sacredness of sanctuaries with their potential to serve as sites of resistance.

2. Davidson (2014) offers an excellent political history of sanctuary, beginning with biblical times and continuing through England's sanctuary laws, the Underground Railroad, churches harboring Vietnam War resisters in the 1970s and Central American refugees in the 1980s and 1990s, and the sanctuary cities of the early 2000s to the present.

3. The last mass extinction event, responsible for the end of the dinosaurs along with more than half of all the earth's species, occurred sixty-five million years ago. A leading hypothesis for its cause is the fallout from a six-mile-wide asteroid that crashed into the area now known as the Yucatán Peninsula.

The First Front: Wild Animal Sanctuaries and the War against Wild Animals

The existential threat to wild animals[4] posed by humans is most visible when it is direct: elephants and rhinoceroses targeted and slaughtered for their tusks; chimpanzees captured in the wild and bred in captivity for laboratory experimentation; goldfinches snared by glue pasted on branches, only one in ten making it alive to market and caged hell. But, importantly, it is also indirect: human destruction of native forest habitat to create cattle grasslands or palm oil plantations, or anthropogenic climate change leading to ocean acidification and the death of coral reefs.

Extinction is the most dramatic and publicized consequence (see chap. 11). Today, just over two hundred years into the unprecedented planetary experiment known as the Industrial Revolution, conservative estimates put the extinction rate at between one thousand and ten thousand species per million species each year (Wilson 2002).

But the permanent erasure of entire species is not all that matters here. Direct and indirect human assaults on wild animals also threaten specific individuals, their families, and their social groupings. A singular focus on species extinction can lead to a disregard for individual animal lives, an approach in which the suffering of the individual is disregarded in favor of a "holistic" approach to the entire ecosystem.[5] Wild animals not (yet) in danger of extinction are the less heralded victims in an ongoing war of attrition that has reduced the number of all vertebrate animals in the wild by more than half in the last forty years alone (McLellan 2014).

How to think wild animal sanctuaries as sites of resistance within this context? Crucially, the fulcrum moment for the emergence of *sanctuary* as a node in the topography of enmity between humans and wild animals occurred when the seemingly immense, untamable, and infinite became understood by *European* colonial and imperial states as finite, fragile, and

4. By wild animals I mean animals belonging to species who have historically thrived without depending on humans. I include in this category wild animals born in captivity, even though, as individuals, these animals are often unable to live in the wild. See Cochrane (2014) for a useful explication of the distinction between biological and sociological domestication.

5. As Lori Gruen (2011, 172) and other ecofeminists have noted, "we can value both collectives and individuals. . . . Species have value that doesn't reduce simply to the cumulative well-being of each individual member of the species, but that value doesn't transcend the members either."

in need of protection.[6] The maw of *Homo esophagus colossus* destabilizes the very category of wild, simultaneously reducing it to a profane object of biopolitical management in the realm of political economy *and* mythologizing it with a racialized fantasy of unspoiled innocence. There are two key historical strands at work here; each deeply shapes the political promises and perils of contemporary wild animal sanctuaries.

The first strand is the repositioning of what is wild as utterly dependent on human paternalism for its survival, as necessitating human-provided *sanctuary*. In the words of Paul Crutzen (Schwägerl 2014), the Nobel Prize–winning atmospheric chemist credited with popularizing the term *Anthropocene*: "it's we who decide what nature is and what it will be. . . . Where wilderness remains, it's often only because exploitation is still unprofitable. Conservation management turns wild animals into a new form of pets."

It is the full-throated scream of the Industrial Revolution that first deploys, in 1879, the word *sanctuary* to connote "an area of land within which (wild) animals or plants are protected and encouraged to breed or grow"[7] when A. P. Vivian writes, in *Wanderings in Western Land*, that "the suggestion . . . of setting apart certain districts as 'sanctuaries,' within which the buffalo should never be molested, is one well worthy of consideration."[8] The connotation catches, and by 1897, *Cornhill Magazine* intones, "The national forests will become, as the new Forest is now in some measure, sanctuaries for all the animals of England." And by 1909 the *Bulletin of the New York Zoological Society* notes, "Around the coast there is gradually being extended a chain of insular bird sanctuaries that means much to the avifauna of North America."

And today, not even 150 years from the first English-language reference to sanctuary in relation to animals, is it an exaggeration to say that all wild animals live in a sanctuary of one kind or another? These may be sanctuaries of commission, one of approximately 120,000 protected areas making up about 13 percent of the earth's surface in the form of preserves,

6. Note the dualism in this formulation. Wild nature pivots from that which must be tamed to that which must be protected. Excluded entirely are non-Western cosmologies that locate humans as *a part of* nature rather than as its aggressor *or* its protector (TallBear 2011, 2013).

7. This quotation and all subsequent citations in this paragraph are from "Sanctuary, II.5.d and II.7" in the *Oxford English Dictionary*, Volume XIV, p. 443.

8. By 1879, of course, buffalo are already close to decimation, their systematic massacre a key tactic in the white extermination or permanent displacement of Native Americans (Cronon 1991).

wilderness areas, national parks, green corridors, or community controlled forests (International Union for Conservation of Nature and United Nations Environment Programme-World Conservation Monitoring Centre 2010),[9] or they may be sanctuaries of omission, those increasingly rare places where capitalism has yet to "develop" sufficient motivation or means for overwhelming invasion.[10] Either way, they are thinkable as *sanctuaries* only because of the vast negative space of destruction surrounding them, and in this sense they are markers of tragedy, not avatars of progress.

In addition to these sanctuaries of omission and commission, there exist numerous specialized, species-specific sanctuaries for orphaned and injured wild animals, some with the goal of one day reintroducing these animals into wild sanctuaries of commission. Still other specialized sanctuaries take in wild animals, such as chimpanzees, elephants, and carnivores, who have been kept or bred in captivity and used in laboratories, the entertainment industry, or as roadside attractions and exotic pets. There is an inherently tragic quality to these wild animal sanctuaries, which recognize that the creatures in their care, while far better off than before, will never know truly free or wild lives (Emmermen 2014).

Philosopher Lori Gruen[11] writes powerfully about the unavoidable ethical paradoxes of wild animal sanctuaries:

> Chimp Haven and genuine wild animal sanctuaries around the world are making a huge difference for captive animals. . . . Yet, even when captive animals have their futures secured . . . they remain captives. . . . Keeping them captive is wrong and releasing them from captivity is wrong. . . . *There may be no ethical way to rectify the wrong we have done.* (Gruen 2011, 161–62, emphasis mine)

Gruen writes specifically about those wild animals most obviously in captivity in sanctuaries such as Chimp Haven, explicitly contrasting them with

9. Marine sanctuaries are not included in this calculation.

10. A third alternative, not explored here, posits the possibility of life in the capitalist ruins (Tsing 2015).

11. Gruen is the originator of The First 100 Project and The Last 1,000 Project. The first creates individual biographies for each of the first one hundred chimpanzees used for scientific research in U.S. laboratories, and the second tracks the ongoing movement of the last one thousand chimpanzees out of laboratory settings and into sanctuaries such as Chimp Haven in Shreveport, Louisiana. See http://first100chimps.wesleyan.edu/ and http://last1000chimps .com/about.html.

wild animals still living in their native habitats. But as *Homo esophagus colossus* further extends its supremacy, an unavoidable question makes itself felt: Is there a single wild animal in existence today who remains free of the tragic tension at the heart of Gruen's analysis?

The second historical strand shaping the terrain of possibility and peril for today's wild animal sanctuaries is the colonial, patriarchal, and white supremacist context in which "nature," including wild animals, first entered the European imaginary as fragilities in need of (white male) protection. In a richly nuanced history, Richard Grove (1995) meticulously demonstrates that modern Western environmentalism and conservationism originated not in the metropoles of Europe but in its colonies, and in particular its island colonies. Although the colonial political economy subjected these island colonies to a brutal plantation slavery, its libidinal economy constructed the islands as sites of fantasy about noble savages and unspoiled nature. Meanwhile, a nascent scientific economy used the island colonies as laboratories in which the impacts of human deforestation on climate and ecology were first measured.

The uneasy mix of these political, racial, and scientific economies created a situation in which "colonial states increasingly found conservationism to their taste and economic advantage, particularly in ensuring sustainable timber and water supplies and in using the structures of forest protection to control their unruly marginal subjects" (Grove 1995, 15). Some of the world's first modern wilderness sanctuaries were imposed by tyrannical force in British occupied Tobago (1763) and Barbados (1765) via ordinances decreeing that woodlands be set aside and preserved "to prevent that drought which in these climates is the usual consequence of the total removal of the woods" (quoted in Grove 1995, 271). The Tobago Caribs, dispossessed of their land by the ordinances, "were forced to become a labour pool or were ignored and finally disappeared, the victims of random killings, food shortages, and imported diseases" (Grove 1995, 284), the first wave of what would eventually become as many as tens of millions of global "conservation refugees" (Dowie 2009; Duffy 2010).

By contrast, similar British ordinances in Grenada, Dominica, and especially St. Vincent were met with a fierce guerilla war executed by the indigenous black Carib population, who saw British attempts to create forest reserves as part of a broader program of land expropriation. Tellingly, the end result of the indigenous population's successful resistance to the British sanctuaries was *less* deforestation on St. Vincent than in Tobago and Barbados, where the British had successfully imposed their forest sanctuaries (Grove 1995).[12]

The historical legacies linking the imposition of wilderness sanctuaries to racist colonialism signals an important need to disaggregate *Homo esophagus colossus*, to contest an overly simplified view that the war against animals is best understood in terms of undifferentiated humans versus undifferentiated nonhuman animals.

Indeed, a central peril of contemporary wild animal wilderness sanctuaries is their lack of attention to the *intrahuman* relationships of domination and oppression that play a central role in shaping historical and contemporary wars against wild animals and their habitats. No matter how well intentioned, the creation and enforcement of wilderness for animal sanctuaries risks a white- *and* greenwashing of the structural conditions of intrahuman domination and oppression that allows the global North to position itself as righteous cop against the wild animal killing thugs of the global South. They also risk occluding the North's twin legacies of capitalism and colonialism (Dawson 2016) while its economies of consumption create the very wildlife trade that cause the most egregious assaults on wild animals and their habitats (one need not wear ivory earrings or ingest rhinoceros horn medicines: 80 percent of the coltan used in cell phones comes from a single national park in Democratic Republic of Congo [Duffy 2010, 10].)

Thus, a critical question raised by the second key strand in the historical legacy of contemporary wild animal sanctuaries is what might it mean to rethink wild animal sanctuaries more broadly as sites of resistance in the struggle for global social justice, one that recognizes the devastating *continuities* between contemporary wild animal protection efforts and the imperialistic racism at the heart of the creation of the some of the world's first wild sanctuaries?[13]

12. Here, the role of antiblack racism should not be underestimated: "While Indians had frequently found themselves eligible for easier treatment [by the colonizers] as noble savages, the black African affinities of the black Caribs invited cultural comparisons with the groups then being imported to other West Indian islands as plantation slaves at a time when the brutality meted out to slaves in the British Windward aisles was almost unparalleled in severity" (Grove 1995, 288). For a history of Native American dispossession from Yosemite, Yellowstone, and Glacier National Parks and the consequences for subsequent wilderness sanctuaries worldwide, see Spence (1999).

13. For one exploratory answer to this question, see Ashley Dawson's (2016, 88–89) provocative argument for a universal guaranteed income for the inhabitants of the world's current biodiversity hotspots (the vast majority of whom are located in the global South) to be funded by a financial transactions tax on speculative global capital flows.

The Second Front: Farm Animal Sanctuaries and the War against Domesticated Animals

Most contemporary Concentrated Animal Feeding Operations (CAFOs) and slaughterhouses hide behind barbed wire, surveillance cameras, and ag-gag laws. Not Fair Oaks Farms.[14] Years ago, this central Indiana industrialized pig and dairy farm turned its operations into an agritourism destination that now attracts over six hundred thousand visitors each year, generating millions of dollars in visitor revenue and, ostensibly, "educating" large numbers of people about the goodness of factory farms.[15] With a $27 per person "adventure pass," a steady stream of visitors, many of them Chicago-area families with young children, gain full access to Fair Oaks's cheerful tour guides, colorful interactive exhibits, climbing walls, and impressive banks of glass walls overlooking, among other areas, a birthing barn where pregnant cows labor to push out sons and daughters (who, once born, are quickly whisked away to a "nursery"); a "farrowing area" where newborn piglets jostle to get milk from mothers confined by metal bars; and a "milking parlor" merry-go-round where thousands of cows step on and off twenty-three hours a day, seven days a week, the milk meant for their babies emptying into vats for shipment to grocery stores across the United States. An exhibit near the glass walls overlooking the pig "gestation area," where females are routinely forced into narrow barred cells and subjected to the sexual violence of forced impregnation by human workers, contains the following display:

> Reproduction of pigs on our farm is NOT done the old fashioned way. Science NOT romance is how we do it.
>
> Our partnership with a pig genetics company has allowed us to successfully enhance our breed standards through Artificial Insemination.
>
> AI allows for more extensive use of quality sires. Superior sires pass on superior traits.
>
> No different than ears of corn that are bred and harvested for sweetness and a hardy nature.

Fair Oaks Farms exemplifies Yi-Fu Tuan's sense of *domestication as domination*, the second, and less publicized, front of the Anthropocene's war

14. See http://www.fofarms.com. All information on Fair Oaks Farms is drawn from primary fieldwork I conducted in April and May of 2016.

15. By contrast, Farm Sanctuary, the world's first and largest sanctuary for formerly farmed animals, receives approximately 10,000 human visitors per year at its New York and California locations.

against animals. At the same time that anthropogenic changes are accelerating mass extinctions of entire animal species, *Homo esophagus colossus* is also intensifying its control and domination over domesticated species through the intensive breeding, confining, and killing of so-called food, companion, and research animals on a scale of tens of *billions* of sentient individuals per year.[16] It is not the threat of extinction that characterizes the war waged by humans against these domesticated animals but rather a never-ending cycle of genetic manipulation, sexual violence, forced separation of family and social units, physical and chemical mutilation, debilitating confinement, and industrialized killing.

What are the possibilities for sanctuary within this topography of staggering suffering, a suffering that now commodifies cruelty as a sightseeing destination? In the abstract, how one answers this question depends crucially on one's assessment of domestication itself.[17]

Some thinkers, notably Gary Francione (2012), view all forms of animal domestication as *necessarily and inherently* exploitative; by virtue of their ten-thousand-year history of selective breeding by humans, domesticated animals are condemned by birth to lives of genetically determined dependency and servility. In this view, an ethical world depends on the complete *extinction* of all domesticated animals; sanctuaries, *at best*, serve as ameliorative halfway houses on the road to a future in which the bark of a dog, meow of a cat, whinny of a horse, grunt of a pig, bleat of a sheep, guffaw of a goat, moo of a cow, and feather dusting of a chicken shall no more be heard on earth.

Others, notably Sue Donaldson and Will Kymlicka (2011), argue that domesticated animals are capable of living flourishing and fulfilling lives but only if granted co-citizenship rights in the human political community.[18] Drawing on a liberal political theory of citizenship, Donaldson and Kymlicka (2015) critique existing sanctuaries as focusing too much on

16. These categorizations are already problematic insofar as they define animals instrumentally according to their usefulness to humans.

17. My summary draws on Cochrane's (2014) helpful overview of four views on domestication.

18. There is not room here to explore the full implications of Donaldson and Kymlicka's (2011) important work, which divides all animals into one of three categories (wild, liminal, and domesticated) and argues that human responsibilities to animals depends on which of the three categories the animals belong to. Two important interlocutors are Cochrane (2014), who argues against the arbitrariness of the tripartite categorization, and Wadiwel (2015), who argues that their vision of co-citizenship still necessitates an underlying power dynamic of human sovereignty and paternalism.

animal refuge and counterproductive forms of animal advocacy instead of on pioneering intentional communities that showcase how domesticated animals might discover their autonomy through relationship with others and exercise it to the fullest extent possible.[19]

These positions help to clarify the terrain of theoretical possibilities. But, contra both Francione *and* Donaldson and Kymlicka, existing sanctuaries make a mess of the abstract requirements established by the coordinates of philosophy and political theory. Defying simplistic categorization as ameliorative halfway houses or conservative institutions of refuge and counterproductive advocacy, existing sanctuaries might be better understood as spaces for ongoing, and necessarily imperfect, practices of entangled empathy,[20] spaces for the development of always provisional enactments of interspecies possibility written in the messy language of mutual care, affect, and embodiment rather than in the clean analytic prose of academic books and articles.

At their best, existing sanctuaries for domesticated animals operate as points of rupture, staging grounds that make room for a competing notion of domestication as mutual "miracles of attunement" (Despret 2004), miracles with the capacity to "tell a story, but tell it more than one way at once, and tell another underneath it uprising through the skin of it," in the words of novelist Ali Smith (2014, 201). While never fully *displacing* Yi-Fu Tuan's rightful linking of domestication with domination, the very

19. While outlining a laudable theoretical vision, one weakness of Donaldson and Kymlicka's (2015) critique of what they call the "refuge + advocacy" model of sanctuary is that it relies on what the authors themselves acknowledge is superficial empirical research into the practices of existing sanctuaries. There is a risk here of devaluing and misrepresenting a rich and complex body of individual and species-specific knowledge built up over decades of sanctuary practices oriented by the very same values of freedom and autonomy put forward by their critique. Beyond Donaldson and Kymlicka, animal activists and Animal Studies scholars would do well to give much greater attention to embodied practices of mutual attunement, to forms of knowledge that Flyvbjerg (2001) calls *phronesis*. With a few notable exceptions (p. jones 2014; M. Jones 2014), most humans working and living directly with formerly farmed animals in sanctuaries are not theorizing their everyday interspecies relations in theoretical language, creating a dangerous possibility for theorists to "read" their interspecies practices without, ironically, actually becoming attuned to them through careful, long-term observation. See Abrell (2016) for a recent example of an ethnography of animal sanctuaries that represents a hopeful step in the right direction.

20. *Entangled empathy* is Lori Gruen's phrase. "In the case of other animals," Gruen (2012) writes, "to empathize well, one must understand the individual's species-typical behaviors as well as her individual personality, and that is not easy to do without observation, over a period of time. Many current discussions of the claims animals make on us fail to attend to the particularity of individual animal lives."

best sanctuaries for domesticated animals function as active explorations, *through praxis rather than abstract theory*, of a range of possible liberatory futures for human-nonhuman animal relationships. Although there are now over fifty sanctuaries for formerly farmed animals in North America (Donaldson and Kymlicka 2015), two sanctuaries in particular help to evoke this range of possibilities.

Farm Sanctuary, the world's longest-running sanctuary for formerly farmed animals, was cofounded in 1986 by Gene Baur and Lorri Houston. Influenced by Latin American liberation theology and the 1980s sanctuary movement to shield refugees from war-torn El Salvador, Guatemala, and Nicaragua from US immigration laws, Baur and Houston initially founded Farm Sanctuary to expose factory farms.[21] After rescuing a sheep, subsequently named Hilda, from a "dead pile" in a Lancaster, Pennsylvania, stockyard, however, Farm Sanctuary began its transformation into an actual, physical sanctuary. Thirty years later, Farm Sanctuary's three New York and New Jersey locations are home to approximately nine hundred rescued chickens, cows, sheep, goats, turkeys, pigs, ducks, and donkeys (Baur 2008). The organization's ten-million-dollar annual operating budget is funded largely through individual donations with the majority of expenses going toward shelter operations and education and outreach. Indeed, Farm Sanctuary views the animals on its shelters as "ambassadors" for the billions of farmed animals who die each year and actively promotes visitor and internship programs that drew over eleven thousand outsiders to its sanctuaries in 2015 (Farm Sanctuary 2015).

Farm Sanctuary's national shelter director, Susie Coston, has been working closely with formerly farmed animals for twenty years and is responsible for training and mentoring many others who have gone on to start their own farm animal sanctuaries. Coston is, in many ways, the farmed animal sanctuary movement's original "veganic intellectual."[22] Interviewing Coston for this essay, I was struck by the immediacy and embodied nature of Coston's answers to my questions. Although she has acquired a huge body of knowledge on formerly farmed animals, Coston eschews thinking in general, theoretical terms and instead makes constant reference to the wants, desires, and aversions of specific individuals. When I asked Coston an abstract question about how Farm Sanctuary navigates

21. Gene Baur, personal communication, 2016.

22. "The veganic intellectual . . . plays the same role as the organic intellectual, but for a group that includes nonhuman animals. The veganic intellectual does not claim to be 'the voice of the voiceless,' but rather recognizes and listens to animal voices" (p. jones 2014, 105).

the tension between the animals' safety and their autonomy, for example, Coston replied by talking about how the sanctuary is constantly adjusting the social groupings of its pigs in order to accommodate the pigs' desires about whom they want to spend time with and the tendency for certain pigs to become the subject of harassment by other pigs. She also told me about how when a sheep needs to go to the hospital for veterinary care, her entire family group travels with her in order to keep all of them feeling secure. For Coston, entangled empathy and the miracle of mutual attunement are enacted through careful, decades-long observation and interaction with specific animal species and individuals, not through the imposition of abstract theories.[23]

VINE, short for Veganism Is the Next Evolution, was cofounded in 2000 by pattrice jones and Miriam Jones. Unlike Farm Sanctuary, VINE does not aggressively promote a visitor's program. VINE views the animals at its sanctuary less as ambassadors whose role is to educate a larger public than as individuals whose self-expressed preferences determine whether or not they are ever named and whether or not they interact with or are seen by humans.[24] Miriam Jones explains,

> This approach to working with liberated, formerly farmed animals is not necessarily that taken by other sanctuaries. Many advocate and cultivate far more contact-intense relationships between human workers and animal "residents." Everyone in those sanctuary situations has a human name, and none of the chickens are allowed to sleep in trees or otherwise (re)learn wild/feral behaviors. We at VINE do not condemn such approaches, but neither do we feel compelled to adopt them ourselves. (M. Jones 2014)

One unique aspect of VINE, and one that other sanctuaries would do well to emulate, is their deliberate adoption of an ecofeminist, intersectional approach to analyzing oppression. For pattrice jones and Miriam Jones, human oppression of animals cannot be separated from racism, heteronormativity, patriarchy, and ecocide, and it must be thought about and acted on alongside those other oppressions. To take just one embodied example, VINE is one of the few sanctuaries that accepts roosters liberated from cockfighting, viewing them as victims of a sexism that instills

---

23. Interview with Susie Coston, August 9, 2016.

24. See Gruen (2014a) for a discussion of the ethics of making animals available to the human gaze.

constant fear in the animals in order to induce displays of aggression associated with and celebrated by hypermasculinity. Through a graduated series of exposures, VINE reintroduces these former fighters to normal life with other roosters and hens, demonstrating that roosters used in cock-fighting do not inevitably have to be euthanized when they are liberated from the industry. Providing sanctuary to these roosters is also, at VINE, a feminist practice of resistance to sexism (p. jones 2009).

Although they experiment with a range of approaches, Farm Sanctuary, VINE, and many other existing sanctuaries for domesticated animals are unique in forcing and focusing our attention—our attunement—away from the abstract, philosophical, and disembodied terrain of so much Animal Studies and animal liberation discourse and onto the material, embodied actuality of our entanglements with specific individual animal lives and of their entanglements with one another. Faced with a mind-numbing "arithmetic of compassion" (Herbert 2007) that scales in the billions, besieged by shortsighted utilitarian measures of advocacy effi-cacy that create mathematical equations to determine how much animal suffering is averted per dollar expended,[25] sanctuaries insist, paradoxically, on the unquantifiable importance of one.

Of *this* one. *This* one, Hilda the sheep, who was left for dead, discarded like trash on a pile in a Lancaster stockyard (Baur 2008, 20–38). *This* one, Maxine the cow, who jumped the wall of a slaughterhouse in Queens and ran free through New York City's streets (Farm Sanctuary, n.d.). *This* one, Zoop the goat, who was found in a New Jersey snowstorm, a six-pound baby walking on her knees (Baur 2008, 144–46). *This* one, Albert the rooster, who hurt his leg and decided that a life that didn't include spending his nights in the tallest tree around was not a life worth living (M. Jones 2014, 90–91). *This* one, ALFie the lamb, and *this* one, Maddox the calf, who prefer the company of each other to others of their own species (p. jones 2014, 102).

If these are the promises of sanctuaries for domesticated animals, what of their perils? One of the biggest, ironically, is that sanctuaries risk becoming victims of their own success, their potential to serve as sites of resistance blunted by an increasing mainstream acceptance, even popu-larity, that brings with it funding, membership, and attention precisely because it is not perceived as fundamentally threatening to the existing political and economic order. At best, this type of domestication risks

25. See, for example, Animal Charity Evaluators, http://www.animalcharityevaluators.org/.

turning sanctuary into a politically defanged, libidinal object of escapist desire, an ecofriendly, interspecies inversion of Fair Oaks Farms. At worst, it risks packaging sanctuary as gift shop merchandise, selling it as lifestyle choice, and branding it as fundraising icon. The attendant danger for animals is that, despite being rescued from a life as mere property, they nonetheless become recirculated and recommodified as objects within a sanctuary economy where their exchange value is calculated in terms of number of donations brought in or amount of celebrity and media attention garnered (Abrell 2016).[26]

A second, closely related peril is that the overwhelming whiteness and North American centrism of the contemporary domesticated animal sanctuary movement allows it to comfortably work within and even intensify a politics of white, global North privilege. One concrete expression of this privilege occurs when sanctuaries fail to develop and act on broader analyses of power that link animal oppression to other forms of unjust domination (see VINE, above), choosing instead to ignore or downplay other movements for social justice. Because of its size and influence, for example, Farm Sanctuary in particular possesses a unique opportunity to invite alliances with parallel movements for justice around issues of race, immigration, and labor in the agricultural industry and, in doing so, to forcefully counteract a xenophobic and racist tendency in the wider animal protection movement that repeatedly scapegoats and criminalizes those low-level, immigrant workers of color most often featured in undercover video footage from slaughterhouses and factory farms.

Similarly, just as Farm Sanctuary's founders were inspired by the 1980s sanctuary movement on behalf of political refugees from Central America, so, too, could the domesticated animal sanctuaries of the present adopt a multioptic vision (Kim 2015) of what it means to "live the farm sanctuary life" (Baur and Stone 2015) by making it an urgent budgetary, programming, and staffing priority to listen to and learn from movements for liberation from antiblack racism, ableism, misogyny, and heteronormativity in larger society (S. Ko 2017; A. Ko 2016; Jackson 2013; Taylor 2014; p. jones 2014). This is not a matter of superficially or instrumentally incorporating the rhetoric and symbols of other struggles against oppression but of asking, at the deepest levels, about how these forms of oppression are linked to one another. Is genuine animal liberation possible in a world in which

26. Susie Coston, interview with the author, 2016.

other forms of domination and oppression continue to exist? How would an authentic grappling with the other faces of oppression alter, perhaps radically, current conceptions and enactments of sanctuary? In the words of Aph Ko (2016b), founder of *Black Vegans Rock* and *Aphro-ism*,

> The biggest issue the white animal rights movement has is that they can't properly locate WHY animal oppression is happening. They see the aftermath of oppression—they see the victims—but most of these activists have no conceptual clue as to why animals are systemically being hurt. Sometimes it's painful to watch activists from the dominant class try to create campaigns to stop animal oppression (without realizing how they are perpetuating it). . . . White folks don't seem to realize that white supremacy systemically harms animals. White folks don't want to move out of their leadership positions, but they want to stop animal oppression, which basically means they don't want to change behaviors that are discursively hurting animals.[27]

The critical question facing animal sanctuaries for domesticated animals is identical to the one facing wild animal sanctuaries: what might it mean to rethink sanctuary more broadly as a site of resistance in the fight for global social justice, one that recognizes how the human-nonhuman divide authorizes violence against *all* who are deemed to be less than human?

Given the relentless extensification and intensification of the war against animals, there is an urgency to unimagine sanctuaries as sacred utopias—places that are no place—and instead (re)think them as sites of potential rupture and resistance. Because sanctuaries are located within topographies of enmity, they must negotiate a series of critical tensions, including between freedom and management, wildness and imperialistic racism, domination and mutual attunement, and single optic versus multioptic analyses of power and oppression.

Sanctuary's liberatory promise lies not in the resolution of these contradictions but rather in its day in and day out exploration of them in the context of embodied, specific, interspecies relationships of mutual attunement. In so doing, sanctuaries might serve as sites of resistance whose aim, ultimately, is the subversion, rather than the mere crossing, of the human-animal divide, a divide that reproduces and is reproduced by antiblack racism, ableism, heteronormativy, and patriarchy. Only with the

27. See http://www.blackvegansrock.com/ and https://aphro-ism.com/.

subversion of this divide will it be possible for all to live without fear, for birds to wake us, and for us beasts to give each other way and return one another's greetings. Only then, that is, might sanctuary become redundant, irrelevant, unnecessary.

Suggestions for Further Reading

Abrell, Elan L. 2016. "Saving Animals: Everyday Practices of Care and Rescue in the US Animal Sanctuary Movement. PhD diss., CUNY.

Despret, Vinciane. 2004. "The Body We Care For: Figures of Anthropo-Zoo-Genesis. *Body and Society* 10 (2/3): 111–34.

Donaldson, Sue, and Will Kymlicka. 2015. "Farmed Animal Sanctuaries: The Heart of the Movement? *Politics and Animals* 1 (1): 50–74.

Dowie, Mark. 2009. *Conservation Refugees: The Hundred-Year Conflict between Global Conservation and Native Peoples*. Cambridge, MA: MIT Press.

Gruen, Lori. 2014. "Dignity, Captivity, and an Ethics of Sight." In *The Ethics of Captivity*, edited by Lori Gruen, 231–47. New York: Oxford University Press.

References

Abrell, Elan L. 2016. "Saving Animals: Everyday Practices of Care and Rescue in the US Animal Sanctuary Movement." PhD diss., CUNY.

Baur, Gene. 2008. *Farm Sanctuary: Changing Hearts and Minds about Animals and Food*. New York: Simon and Schuster.

Baur, Gene, and Gene Stone. 2015. *Living the Farm Sanctuary Life: The Ultimate Guide to Eating Mindfully, Living Longer, and Feeling Better Every Day*. New York: Rodale.

Broswimmer, Franz J. 2002. *Ecocide: A Short History of Mass Extinction of Species*. Sterling, VA: Pluto.

Cochrane, Alasdair. 2014. "Born in Chains? The Ethics of Animal Domestication. In *The Ethics of Captivity*, edited by Lori Gruen, 156–73. New York: Oxford University Press.

Cronon, William. 1991. *Nature's Metropolis: Chicago and the Great West*. New York: W. W. Norton.

Davidson, Michael J. 2014. "Sanctuary: A Modern Legal Anachronism." *Capital University Law Review* 42 (3): 583–618.

Dawson, Ashley. 2016. *Extinction: A Radical History*. New York: Or Books.

Despret, Vinciane. 2004. "The Body We Care For: Figures of Anthropo-Zoo-Genesis." *Body and Society* 10 (2/3): 111–34.

Donaldson, Sue, and Will Kymlicka. 2011. *Zoopolis: A Political Theory of Animal Rights*. Oxford: Oxford University Press.

———. 2014. Unruly Beasts: Animal Citizens and the Threat of Tyranny. *Canadian Journal of Political Science* 47 (1): 23–45.

———. 2015. "Farmed Animal Sanctuaries: The Heart of the Movement?" *Politics and Animals* 1 (1): 50–74.

Dowie, Mark. 2009. *Conservation Refugees: The Hundred-Year Conflict between Global Conservation and Native Peoples.* Cambridge, MA: MIT Press.

Dubey, Ashutosh. 2005. "The Sanctuary." *Indian Literature* 49 (6): 32–33.

Duffy, Rosaleen. 2010. *Nature Crime: How We're Getting Conservation Wrong.* New Haven, CT: Yale University Press.

Emmerman, Karen. 2014. "Sanctuary, Not Remedy: The Problem of Captivity and the Need for Moral Repair." In *The Ethics of Captivity*, edited by Lori Gruen, 213–30. New York: Oxford University Press.

Farm Sanctuary. 2015. *Farm sanctuary 2015 year in review.* http://www.farmsanctuary .org/wp-content/uploads/2012/03/2015-Fall-Newsletter-YIR-spread.pdf.

———. N.d. "Maxine: Daring Dash around Queens Lands Renegade Cow at New York Shelter. https://www.farmsanctuary.org/the-sanctuaries/rescued-animals /featured-past-rescues/daring-dash-around-queens-lands-renegade-cow-at-new -york-shelter/.

Flyvbjerg, Bent. 2001. *Making Social Science Matter: Why Social Inquiry Fails and How It Can Succeed Again.* New York: Cambridge University Press.

Francione, Gary. 2012. "'Pets': The Inherent Problem of Domestication." *Animal Rights: The Abolitionist Approach* (blog), July 31. http://www.abolitionistapproach.com /pets-the-inherent-problems-of-domestication/

Grove, Richard. 1995. *Green Imperialism: Colonial Expansion, Tropical Island Edens, and the Origins of Environmentalism, 1600–1860.* Cambridge: Cambridge University Press.

Gruen, Lori. 2011. *Ethics and Animals: An Introduction.* Cambridge: Cambridge University Press.

———. 2012. "Navigating Difference (Again): Animal Ethics and Entangled Empathy." In *Strangers to Nature: Animal Lives and Human Ethics*, edited by Gregory Smulewicz-Zucker, 213–33. New York: Lexington Books.

———. 2014a. "Dignity, Captivity, and an Ethics of Sight." In *The Ethics of Captivity*, edited by Lori Gruen, 231–47. New York: Oxford University Press.

———. 2014b. *The Ethics of Captivity.* New York: Oxford University Press.

Herbert, Zbigniew. 2007. *The Collected Poems, 1956–1998.* Translated by Alissa Valles. New York: Ecco.

International Union for Conservation of Nature and United Nations Environment Programme-World Conservation Monitoring Centre. 2010. *World Database on Protected Areas.* Switzerland: World Commision on Protected Areas; Cambridge: UNEP-WCMC.

Jackson, Iman Zakiyyah. 2013. "Animal: New Directions in the Theorization of Race and Posthumanism." *Feminist Studies* 39 (3): 669–85.

Jones, Miriam. 2014. "Captivity in the Context of a Sanctuary for Formerly Farmed

Animals. In *The Ethics of Captivity*, edited by Lori Gruen, 90–101. New York: Oxford University Press.

jones, pattrice. 2009. "Rehabilitating Fighting Roosters: An Ecofeminist Approach." *The Abolishionist-Online*. http://sanctuary.bravebirds.org/wp-content/uploads /2009/05/rehab2.pdf.

———. 2014. "Eros and the Mechanisms of Eco-defense." In *Ecofeminism: Feminist Intersections with Other Animals and the Earth*, edited by Carol J. Adams and Lori Gruen, 91–106. New York: Bloomsbury Academic.

Kim, Claire Jean,. 2015. *Dangerous Crossings: Race, Species, and Nature in a Multi-cultural Age*. Cambridge: Cambridge University Press.

Ko, Aph. 2016. "Aph ko: Anti-Racist Activist Fighting for Animal Liberation." *Striking at the Roots* (blog), May 31. https://strikingattheroots.wordpress.com/2016/05/31/aph -ko-anti-racist-activist-fighting-for-animal-liberation/.

Ko, Syl. 2017. "Notes from the Border of the Human-Animal Divide: Thinking and Talking about Animal Oppression When You're Not Quite Human Yourself." In *Aphro-ism: Essays on Pop Culture, Feminism, and Black Veganism from Two Sisters*. New York: Lantern Books.

Kolbert, Elizabeth. 2014. *The Sixth Extinction: An Unnatural History*. London: Blooms-bury.

Leakey, Richard E., and Roger Lewin. 1995. *The Sixth Extinction: Patterns of Life and the Future of Humankind*. New York: Doubleday.

McLellan, Richard, ed. 2014. *Living Planet Report, 2014: Species and Spaces, People and Places*. Gland: World Wildlife Fund for Nature.

Merchant, Carolyn. 1980. *The Death of Nature: Women, Ecology, and the Scientific Revolution*. San Francisco: Harper and Row.

Oxford Essential Dictionary of the U.S. Military. 2001. New York: Oxford University Press.

Schwägerl, Christian. 2014. *The Anthropocene: The Human Era and How It Shapes Our Planet*. Santa Fe, NM: Synergetic.

Smith, Ali. 2014. *How to Be Both*. New York: Pantheon.

Spence, Mark D. 1999. *Dispossessing the Wilderness: Indian Removal and the Making of the National Parks*. New York: Oxford University Press.

TallBear, Kim. 2011. "Why Interspecies Thinking Needs Indigenous Standpoints: Theo-rizing the Contemporary." *Cultural Anthropology*, April 24. https://culanth.org/field sights/260-why-interspecies-thinking-needs-indigenous-standpoints.

———. 2013. "An Indigenous Approach to Critical Animal Studies, Interspecies Think-ing, and the New Materialisms." Lecture presented at the University of Washing-ton, "Borders of Kinship: Species/Race/Indigeneity," Latin American and Carib-bean Studies Program, the Jackson School of International Studies, the Simpson Center for the Humanities, and the Institute for the Study of Ethnicity, Race, and Sexuality.

Taylor, Sanaura. 2014. "Interdependent Animals: A Feminist Disability Ethics of Care." In *Ecofeminism: Feminist with Other Animals and the Earth*, edited by Carol J. Adams and Lori Gruen, 109–26. New York: Bloomsbury Academic.

Tsing, Anna Lowenhaupt. 2015. *The Mushroom at the End of the World: On the Possibility of Life in Capitalist Ruins*. Princeton, NJ: Princeton University Press.

Tuan, Yi-Fu. 1984. *Dominance and Affection: The Making of Pets*. New Haven, CT: Yale University Press.

Wadiwel, Dinesh Joseph. 2015. *The War against Animals*. Leiden: Brill.

Wilson, Edward O. 2002. *The Future of Life*. New York: Random House.

24 SENTIENCE

Gary Varner

In a memorable episode of *Star Trek: The Next Generation* titled "The Measure of a Man," scientist Bruce Maddox arranges to get the android Commander Data transferred to his command in order to disassemble him as part of his work on building more androids with Data's superhuman abilities. The district's JAG (judge advocate general) rules that Data is the property of Star Fleet and can't refuse the procedure any more than the ship's computer could "refuse a refit." When Captain Picard objects, a hearing is scheduled with Picard as Data's advocate. Maddox argues that Data can't have rights because "it is not a sentient being." When Picard asks, "What is required for sentience?," Maddox replies, "Intelligence, self-awareness, consciousness." Picard then proceeds to demonstrate that Data is both self-aware and intelligent, and that suffices to convince the JAG to "rule that Lieutenant Commander Data has the freedom to choose."[1] Although the JAG doesn't add "having freedom to choose" to Maddox's characterization of sentience, the JAG's statement at least suggests an association with self-determination, autonomy, or maybe personhood.

That *sentient* is sometimes used as a synonym of *person* is confirmed by another episode, this time from *Star Trek: Enterprise*. In "Rogue Planet," the crew encounters a solitary planet with an alien space ship on the surface. When an "away team" investigates, they find that the alien ship belongs to the Eska, a humanoid race that has been using the planet for recreational hunting for nine generations. The hunters explain that "there are higher

---

1. The episode is "The Measure of a Man" from season two. The court scene dialog is from 36 to 42 minutes into the 45½ minute episode (sans commercials). The reference to a "refit" is from 20 minutes in.

primates here" but say that they "don't touch" those. When asked why the Enterprise's sensors didn't detect the Eska hunters on the surface, they explain that they "use sensing cloaks [to] keep the wildlife from spotting us." At first, the Eska seem to be hunting only a species of wild boar, but later the Enterprise crew discovers that they are also using their cloaking devices to hunt "the Wraiths," a species of "shape-shifters" who can read the thoughts of humanoids. The Wraiths can take any form, and in the course of the episode, one of them appears to Captain Archer and talks to him as an adult human being. After the Wraith tells him that the Eska are hunting her species, Archer is incensed; as he puts it to his crewmates, "Hunting wild boar is one thing, but they're killing a sentient species," and "I'll be damned if I'm going to let anyone shoot her."[2] Adult humans are the paradigm case of what counts as a *person* when the term is used in the specialized sense that is distinct from the sense of "member of the species *Homo sapiens.*" So when Archer reacts to the Wraith's interacting with him as a human being by saying that "Hunting wild boar is one thing, but they're killing a sentient species," he seems to be using the term *sentient* as analogous to *person* in the sense of that term that identifies persons with individuals who have some complex of cognitive abilities (see chap. 18).

From these two popular culture examples, it is clear that the concept of sentience, or being sentient, is both pervasive and contested. The *Oxford English Dictionary* (*OED*) places the term *sentient* in its "Band 5," referring to "words which occur between 1 and 10 times per million words in typical English usage." This includes "words which would be seen as distinctively educated, while not being abstruse, technical, or jargon."[3] Thus, the word is used fairly freely, both by academics writing in diverse disciplines and by authors of novels, screenplays, and televisions series. As illustrated in the above examples, however, users often do not explicitly define *sentience*, and they implicitly give it a wide variety of significantly different meanings, including

1. conscious,
2. intelligent,
3. self-aware,

2. The episode is "Rogue Planet" from season one. The initial discussion of hunting is 10–12 minutes in. The Archer quotes are at 36 and 38–39 minutes in. Relatedly, at about 33 minutes in, Commander T'pal says that the Wraiths "sound like intelligent, sentient beings."

3. "Key to frequency: Calculating frequency," http://public.oed.com/how-to-use-the-oed /key-to-frequency/, and the entry on "sentient."

4. having freedom of choice or autonomy, and

5. qualifies for personhood.

That the term has assumed this broad range of meanings is not surprising given its etymology.

The English term *sentient* is derived from the Latin *sentient-em*, which is the present participle of *sentīre*, which means simply "to feel." One can "feel" in substantially different ways, however. When you touch a hot stove you "feel" heat on your skin, and if you burn yourself you will "feel" pain in your hand. You can also "feel" emotions such as relief, fear, anticipatory dread, elation, and so on. We sometimes describe more cognitive states or thoughts—such as believing something, making a judgment, and so on— as "feeling" a certain way. And we even describe desires as feelings, as in "I feel like having something to eat."

The Core Meaning of *Sentience*

Although in everyday language the term *sentient* has a variety of importantly different meanings, in Animal Studies one definition of the term has become the "core" or default meaning of the term, specifically (6) Capable of conscious suffering and/or enjoyment.

This is the stipulative definition that Peter Singer gave the term in *Animal Liberation*, where he argued that

> If a being is not capable of suffering, or of experiencing enjoyment or happiness, there is nothing to be taken into account. So the limit of sentience (using the term as a convenient if not strictly accurate shorthand for the capacity to suffer and/or experience enjoyment) is the only defensible boundary of concern for the interests of others. (Singer 1975, 8–9)

Singer's caveat that "sentience" is "a convenient if not strictly accurate shorthand" is a nod to the fact that in everyday usage, "sentience" is assigned a number of other, importantly different meanings.

When *sentience* is defined in this way, the term bears an important relationship to physical pain (see chap. 17). On the one hand, being capable of feeling pain suffices to make an individual sentient. This is because, in the context of clinical medicine, *pain* is defined as an *aversive* conscious experience. There are anomalous cases, of course: masochists are described as *liking* physical pain, and under certain circumstances (e.g., in the heat of combat), people report not feeling any pain when injured in ways that would normally cause intense physical pain. Nevertheless, physical pain is normally a type of conscious *suffering*, and therefore, if animals of a given species are capable of feeling pain, that suffices to make them sentient even if they lack various cognitive capacities associated with being self-aware, autonomous, or a person.

While the ability to feel pain is a *sufficient* condition for being sentient, it is not a *necessary* condition. For we can imagine individuals who are incapable of suffering from physical pain who are nevertheless capable of suffering and enjoyment of various other kinds. The android in *Star Trek*, Commander Data, is represented as never feeling physical pain, but he nevertheless seems to enjoy various pastimes, including playing the violin and poker. And in the real world, there are a very small number of humans who are incapable of feeling physical pain. These are patients with CIP, congenital insensitivity to pain. As depicted in the documentary film *A Life without Pain* (2005), patients with CIP have dysfunctional nociceptors (peripheral elements of the nervous system involved in feeling physical pain), and so they feel only pressure on their skin when, for instance, they are pricked with a pin in a way that would cause pain (if only minor) to a person with functional nociceptors. But in the film, the three young people with CIP are shown enjoying activities typical of youngsters, and the typical range of human emotions, including joy, humor, anger, fear, depression, and disappointment. Having any of those emotions would by itself suffice for sentience; for instance, joy is a form of conscious enjoyment, and disappointment is a form of conscious suffering, so having the capacity to experience either emotion is, like the capacity for feeling physical pain, a sufficient condition for being sentient.

However, the film *A Life without Pain* also depicts how difficult it is for parents to raise a child with CIP. One of the children in the film chewed her fingers so badly that her parents had to restrain her arms in braces, and later she rubs her eyes excessively until they made her wear goggles. Another child with CIP tried to smooth out some wrinkles on her hand with a hot iron, and when a third broke her leg in play, it wasn't discovered for two weeks. Not surprisingly, only approximately one hundred people worldwide are diagnosed with CIP, and many of them don't survive long enough to have children themselves.

So in the real world—apart from science fiction—we don't normally encounter any individuals who are plausible candidates for sentience of any kind that are not also capable of feeling physical pain. For that reason, the capacity for feeling physical pain is commonly treated as matching the extent of sentience, and scientific research concerning which animals are sentient has focused specifically on the question of which animals can feel physical pain.

How Widespread Is Sentience in the Animal Kingdom?

Discussions of which animals can feel pain inevitably involve "arguments by analogy," because the "subjective feel" of pain is not observable in the way that the physical movements and physiological reactions associated with feeling pain are observable. That is, we can see an animal withdrawing a limb from a burning-hot stimulus and subsequently avoiding that stimulus, and scientists can measure physiological reactions associated with consciousness of pain, such as activity in certain areas of the brain or secretion of endogenous opioids. But the conscious *suffering* itself—which is what makes the animal sentient—cannot be observed or measured scientifically. Instead, an "argument by analogy" can be constructed by collecting various kinds of scientific research on pain and comparing what is known about humans with what is known about various taxa of nonhumans.

Arguments by analogy are used in various contexts regarding properties that, like the subjective feel of physical pain, cannot be directly observed. The general structure of an argument by analogy regarding pain in a given species of animals can be represented as follows:

1. We know that both *human beings* and *animals of species S* have properties *a, b, c, . . .* and *n*.
2. We know that *human beings* consciously feel pain.
3. Therefore, *animals of species S* probably also can feel pain.[4]

Such arguments are *inductive* rather than deductively *valid*, meaning that the premise's being true does not rule out the conclusion's being false. How much confidence we should have in the conclusion that animals of species S can feel pain depends on how relevant the properties *a, b, c, . . .* and *n* are to the ability to feel pain.

4. This schema of the argument is adapted from Varner (2012, 108).

Based on studies of human beings' reports about their experiences of pain, it is believed that the following are relevant to the ability to feel pain:

- Having nociceptors (as noted above, humans with dysfunctional nociceptors report not feeling physical pain)
- Having nociceptors connected to the brain (humans with broken spinal cords report not feeling pain in the parts of their bodies from which nociceptive signals can no longer reach their brains even though spinally mediated reflexes in an affected limb still occur.
- Endogenous opioids are secreted when painful stimuli are encountered (in humans, this functions to mute the aversive "feel" of pain and is involved in the "no pain in the heat of combat" effect mentioned above).
- Responses to damaging stimuli are analogous to humans in pain, including things like avoiding the stimuli, favoring injured limbs, etc.
- Responses to damaging stimuli are modified by known analgesics.

Roughly speaking, the available evidence suggests that the above comparisons between human beings and various animals hold for vertebrates but not invertebrates (with the exception of cephalopods). Elsewhere (Varner 2012, § 5.3) I have described this as "the standard argument by analogy regarding pain" because it has convinced so many people that vertebrate animals probably can feel pain but invertebrates (with the exception of cephalopods) probably cannot.

There is still significant controversy over the scope of conscious pain in the animal kingdom, however. For instance, a prominent advocate of the view that fish do not feel pain (Rose 2002) argues that only *mammals* are conscious, based on claims about how the mammalian neocortex supports consciousness in humans. In response, others have argued that evidence supports convergent functional evolution among the mammalian neocortex, the paleocortex of fish and reptiles, and the avian hyperstriatum (see, e.g., Braithwaite 2010 [on fish] and Güntürkün and Bugnyar 2016 [on birds]).

Another reason that the "standard" argument by analogy is controversial is that relevant research is ongoing, and as Colin Allen has pointed out, "The direction of discovery here seems uniformly towards identifying *more* similarities between diverse species." For instance, when I first published a table summarizing the "standard" argument in the mid-1990s (Varner 1998, 53), there was no evidence of functional nociceptors in invertebrates or, for that matter, fish. But when I updated the same table about twenty years later (Varner 2012, §§ 5.3–5.4), evidence of functional nociceptors had emerged in various invertebrates, including leeches and marine snails. Allen therefore suggests that "the indicators in Varner's table noting the lack of evidence for nociceptors [in various taxa] represent just that— namely a lack of evidence, not a lack of nonciceptors." Allen adds that "in the absence of a guiding theory" of what consciousness does for an organism, "it is virtually impossible to decide how to weight" the various comparisons listed (Allen 2004, 623) in the "standard" argument by anal- ogy, that is, to decide which comparisons are most relevant to the ability to consciously feel pain.[5]

Why Does Sentience Matter, Morally Speaking?

It is easy to say why the capacity for feeling physical pain matters, morally speaking: under normal circumstances, pain *hurts*. This is why, as noted above, clinicians define "pain" as an *aversive* conscious experience.

With regard to the other, more cognitive capacities that are sometimes associated with sentience—(1) consciousness, (2) intelligence, (3) self- awareness, (4) having freedom of choice or autonomy, and (5) being a person—why would these capacities be thought to matter, morally speak- ing? The answer is that if these capacities allow an individual to have expe- riences of suffering and/or enjoyment that they would otherwise lack, then these capacities would be relevant from a sentientist perspective.

First, consider *intelligence*. This is, of course, a word with very broad con- notations. The *OED* identifies eight basic meanings of the noun, including "the action or fact of mentally apprehending something; understanding, knowledge, comprehension (*of* something)." So construed, intelligence can allow an individual to experience various kinds of suffering and enjoy- ment. Consider the difference that having some mathematical intelligence or comprehension of music theory makes to an individual: those forms of intelligence will enable the individual to enjoy solving problems in mathe- matics and appreciating various things about music. Without those capac- ities, the individual would be unable to experience those enjoyments. And at the same time, various forms of intelligence enable individuals to suffer in parallel ways. Only a "mathematically literate" person can feel the frus- tration of working a difficult math problem, and only a music aficionado

5. For a fuller treatment of the standard argument by analogy and Allen's criticisms, see Varner (2012, chap. 5).

can have certain kinds of aesthetic sensibilities offended. In such ways, intelligence of various kinds can enable a sentient being to both enjoy and suffer in ways that beings lacking those kinds of intelligence cannot.

For that reason, many people believe that while all sentient creatures are *morally considerable*, the lives of some sentient creatures have greater relative *moral significance* than others. That is, they distinguish between what qualifies an individual for moral consideration at all and what justifies establishing priorities among interests in cases of conflict or triage.

Prominent among the capacities that are claimed to enhance the moral significance of an individual's life is autonomy. The concept of autonomy has been extensively discussed by moral and political philosophers, especially (but not exclusively) by those working in the Kantian tradition. Famously, Immanuel Kant held that to act autonomously is to act according to laws that one gives oneself (Kant [1785] 1948, 98–100). Others have described autonomy differently, but a common theme is that to achieve autonomy, one must understand that who one is, what kind of a person one is, is in some significant measure a matter of one's own choosing. Here "person" is understood in a specialized sense that is different than "being a member of *Homo sapiens*," but to achieve that kind of perspective on one's own life requires a degree of cognitive sophistication that is, presumably, beyond that of most if not all nonhuman animals.

Various philosophers who argue that there is special value in such autonomy will be inclined to say that the lives of human beings have special moral value by virtue of this complex cognitive capacity that we have. In such a view, a human's life is more morally significant than that of an animal who is sentient by virtue of feeling physical pain but who lacks the capacity for autonomy. In such a view, Captain Archer was right to think that the Eskas' hunting of the very humanlike Wraiths fell into a different moral category than their hunting of wild boars.

This kind of complication within a sentientist system of ethics is what Singer had in mind late in his widely reprinted chapter "All Animals Are Equal" from *Animal Liberation*. There he argued that "The wrongness of killing a being is more complicated" than the wrongness of inflicting pain and that

> a rejection of speciesism does not imply that all lives are of equal worth. While self-awareness, intelligence, the capacity for meaningful relations with others, and so on are not relevant to the question of inflicting pain—since pain is pain, whatever other capacities, beyond the capacity to feel pain, the being

may have—these capacities may be relevant to the question of taking life. It is not arbitrary to hold that the life of a self-aware being, capable of abstract thought, of planning for the future, of complex acts of communication, and so on, is more valuable than the life of a being without these capacities. (Singer 1975, 18, 21–22)

Thus, a sentientist can hold that while the ability to feel physical pain suffices to give an individual moral *standing*, various cognitive capacities can give an individual the ability to experience various other kinds of conscious suffering and enjoyment. And when the question is not about inflicting physical pain but, rather, the value of an individual's life, a sentientist can consistently hold that the lives of individuals with certain cognitive abilities are more morally *significant* than the lives of individuals lacking the cognitive capacities in question.

Philosophers working within the sentientist tradition have defended such a view, and in doing so they often invoke the concepts of personhood and some form of autonomy. For instance, in subsequent work, Singer (1979) defined *persons* as "rational and self-conscious" beings (76), sometimes glossing "rational and self-conscious" as having "a life that is biographical" (1993, 126), which suggests a conception of autonomy that consists of choosing and striving to live a life "story." In various publications, Singer has described persons so defined as "not replaceable" in contrast to individuals lacking "self-consciousness" and "a life that is biographical," although he has given various different reasons for thinking that persons' lives have special moral significance.[6]

Sentientism and The Rest of Nature

An important implication of taking a sentientist stance in ethics is that the sphere of moral consideration will be limited to conscious beings. Above, under the heading "How Widespread Is Sentience in the Animal Kingdom?," we noted that the scope of sentience is usually assumed to match the scope of physical pain, and we noted that "the standard argument by analogy regarding pain" concludes that consciousness of pain is, with few exceptions, limited to vertebrate animals. That means that a sentientist

6. For an overview of Singer's reasoning on personhood and replaceability, see Varner (2012, § 9.4).

must treat the large majority of living organisms as having only *instrumental* value, morally speaking.

Many environmental ethicists have rejected sentientism for just this reason: environmentalists typically think that many nonsentient entities have *intrinsic* value, including nonsentient organisms as well as "holistic" entities such as species and ecosystems. For that reason, the term *sentient* was made into "an '-ism'" by environmental philosophers. John Rodman introduced the term *sentientism* in a review of Singer's *Animal Liberation*. Criticizing Singer for not extending moral consideration to more of nature, Rodman (1977) wrote,

> In the end, Singer achieves "an expansion of our moral horizons" just far enough to include most animals. . . . The rest of nature is left in a state of thing-hood, having no intrinsic worth, acquiring instrumental value only as resources for the well-being of an elite of sentient beings. Homocentrist rationalism has widened out into a kind of zoöcentrist sentientism. (Rodman 1977, 91)

With many thinking that sentientist ethical thinking is at odds with sound environmental ethics, environmental philosophers commonly use the term *sentientism* derisively, to mean any view that "*arbitrarily* favors sentients over nonsentients" (Linzey 1998, 311, emphasis in original).[7] Related claims against sentientists include that they should support the removal of large predators that cause suffering to sentient animals in natural environments and that they should oppose the use of hunting to control wildlife populations, even when, because of local extinction of natural predators, those wildlife populations threaten the integrity of their ecosystems.

On the other hand, some philosophers are glad to "own" the term *sentientism* and offer various responses to such objections. Following Singer (1975), who argues that sentience is the only *nonarbitrary* cutoff for morally significant interests (8–9), some sentientists have argued that they can consistently endorse hunting when it is necessary to maintain ecological integrity (see, e.g., Varner 2010). And some have argued that on sentientist grounds we *should* leave natural predators in place, either because human interventions in natural systems have various unintended side effects or

---

7. Linzey there claims to have coined the term in 1980, but Rodman's essay appeared three years earlier.

because, with limited resources available, we can do more to improve animal welfare by focusing on our relationships with domesticated species (see, e.g., Sapontzis 1987, 247, and Singer 2009, 461).

Sentience and the Greatest Happiness Principle

Sentience has also played a key role in the utilitarian tradition in ethics. Although the goal of utilitarianism is sometimes described as producing "the greatest good for the greatest number," *utilitarianism* is best defined as the view that (at least *ultimately*) the right thing to do is to arrange things so as to maximize aggregate happiness.[8] Although this "greatest happiness" theory of morality has had proponents at least since ancient Greece, the label *utilitarianism* took hold during the nineteenth century with the writings of Jeremy Bentham (1748–1832) and John Stuart Mill (1806–1873). Bentham and Mill, along with fellow Englishman Henry Sidgwick (1838–1900), are now known as "the classical utilitarians."

Bentham's ([1780] 1948) frequently cited quote—"the question is not, Can they *reason*? nor, Can they *talk*? but, Can they *suffer*?" (412n, italics in original)—is one of the clearest endorsements of sentientism. In this footnote, he also points to parallels between popular attitudes toward nonhuman animals and discredited views that once justified human slavery.[9] While he doesn't use the term *sentience*, Bentham's reference to "suffering" suggests the "core sense" of the term described above. He is saying, in effect, that utilitarians' commitment to maximizing aggregate happiness commits them to being concerned with the suffering of all nonhuman animals who are capable of suffering (or, presumably, conscious enjoyment) regardless of whether or not they have the various other more cognitive capacities characteristic of adult humans.

Mill similarly refers to "all of sentient creation" in his *Utilitarianism* ([1861] 1957). In a passage emphasizing that in a modern democratic society there is no reason that "an existence exempt as far as possible from

8. The "at least ultimately" qualification is necessary because some utilitarians endorse a "two-level" view that emphasizes that utilitarian reasons can be given for not thinking about maximizing happiness all of the time. The reference to "arrang[ing] things so as to maximize happiness" is necessary for related reasons (see Varner 2012).

9. Boddice (2010) emphasized that Bentham's famous "Can they suffer?" quote is commonly taken out of context and that unlike contemporary animal rights advocates, Bentham endorsed both agricultural and medical uses of animals.

pain, and as rich as possible in enjoyments, both in point of quantity and quality" should not be "secured to all mankind," Mill adds, "and not to them only, but, so far as the nature of things admits, to the whole sentient creation" (chap. 2, ¶ 10). For his part, Sidgwick ([1874] 1907) noted, in his *Methods of Ethics*, that not only Bentham and Mill but "the Utilitarian school generally" are committed to considering not just human happiness but "any pleasure [and, presumably pain] of any sentient being" (414).

In the preceding section we noted that sentientists commonly claim that while the ability to feel physical pain suffices to give an individual moral standing, certain cognitive capacities can add relative moral *significance* to the lives of the individuals who have them. A famous passage in Mill's *Utilitarianism* can be understood as an example of this sort of claim. In chapter 2 Mill notes that "utilitarian writers in general" have recognized "the superiority of mental over bodily pleasures." Mill ([1861] 1957) argues that this is because intellectual pleasures are *qualitatively* superior to bodily pleasures, and he draws the conclusion that "it is better to be a human being dissatisfied than a pig satisfied" (12, 14). This famous quotation suggests that in Mill's view, humans' cognitive capacities give their lives special moral significance.

In Animal Studies, *sentience* is standardly defined in terms of the capacity for conscious suffering and/or enjoyment. So defined, the scope of sentience in the animal kingdom is commonly assumed to match the scope of physical pain, because having that capacity is a sufficient condition for being sentient. At the same time, various sentientists endorse a hierarchical view about the value of animals' lives, arguing that various cognitive capacities enhance individuals' capacities for enjoyment and suffering in ways that give their lives special moral significance. Such a sentientist stance has historical precedents in the writings of utilitarian philosophers, although contemporary environmental ethicists often dismiss sentientism as a basis for an *environmental* ethics.

Suggestions for Further Reading

Allen, Colin, and Michael Trestman. 2016. "Animal Consciousness." *Stanford Encyclopedia of Philosophy*, edited by Edward N. Zalta. Plato.stanford.edu, October 24. https://plato.stanford.edu/archives/win2016/entries/consciousness-animal/.

Braithwaite, Victoria. 2010. *Do Fish Feel Pain?* New York: Oxford University Press.

Gruen, Lori. 2017. "Conscious Animals and the Value of Experience." In *The Oxford*

Handbook of Environmental Ethics, edited by Stephen Gardiner and Allen Thompson, 91–100. New York: Oxford University Press.

Proctor, Helen S., Gemma Carder, and Amelia R. Cornish. "Searching for Animal Sentience: A Systematic Review of the Scientific Literature." *Animals* 3 (3): 882–906, doi:10.3390/ani3030882. https://www.ncbi.nlm.nih.gov/pmc/articles/PMC449 4450/.

Varner, Gary. 2001. "Sentientism." In *A Companion to Environmental Philosophy*, edited by Dale Jamieson, 192–203. Malden, MA: Blackwell.

———. 2012. "Which Animals Are Sentient?" In *Personhood, Ethics, and Animal Cognition: Situating Animals in Hare's Two-Level Utilitarianism*, 105–132. New York: Oxford University Press.

References

Allen, Colin. 2004. "Animal Pain." *Nous* 38: 617–43.

Bentham, Jeremy. (1780) 1948. *Introduction to the Principles of Morals and Legislation*. Edited by Willfrid Harrison. Oxford: Basil Blackwell.

Boddice, Rob. 2010. "The Moral Status of Animals and the Historical Human Cachet." *JAC* 30 (3/4): 457–89.

Braithwaite, Victoria. 2010. *Do Fish Feel Pain?* New York: Oxford University Press.

Güntürkün, Onur, and Thomas Bugnyar. 2016. "Cognition without Cortex." *Trends in Cognitive Science* 20 (4): 291–303.

Kant, Immanuel. (1785) 1948. *The Moral Law: Kant's Groundwork of the Metaphysics of Morals*. Translated by H. J. Paton. New York: Hutchinson's Library.

Linzey, Andrew. 1998. "Sentientism." In *Encyclopedia of Animal Rights and Animal Welfare*, edited by Marc Bekoff. Westport, CT: Greenwood Press.

Mill, John Stuart. (1861) 1957. *Utilitarianism*. Indianapolis: Bobbs-Merrill.

Rodman, John. 1977. "The Liberation of Nature?" *Inquiry* 20: 83–145.

Rose, James. 2002. "The Neurobehavioral Nature of Fishes and the Question of Awareness and Pain." *Reviews in Fisheries Science* 10: 1–38.

Sapontzis, Steve. 1987. *Morals, Reason, and Animals*. Philadelphia: Temple University Press.

Sidgwick, Henry. (1874) 1907. *The Methods of Ethics*. 7th ed. London: Macmillan.

Singer, Peter. 1975. *Animal Liberation: A New Ethics for Our Treatment of Animals*. New York: Harper Collins.

———. 1979. *Practical Ethics*. New York: Cambridge University Press.

———. 1993. *Practical Ethics*. 2nd ed. New York: Cambridge University Press.

———. 2009. "Reply to David Schmidtz." In *Singer under Fire: The Moral Iconoclast Faces His Critics*, edited by J. A. Schaler, 455–62. Chicago: Open Court.

———. 2011. *Practical Ethics*. 3rd ed. New York: Cambridge University Press.

Varner, Gary. 1998. *In Nature's Interests? Interests, Animal Rights, and Environmental Ethics*. New York: Oxford University Press.

———. 2010. "Environmental Ethics, Hunting, and the Place of Animals." In *The Oxford Handbook of Animal Ethics*, edited by Tom L. Beauchamp and R. G. Frey, 855–76. New York: Oxford University Press.

———. 2012. *Personhood, Ethics, and Animal Cognition: Situating Animals in Hare's Two-Level Utilitarianism*. New York: Oxford University Press.

25 SOCIALITY

Cynthia Willett and Malini Suchak

Tara and Borie were working hard at a task that required both of them to cooperate to get rewards. Midway through, Katie elbowed her way in, pushing Tara out of the way. Borie was not really keen on working with Katie and walked away shortly thereafter, leaving Katie by herself at a task that required two individuals. Katie looked around and realized that Tara was still nearby but would take some convincing to return. Katie approached Tara with her hand out. Tara accepted her hand; then the two embraced and enjoyed a brief play tussle before returning to work on the cooperative task. This anecdote sounds like a story a parent might tell a child to illustrate the importance of playing nice with others, but in fact, Tara, Borie, and Katie were all chimpanzees participating in a cooperation study at the Yerkes National Primate Research Center, where one of us (Malini Suchak) made these observations. Interactions such as this one are undeniably reminiscent of our own interactions with each other and challenge the notion that we humans are alone in our complex sociality.

Human exceptionalism and a persistent stream of positivism in the sciences inherited from a legacy of Cartesianism have limited our understanding of animal sociality. As intertwined ideologies, human exceptionalism and Cartesian positivism raise up a narrow range of cognitive capacities—centered on abstract thinking—as the pinnacle of evolution. The reliance on this narrow focus of achievement reinforces "prejudicial barriers" against the often complicated and poorly understood social dynamics of nonhuman animals and perhaps the human species as well (Pribac 2016).

Phenomenologists such as Maurice Merleau-Ponty have shown that an embodied mode of engaging the world is missed by the Cartesianism that pictures a disembodied mind using the body as a mechanical tool to gain sensory information for mental representations. Phenomenology reminds us that representational forms of thinking are derivative artifacts of this ordinary way of being in the world. Rather than rely on propositional knowing that something is the case, they draw on an embodied capacity for knowing how to respond in a particular situation. This embodied model better accounts for ordinary human interactions while allowing for considerable continuity between human and nonhuman animals. The phenomenological approach weakens that Cartesianism in science that undergirds human exceptionalism by assuming prejudicially an ontological gap between human and nonhuman animals based on capacities for abstract thought. Abstract thought, assumed to separate human minds from animal bodies, may be derivative and not essential for ordinary forms of interaction.

Our approach takes a step further toward understanding species variations in sociality by shifting from the phenomenology of perception to psychological studies of emotions and affect (see chaps. 8, 9). In effect, this means shifting the emphasis in targeted helping and other aspects of sociality in animals from perceptual models to models of affect attunement (Stern 1985; Willett 2014). The phenomenological approach offers an important challenge to Cartesianism and a useful model for understanding the perception of objects through embodied experience. But theories of affect attunement have the advantage of developing directly out of a focus on socio-affective experience and intersubjectivity (Willett 1995). Moreover, rich and complex capacities for affect attunement across species strongly support a layered, horizontal model of sociality.

Affect Attunement

Affect theory grew out of research on communication between human caregivers and their preverbal children. Psychologist Daniel Stern, who introduced the term *affect attunement*, presented evidence suggesting that a layered continuity model better characterizes childhood development and maturation than models that posit radical discontinuities between irrational children and rational adults. Capacities for nonverbal, affect-oriented "dialogues" originally occurring between the infant and his or her mother continue to feature significantly as layers of adult communication.

Thus, it is not surprising that affect theory can be used to understand how meaningful encounters occur within and across nonhuman animals species (Willett 2014).

Affect attunement appears as a holistic, intuitive form of communication in contrast with highly intentional, abstract, and analytic modes of expression. However, like the latter, affect attunement can be precisely described if not conceptualized, and it allows for correction. Stern categorizes these prereflective feelings as either hedonic or vitality affects. The former register degrees of pain or pleasure, and the latter trace the sheer energy associated with degrees of arousal. Examples of the latter are especially salient in the dynamic and rhythmic patterns of music and of corporeal movement such as dance. A movement or musical piece can be fleeting or heavy; it may take the form of a crescendo or a syncopation. When hedonic tones express beliefs and/or consciously intentional desires, they take on the more complex layers of emotions such as shame and anger or love and pride. Animals express layers of affects and even complex emotions in nonverbal interactions. For example, shame can appear for a human or chimp as an avoidance of eye contact.

The communication of hedonic and vitality affects can operate across diverse sensory modalities, as for when one creature responds with sound to another creature that has signaled an affect through movement or gesture, composing multimodal styles of communication. *Multimodal* refers to any two or more modalities (vocal or other sources of sound, visual through facial expression, gestural, olfactory, etc.) that are used at the same time. This capacity for *cross modal correspondence*, to use Stern's term, signals mutual understanding and a recognition of intersubjectivity in a way that sheer imitation does not. Given this intersubjective element, a response may take on a dissonant rather than a harmonious quality, expressing disagreement (irritation or anger) or a noncomplementary affect (e.g., a sense of fear in contrast with the other's sense of calm). The aim may be to alter rather than to share the mind-set or perspective of the other.

Affect attunement as originally proposed articulates a preverbal social bond between infant and caregiver based on a nonegoistic, preconscious immersion in the rhythms and tones of life. However, it seems to underlie communication in nonverbal interactions for nonhumans as well. Consider the remarkable report of a cat in a nursing home who curls up around dying elderly patients and remains steadily by their side as they, unbeknown to medical personnel, pass away (de Waal 2009, 73). Or of a sacred moment of silent wonder by a stream for baboons marching in synchronized procession across their homeland in East Africa (Smuts 2001). Or of the sense of reciprocity and of fairness that spontaneously emerges among wolves and other carnivores at play (Bekoff 2007, 85–110). Or grieving rituals for the dead performed by corvids (a family of birds that includes magpies, crows, and ravens: Bekoff 2007, 1–2) and mammals (Bradshaw 2009). To top it all off, a sense of humor has been observed in a number of corvids and mammals (Bekoff 2007, 59–60). Such anecdotes are far from rare. Perhaps nonhuman animals fail to have a capacity for abstract reasoning, but our furry and feathery kin nonetheless demonstrate rich elements of intersubjective response, and they can extend their social encounters to creatures beyond their own species.

Interspecies Communication

There are an unknown range of modes of vitality and hedonic attunements operating within and across species. The musical metaphor of attunement with its emphasis on sound and movement may not seem suited to capture, say, the olfactory sense important for many species, yet the attunement trope does serve to remind us of the importance of nonrepresentational sources of communication even for our own species. Steven Mithen, author of *The Singing Neanderthals* (2006), provides evidence from ethology and evolutionary theory for the importance of sound in affect-based communication. He cites Bruce Richman's study of "a profusion of rhythmic and melodic forms" in the movements and vocalizations among gelada monkeys: "As they approach one another, walk past one another or take leave of one another, as they start or stop social grooming, as they threaten someone because he is too close to a partner, solicit someone's support or reassurance, in fact as they do the infinite variety of different social actions that make up the minute-to-minute substance of their social lives they always accompany these actions with vocalizing" (Mithen 2006, 110, quoting Richman 1987, 199). Unlike other baboon species, these "humming baboons" eat grass, which means (like cows) they are constantly ripping up grass and eating. They keep their hands busy, which is a problem for a primate since the primary method of social bonding is grooming. It has been suggested that their constant vocalizations, mostly humming, are a sort of "vocal grooming." When their hands are busy, they use voice to maintain social bonds. Their vocal grooming also seems to fall into the musical pattern of call and response, as close social partners will hum back and forth to each other (much as close social partners of other primates reciprocally groom each other). Reciprocity as demonstrated through these call and response attunements is central to sociality, but so is competition and predation.

The capacity for communication across species orients creatures to others in accordance with biological needs to propagate and survive. For this reason the attunement of predator and prey may be the clearest example of synchronization. Noctuid moths have evolved the capacity to detect bat sonar to escape capture. The development of analogous capacities and corresponding affects may enable a significant range of communication-based interactions overlooked by modern humans. Until recently, humans may have possessed profound familiarity with other animals. Anthropologist and psychologist Barbara Smuts (Smuts 2001) speculates that "Paleolithic hunters learned about the giant bear the same way the bear learned about them: through the intense concentration and fully aroused senses of a wild animal whose life hangs in the balance. Our ancestors' survival depended on exquisite sensitivity to the subtle movements and nuanced communication of predators, prey, competitors, and all the animals whose keener sense of vision, smell, or hearing enhanced human apprehension of the world" (294). This enhanced apprehension has enabled Smuts, after extensive fieldwork, to learn how to live with baboon communities. Smuts explains that while she did not "literally [move] like a baboon—my very different morphology prevented that . . .—I was responding to the cues that baboons use to indicate their emotions, motivation, and intentions to one another, and I was gradually learning to send such signals back to them" (295). The exchange of gestures between baboons and their human guest communicated a range of meanings from polite acknowledgment to the need for privacy and respect, lending evidence for our human capacity to engage other species not as objects or dependents but as competitors and companions.

Reciprocity

As mentioned above, wolves at play may provide evolutionary insight into the origins of reciprocity and even a sense of fairness in prosocial behavior. Marc Bekoff writes,

> In my long-term research projects studying the details of social play in domestic dogs and wild canids, I learned that although play is fun, it is also serious business. When canids—members of the dog family—and other animals

play, they use actions such as vigorous biting, mounting, and body slamming that could be easily misinterpreted. So it is important for them to communicate clearly what they want and expect. Animals at play are constantly working to understand and follow the rules and to communicate their intentions to play fairly. They fine-tune their behavior on the run, carefully monitoring the behavior of their play partners and paying close attention to infractions of the agreed-upon rules. Four basic aspects of fair play in animals are: ask first, be honest, follow the rules, and admit you're wrong. (Beckoff 2015, viii–ix)

Interestingly, play teaches various species rules of companionship and belonging by temporarily suspending vertical hierarchies endemic to their species. Higher status, more powerful animals learn to use soft bites and self-handicap when play fighting. When animals violate the rules, they may be excluded from the group. The leveling of hierarchies in play behavior resonates with a horizontal model of sociality. Play may well provide for otherwise hierarchal interaction an egalitarian model of exchange necessary for some forms of cooperation. Moreover, play is found in a surprisingly large range of animal species, including birds and fish, and hence may account for aspects of sociality in nonhumans that are more sophisticated than often assumed. Note that understanding the rules of fair play does not seem to require capacities for abstract, representational thought.

The simplest form of reciprocity is called *mutualism* and is found in animals who may lack a sophisticated sense of play as described above. However, even these animals may exhibit more complex forms of sociality than generally assumed. Mutualism, or "you scratch my back, and I'll scratch yours," is an exchange of benefits that displays cooperative, egoistic behavior. It is evolutionarily simple to explain in that each of the parties experiences immediate benefits. Mutualism has been attributed to monkeys such as baboons, who are viewed as lacking a narrative sense of time. Unlike humans, baboons are said to live with a spatialized sense of the here and now, and as a consequence, they exchange favors only for immediate payback (Barrett, Henzi, and Rendall 2007). In contrast, humans are known to engage in more complicated forms of cooperation based on a time lag between acts of exchange, displaying a more generalized sense of reciprocity. This may be due to our ability to keep records and use language to document interactions rather than an advanced sociality that is uniquely human (Basu et al. 2009). Mutualism may be evolutionarily simple (in that everyone benefits) but more complex on the proximate level, where emotions, decision making, relationships, and learning occur. Further, complicated features of baboon interaction, including playful encounters, long-term friendships and competitive rivalries, and even a sense of wonder suggest a more complicated social life than one implied by a narrow concept of mutualism. Barbara Smuts's observations of baboons' nuanced use of social and personal space to indicate changes in status and degrees of intimacy and competition demonstrate that even these monkeys may experience nuanced social emotions without drawing on narrative-style representations. Clearly, a traditional approach, examining relationships as a series of emotionless exchanges, is not sufficient to understand the complexity of other-than-human social interactions.

Multilayered Sociality

Our layered, horizontal approach to sociality acknowledges the considerable variation within and across species. The spectrum includes more solitary animals—such as highly intelligent owls and bears—at one end and more group-oriented creatures—such as some of the great apes, ultrasocial insects, and some corvids—at other end. Following new research critical of the social intelligence hypothesis and simple measurements of social complexity, this approach does not assume that cognitive or technological intelligence lines up with social intelligence (Holekamp 2007) or that social complexity depends on group size (Bergman and Beehner 2015). In this spectrum, relationality appears as multiply layered (Stern 1985) and messy (Hinde 1976) rather than as culminating in human cognitive capacities or prejudicially reasserting an ontological gap between the human and nonhuman animal. This spectrum is obscured by a narrowly construed scientific methodology where the legacy of Cartesian positivism discourages holistic science. Holistic approaches are necessary for correcting overly mechanistic models that often include limited notions of what counts as evidence and prejudicial assumptions about the value of abstract reasoning and representational thought for sociality.

We develop our approach through key examples that sketch out significant markers for variations of sociality. Initial research suggests that within species, gender features as an important factor for various species. Socially attuned, empathy-related behaviors such as consolation appear to have a female bias, at least in some species. For example, female bystanders show more consolation behavior than male bystanders in chimpanzees (Romero, Castellanos, and de Waal 2010), and human females score higher than males on self-reported empathy scores (Eisenberg and Lennon 1983). This pattern is evident in children as young as one to two years of age, where girls show more prosocial behavior than boys (Zahn-Waxler et al. 1992). (However, note that others, such as Roth-Hanania, Davidov, and Zahn-Waxler 2011, have not found gender differences.) Findings about the ways social expectations affect children early on make it impossible to separate nature and nurture.

Some suggest that the root of gender differences comes from a common origin in parental concern and the honing of affect attunement skills found in maternal care. But the multilayered model of socio-affective attunement allows us to capture subtle features of caregiving that manifest an organized subjective presence of self and self-other distinction on the part of the infant. This attunement goes both directions. It also manifests in the mother's prereflective and yet socially skilled and culturally inflected response to her offspring. Neither mother nor child should be viewed as experiencing a presubjective fusion with the other. The infant may look to the mother's emotional reactions to assess dangers (social referencing), and the mother may prereflectively respond to emotions elicited by the infant's cries to increase caregiving. Thus, in most mammalian species, behaviors at the often poorly understood core of socio-affective response may have a gendered basis due to the disproportionate amount of parental care provided by females across species. A different pattern might be expected in most birds, who exhibit biparental care, or species who exhibit alloparenting behavior. However, as empathy research is extremely mammal-centric, there are not yet the data to test out such hypotheses.

Bonding between individuals is a basic form of prosocial interaction that is widespread across species. Although solitary species may not develop close bonds with peer conspecifics, a key feature of mammalian development is a high level of maternal care, which is true across all mammals, even the less social species. Therefore, all mammals form mother-offspring bonds, often for a period of several years. Bears are often considered the least social of carnivores. However, cubs go through an extended period of development, staying with the mother for up to three years (MacDonald 2006). This period is important both for the bears to gain enough body mass to survive on their own as well as for the bears to learn how to hunt by going with the mother when she hunts. Mothers are known for fiercely protecting their young, as adult males often engage in infanticide. Polar bear mothers even have a specific "chuffing" call that serves to promote contact between mother and offspring (Wemmer, von Ebers, and Scow 1976). Thus, this small social unit serves an important function, and altogether bears can spend much of their lives with other individuals.

Of course, for many species, social interactions extend to peers or individuals of the same age group, and these relationships are frequently reinforced through affiliative behaviors. Social animals are known for the wide degree of affiliative behaviors ranging from grooming in nonhuman primates to trunk to mouth interactions in elephants. Less widely acknowledged and understood are bonding behaviors that occur among animals typically viewed as solitary. Our house cats were originally domesticated from the solitary Near Eastern wildcats about eight thousand years ago (Driscoll et al. 2009). During domestication, house cats seem to have developed an ability to tolerate the presence of other cats if they are socialized with others early in life. Despite this limited evolutionary history of sociality, feral cats will readily form bonds and live in groups of ten to thirty individuals (Crowell-Davis, Curtis, and Knowles 2004). They show a number of different affiliative behaviors, including allorubbing, grooming, and sleeping in close proximity to each other. They also have developed specific communicative signals to indicate friendly intentions, notably, approaching others with their tail up (often with a slight kink at the end). They use this with other cats as well as when approaching humans with whom they wish to affiliate (Bradshaw and Cameron-Beaumont 2000). This signal is so powerful that in experimental studies where cats are presented with a silhouette of a cat approaching with its tail up or down, they will respond in kind.

Sociality is by no means limited to prosocial, affiliative, and cooperative behavior. Group living often entails competition over resources and in many species flexibly balancing the need for cooperative behavior with competition. Hyenas live in complex societies called clans, which range from six to ninety individuals (Holekamp 2007). Females are dominant over males, and they form very strong, linear hierarchies that are not determined by strength or size but rather their social network. They quite readily engage in mutualism when hunting as a group for large prey or defending their prey from other large predators such as lions. When participating in an experimental task requiring cooperation, hyenas well versed in the task will even modify their behavior to help naive individuals. However, they are also characterized by intense competition. Dominant females maintain their position through aggression toward lower-ranking individuals and will displace lower-ranking hyenas around carcasses to take the first share of the prey. Thus, hyena life, like that of so many other social species, requires a delicate balance between competition and cooperation to survive.

A sense of fairness may help various species strike a balance between competition and cooperation by creating social norms to keep competition in check. For example, chimpanzees who are engaging in cooperation will allow for relatively normal expressions of dominance where a dominant individual takes a subordinate's place, often at a resource, but take action against freeloaders (who violate norms of fairness; Suchak et al. 2016). This behavior may help mitigate any selfish tendencies of individuals who might prefer to freeload over contributing to the group effort. Initial work on fairness was done by Sarah Brosnan, who found that capuchin monkeys would go on strike if their partner received a higher quality reward than they did. Since that time, similar responses have been demonstrated in a number of species ranging from dogs to chimpanzees to ravens (Brosnan and de Waal 2003; Range et al. 2009; Wascher and Bugnyar 2013). An important note, however, is that with the exception of chimpanzees, most individuals only show a negative response if they receive less than their partner or they put in more effort than their partner for the same outcome. In contrast, both chimpanzees and humans will show a negative response for getting more, demonstrating, perhaps a more robust, other-regarding sense of justice.

Although the above examples illustrate the many ways in which sociality manifests across species, it is important to note that none of the above examples require that the participants in the social exchange have a sense of what the other knows or perceives. For example, a mother polar bear chuffing to establish contact with her young does not need to imagine what the young must be thinking in order to achieve the desired end—a reunion. Similarly, Kay Holekamp describes a simple rule of thumb that may account for the sophisticated hunting skills of hyenas: "Move whenever you need to in order to keep the selected prey animal between you and another hunter" (Holekamp 2007). These heuristics or rules of thumb drive minute-to-minute decision making in the social context. That many species, including humans in many cases, can achieve such sophisticated ends without abstract capacities for developing and representing ideas illustrates how the narrower view of sociality as requiring abstract cognitive capacities truly limits our abilities to understand sociality across species. Here to some extent we are setting aside the vertical line from affect to cognition.

With this approach we may be also be setting aside other binaries implicated in human exceptionalism. For example, humans are often pointed to as the only species to possess true altruism, with some going so far as to say chimpanzees, our closest relatives, are "indifferent to the welfare of unrelated group members" (Silk, et al. 2005). Scientists have created a hierarchy of cognitive abilities required for true altruism that no nonhuman animal could possibly achieve. In fact, no human animal could possibly design an experiment to adequately test such abilities (see chap. 3). A horizontal approach such as the one outlined here could benefit an exploration of many other areas of sociality as it opens up the possibility of a vast richness of interactions that we are currently ignoring in favor of a priority given to cognitively complex mechanisms.

Our horizontal approach registers messy, layered emotions that are not easily quantifiable and variations within species that are not easily replicated. Socio-affective dimensions are emphasized over abstract cognition that may pertain only to humans but otherwise is only one skill among many and may not be central for understanding sociality in either humans or nonhumans. Affects communicate within but also across species boundaries. Animals synchronize through multiple sensory modalities to forge social fields where cooperation, competition, and predation are at stake. At the most basic layer of these social fields, hedonic tones and vital rhythms express multiple dimensions of meaning. Felt relationships can be pleasant or painful, intense or mild, joyful or outrageous, or boring. At more complex layers, animals may grieve, feel shame, enjoy humor and play, or express wonder. For the most part, we humans will probably never know. Still, the range of value and meaning in fields of affect points toward a wider scope for communication and sociality within and across species.

Suggestions for Further Reading

Barrett, Louise, Peter Henzi, and Drew Rendall. 2007. "Social Brains, Simple Minds: Does Social Complexity Really Require Cognitive Complexity?" *Philosophical Transactions of the Royal Society of London B: Biological Sciences* 362 (1480): 561–75.

Bekoff, Marc. 2007. *The Emotional Lives of Animals.* Novato, CA: New World Library.

Bradshaw, G. A. 2009. *Elephants on the Edge: What Animals Teach Us about Humanity.* New Haven, CT: Yale University Press.

de Waal, Frans. 2006. *Primates and Philosophers: How Morality Evolved.* Princeton, NJ: Princeton University Press.

Smuts, Barbara. 2001. "Encounters with Animal Minds." *Journal of Consciousness Studies* 8: 293–309.

Willett, Cynthia. 2014. *Interspecies Ethics.* New York: Columbia University Press.

References

Barrett, Louise, Peter Henzi, and Drew Rendall. 2007. "Social Brains, Simple Minds: Does Social Complexity Really Require Cognitive Complexity? *Philosophical Transactions of the Royal Society of London B: Biological Sciences* 362 (1480): 561–75.

Basu, S., J. Dickhaut, G. Hecht, K. Towry, G. Waymire. 2009. "Recordkeeping Alters Economic History by Promoting Reciprocity. *Proceedings of the National Academy of Sciences* 106 (4): 1009–14.

Bekoff, M. 2007. *The Emotional Lives of Animals.* Novato, CA: New World Library.

———. 2015. Preface to *Entangled Empathy: An Alternative Ethic for Our Relationships with Animals*, by Lori Gruen. vii–ix. New York: Lantern Books.

Bergman, T. J., and J. C. Beehner. 2015. "Measuring Social Complexity." *Animal Behaviour* 103: 203–9.

Bradshaw, G. A. 2009. *Elephants on the Edge: What Animals Teach Us about Humanity.* New Haven, CT: Yale University Press.

Bradshaw, J., and C. Cameron-Beaumont. 2000. "The Signalling Repertoire of the Domestic Cat and Its Undomesticated Relatives. In *The Domestic Cat: The Biology of its Behaviour*, edited by Dennis C. Turner, 67–94. Cambridge: Cambridge University Press.

Broesch, T. L., T. Callaghan, J. Henrich, C. Murphy, and P. Rochat. 2010. "Cultural Variations in Children's Mirror Self-Recognition." *Journal of Cross-Cultural Psychology* 42: 1018–29.

Brosnan, S. F., and F. B. de Waal. 2003. "Monkeys Reject Unequal Pay. *Nature* 425 (6955): 297–99.

Crowell-Davis, S. L., T. M. Curtis, and R. J. Knowles. 2004. "Social Organization in the Cat: A Modern Understanding." *Journal of Feline Medicine and Surgery* 6 (1): 19–28.

de Waal, F. 2006. *Primates and Philosophers: How Morality Evolved.* Princeton, NJ: Princeton University Press.

———. 2009. *The Age of Empathy.* New York: Random House.

Driscoll, C. A., J. Clutton-Brock, A. C. Kitchener, and S. J. O'Brien. 2009. "The Taming of the Cat." *Scientific American* 300 (6): 68–75.

Eisenberg, N., and R. Lennon. 1983. "Sex Differences in Empathy and Related Capacities." *Psychological Bulletin* 9 (1): 100–131.

Gruen, L. 2015. *Entangled Empathy: An Alternative Ethic for Our Relationships with Animals.* New York: Lantern Books.

Hinde, R. A. 1976. "On Describing Relationships." *Journal of Child Psychology and Psychiatry* 17 (1): 1–19.

Holekamp, K. E. 2007. "Questioning the Social Intelligence Hypothesis." *Trends in Cognitive Sciences* 11 (2): 65–69.

MacDonald, D. 2006. "The Bear Family." In *The Encyclopedia of Mammals*, edited by D. MacDonald, 574–89. Oxford: Oxford University Press.

Mithen, S. 2006. *The Singing Neanderthals: The Origin of Music, Language, Mind, and Body*. Cambridge, MA: Harvard University Press.

Pribac, T. B. 2016. "Someone Not Something: Dismantling the Prejudicial Barrier in Knowing Animals (the Grief That Follows)." *Animal Studies Journal* 5 (2): 52–77.

Range, F., L. Horn, Z. Viranyi, and L. Huber. 2009. "The Absence of Reward Induces Inequity Aversion in Dogs. *Proceedings of the National Academy of Sciences* 106 (1): 340–45.

Richman, B. 1987. "Rhythm and Melody in Gelada Vocal Exchanges." *Primates* 28: 199–223.

Romero, T., M. A. Castellanos, and F. B. de Waal. 2010. "Consolation as a Possible Expression of Sympathetic Concern among Chimpanzees." *Proceedings of the National Academy of Sciences* 107 (27): 12110–15.

Roth-Hanania, R., M. Davidov, and C. Zahn-Waxler. 2011. "Empathy Development from 8 to 16 Months: Early Signs of Concern for Others." *Infant Behavior and Development* 34 (3): 447–58.

Silk, J. B., S. F. Brosnan, J. Vonk, J. Henrich, D. J. Povinelli, A. S. Richardson, S. P. Lambeth, J. Mascaro, and S. J. Schapiro. 2005. "Chimpanzees Are Indifferent to the Welfare of Unrelated Group Members." *Nature* 437 (7063): 1357–59.

Smuts, B. 2001. "Encounters with Animal Minds." *Journal of Consciousness Studies* 8: 293–309.

Stern, D. N. 1985. *The Interpersonal World of the Infant: A View from Psychoanalysis and Developmental Psychology*. Mineola, NY: Basic Books.

Suchak, M., T. M. Eppley, M. W. Campbell, R. A. Feldman, L. F. Quarles, and F. B. M. de Waal. 2016. "How Chimpanzees Cooperate in a Competitive World." *Proceedings of the National Academy of Sciences* 113 (36): 10215–20.

Wascher, C. A., and T. Bugnyar. 2013. "Behavioral Responses to Inequity in Reward Distribution and Working Effort in Crows and Ravens." *PLoS ONE* 8 (2): e56885.

Wemmer, C., M. von Ebers, and K. Scow. 1976. "An Analysis of the Chuffing Vocalization in the Polar Bear (*Ursus maritimus*)." *Journal of Zoology* 180 (3): 425–39.

Willett, C. 1995. *Maternal Ethics and Other Slave Moralities*. New York: Routledge.

———. 2014. *Interspecies Ethics*. New York: Columbia University Press.

Zahn-Waxler, C., M. Radke-Yarrow, E. Wagner, and M. Chapman. 1992. "Development of Concern for Others." *Developmental Psychology* 28 (1): 126–36.

26 SPECIES

Harriet Ritvo

Concern with kinds runs through the literature of the nineteenth century. Sometimes such concern was existential, as when Tennyson famously worried that Nature was not "careful of the type" since "scarped cliff and quarried stone" showed that "a thousand types are gone . . . [and] all shall go"(*In Memoriam*, 1850). But more often it reflected puzzlement (or perhaps more violent emotions) about where particular kinds began and ended or about how to tell one kind from another. Lewis Carroll, for example, obliquely broached many of the underlying scientific and philosophical problems. In chapter 5 of *Alice in Wonderland* (1865) the Pigeon identifies Alice as a serpent, first on anatomical grounds (her long neck) and then on behavioral ones (her confessed predilection for eating eggs). When Alice protests this taxonomic placement, the Pigeon firmly asserts the precedence of function over form: "You're looking for eggs; I know *that* well enough; and what does it matter to me whether you're a little girl or a serpent?" Alice, however, clings to anatomy as the determinant of correct classification, as she demonstrates in the following chapter, when she warns the Duchess' baby, "If you're going to turn into a pig, my dear . . . I'll have nothing more to do with you." Charles Kingsley spun social exclusion a bit differently in *Water Babies* (1863), where a snobbish salmon reluctantly acknowledges the relation of her kind to the lowly trout, whom she critiques for their laziness in failing to migrate, which has caused them to become "ugly and brown and spotted and small." But even this remote and disparaged kinship carries threats that go well beyond that of mere mortification. She is particularly

I am grateful to Alison Laurence for her help with the research for this article and to Janet Browne for her comments on an earlier draft.

horrified by the possibility of a reproductive breach of the species barrier. She learns that "one of them propose[d] to a lady salmon," although she is reassured (perhaps unrealistically) to hear (from a gentleman salmon) that "there are very few ladies of our race who would degrade themselves by listening to such a creature" (chap. 3). In *The Island of Doctor Moreau* (1896), H. G. Wells's protagonist uses surgery to overcome any such reluctance to hybridize on the part of his experimental subjects.

Of course, the producers (and consumers) of literary works were not alone in their fascination with such topics. Although novelists and poets may have presented the issues raised by taxonomy in their most attractive and compelling form, they did not make the most authoritative pronouncements on such controversies. Scientists and naturalists had been worrying about how to define species in the abstract and how to recognize and delimit them in the flesh—and fighting about these issues—for a long time, although the Google Ngram viewer suggests that the phrases "species problem" and "species question" only emerged in a statistically significant way during the nineteenth century.

The reasons that "species" has been a persistent source of trouble are not too difficult to discern. Species is the anchor of all the more or less elaborate taxonomic systems that have been devised to arrange, organize, and explain the diversity and number of kinds of living organisms. Their early variations reflected differences of more or less learned opinion, and these systems have continued to alter in response to the increase of biological knowledge. Thus, the stripped-down set of nested taxonomic categories listed on the title page of Linnaeus's *Systema naturae* (the 1758 edition, which is still considered the starting point for animal taxonomy and nomenclature; the starting point for plants is his *Species plantarum* of 1753)—kingdom, class, order, genus, species—is readily recognizable as the forerunner of the standard list promulgated in twentieth-century biology classrooms—kingdom, phylum, class, order, family, genus, species. But in recent decades its core meaning has been modified by the replacement of the traditional phylogenetic tree by the cladogram, which looks more like a semisupine toothbrush. The relation among categories has been complicated by the proliferation of taxa at all levels, including, but not limited to, clade, suborder, infraorder, parvorder, superfamily, and subfamily. Systematic schemas have proliferated along with categories and subcategories; no consensus exists about a single replacement for the seven-step schema of mid-twentieth-century textbooks.

In all these alternative schemas, however, the species has remained the foundational taxon, and popular appreciation of its significance has been heightened by the emergence of "endangered species" as the focus of concerns about extinction (see chap. 11). While all the higher categories (along with such lower ones as subspecies and variety) are inherently relational—that is, for example, "order" and "family" are not normally defined except with reference to the categories that they contain or are contained by—the species ostensibly refers to a group of organisms that exists independently of any taxonomic system. *Ostensibly*, however, should be emphasized. It is not necessary to go further than Charles Darwin to see the difficulties inherent in such a bald formulation—one that assumes that species are easy, or even possible, to delimit. His theory of evolution by natural selection (like any evolutionary theory) dissolved the chronological boundaries between parent species and their offspring. And he also addressed the porousness of synchronic boundaries. The second chapter of *On the Origin of Species* is devoted to "Variation under Nature," and it includes an extended account of the difficulty of distinguishing between species and varieties (that is to say, of deciding whether a particular form deserves to be ranked as a species). After weighing the criteria for such decisions and citing the divergent opinions of learned experts, Darwin opted for an explanation that privileged social construction over biological analysis:

> Close investigation, in most cases, will bring naturalists to an agreement how to rank doubtful forms. Yet it must be confessed that it is in the best-known countries that we find the greatest number of forms of doubtful value. I have been struck with the fact that, if any plant or animal in a state of nature be highly useful to man, or from any cause attract his attention, varieties of it will almost universally be found recorded. These varieties, moreover, will often be ranked by some authors as species. (Darwin [1859] 1964, 50)

In other words, to quote an old movie, "If you build it, they will come" (*Field of Dreams*, 1989).

A History of Trouble

But these abstract boundary issues are not the concerns that have been most troublesome historically as the people who needed to deploy the species category have attempted to define it. (It is worth noting that anyone who studies particular organisms must assume, at least for the purposes of discussion, that those individual species exist or existed, no matter how nuanced is their understanding of the problems implicit in that assumption.) The *Oxford English Dictionary* (*OED*) locates the first use of the term *species* with reference to animal or plant kinds in the early seventeenth century in Edward Topsell's discussion of crocodiles in *The History of Serpents*, although this clearly overlaps its somewhat earlier usage to signify "kinds" in many other contexts. The *OED* also notes, very prudently, that: "the exact definition of a species, and the criteria by which species are to be distinguished (esp. in relation to genera or varieties), have been the subject of much discussion."

Unlike the *OED*, most dictionaries are less circumspect, defining zoological and botanical species as groups of organisms that can produce fertile offspring, and this is also the definition offered by many naturalists and introductory biology texts even if modified to accommodate species separated by geography or behavior rather than by reproductive physiology or to acknowledge that there are some obvious exceptions—that is, distinct species that produce fertile offspring, such as the wolf and the coyote or the bison and the domestic cow. For example, according to the *Oxford Dictionary of Zoology*, a species is "an interbreeding group of biological organisms that is isolated reproductively from all other organisms. . . . Most species cannot interbreed with others; a few can, but produce infertile offspring; a smaller number may actually produce fertile offspring"(Allaby 2009, 590).

But the more intensely people are inclined to think about these matters, the more complicated they tend to find them. Among biologists and allied academics, the species question or problem is still a very live issue or set of issues. Recent decades have seen the publication of books with titles such as *The Species Problem: Biological Species, Ontology, and the Metaphysics of Biology*; *Do Species Exist? Principles of Taxonomic Classification*; and *Species: A History of an Idea*. And these are just the tip of an iceberg primarily composed of technical articles in specialized journals. While the authors of these books and articles agree on the importance of their topic, they approach it from divergent perspectives and arrive at very distinct conclusions. Some, for example, are optimistic and pragmatic: "Should the day ever come when it is generally agreed that the species problem has indeed been solved—and there are really no theoretical reasons preventing this—the Modern Synthesis will then be complete" (Stamos 2003, 356). Others are neither optimistic nor pragmatic: "The species problem is not solved and it cannot be ignored"(Kunz 2012, xvi), and "The more a species concept is used as an element of order and the more it is suited for application in practical taxonomy, the more it is vulnerable due to lacking theoretical consistency"(xvi, 218). And some are simply exasperated: "We might just value conceptual clarity and stop trying to employ the dead in support of modern views, while not overvaluing modern views at the expense of a strawman of the past"(Wilkins 2009, 234).

Wild or Not

If scientific and philosophical arguments about such matters continue to be strenuous, the definition of species also has consequences for nonspecialists as well as for the organisms it classifies. When they are transposed from the journals and conferences of experts to the less rarefied discourse of the larger culture in which science is embedded, arguments about species begin to overlap with taxonomies based on different criteria. For example, the categories of "wild" and "domesticated" have been taxonomically potent at least since the emergence of modern classification systems in the eighteenth century, and they were socially and economically potent for centuries and millennia before then. Most versions of modern systematic taxonomy have enshrined these categories in the form of nomenclature, emphasizing the value added by domestication with Latinate binomials, so *Bos taurus* is the offspring of the extinct ancestral *Bos primigenius* (i.e., the aurochs), and *Canis familiaris* is the offspring of the extant *Canis lupus* (i.e., the wolf). Two hundred years ago, in the freewheeling early days of systematic zoology, domesticated animal kinds were frequently allotted their own genera, with breeds of dogs or cattle consequently elevated to the level of species or subspecies (see Ritvo 1997, chap.2) Thus, in his *General History of Quadrupeds*, Thomas Bewick identified the mastiff as *Canis Molossus* and the old English hound as *Canis Sagax* (rather than lumping them with the rest of *Canis domesticus*; Bewick 1824, 336, 342).

But, of course, potency does not necessarily produce or require clarity. Although the categories of "wild" and "domesticated" are implicitly opposed, drawing the line between them—or, to put it another way, establishing mutually exclusive definitions—has never been easy. Many animals (and even more plants) have inevitably remained tantalizingly ambiguous or ambivalent. Several factors have contributed to this persistent imprecision. Some are scientific, deriving ultimately from the elusiveness of an abstract definition of *species* (and, even more, of both higher and lower taxa). Others, at least equally influential, reflect cultural notions about categories and relative value. As a result, the increasingly sophisticated analytic tools of modern biological science have not always made things much clearer, especially since additional information does not inevitably indicate how that information is to be prioritized (i.e., which characteristics are the most significant ones with regard to classification.)

In particular, although domesticated animal kinds are routinely denominated as species separate from their wild ancestors, the theory behind this widespread practice has been elusive. The guidance offered on this point by the website of the International Commission on Zoological Nomenclature, which, by its own declaration, "acts as adviser and arbiter for the zoological community by generating and disseminating information on the correct use of the scientific names of animals," is not particularly firm:

> Wild vs. domestic animal names. The majority of domestic animals and their wild ancestors share the same name but in a few cases the two forms were named separately, which has created confusion. It was proposed that the first available specific name based on a wild population be adopted. Therefore, despite the fact that these names post-dated or were contemporary with those based on domestic derivatives, the Commission recently conserved, as valid, the usage of 17 species names based on wild species.[1]

Nevertheless, taxonomists continue to stress the importance of maintaining separate binomials not only for reasons of intellectual clarity but also because in many cases both the lived experience and the legal status of the two forms are very different (Gentry, Clutton-Brock, and Groves 2004, 645–51). Such decisiveness prescribes a clear course of action, but it leaves the underlying question unanswered.

Or perhaps its implied answer is based on surprising grounds. For example, in a recent article, three distinguished taxonomists argued that "since wild species and their derivatives are recognizable entities, it is desirable to separate them nomenclaturally when distinct names exist." In this formulation the key term—*recognizable*—refers to judgments that interested laypersons can make as confidently (or as provisionally) as can specialists. The "four main characteristics" of domesticated animals that they specify allow plenty of room for individual judgment—or for argument: breeding controlled by humans, provision of a useful product or service, tameness, selection away from the wild type (Gentry, Clutton-Brock, and Groves 2004, 649, 645). (It is worth noting that they do not mention the conventional, if problematic, criterion: the ability or inability for

---

1. http://iczn.org/content/biodiversity-studies.

crosses to produce fertile hybrid offspring.) One of the commonest kinds of pets thus provides an example of the definitional difficulties that may remain (or emerge). Most people would not think twice before characterizing house cats as domesticated, but the authors of an article adding 5,000 years to their historical association with humans (based on both DNA and archaeological evidence) hedge their bets. They answer the question, "Are today's cats truly domesticated?" with notable restraint: "Although they satisfy the criterion of tolerating people, most domestic cats are feral and do not rely on people to feed them or to find them mates. . . . The average domestic cat largely retains the wild body plan" (Driscoll et al. 2009, 75).

These quotations have been taken from articles published in scientific journals, and so their authors do not commit themselves with regard to whether this ambiguous status is a good thing or a bad thing. Such ostensible objectivity has not, however, characterized everyone with an interest in whether a particular animal or group of animals belongs to a domesticated species or a wild one. Over time, while the desire to distinguish between wild forms and their domesticated relatives has remained constant, the valence of this distinction has shifted significantly. The eighteenth-century practice of labeling breeds as species simultaneously celebrated and reified the power of domestication; it might also enhance the cash value of breeds whose unique qualities were deemed to merit such recognition. But an alternative to the traditional preference for domestication was already emerging; along with the first blush of the romantic movement, wildness gained in cachet, at least from some privileged perspectives. Thus, the aristocratic proprietors of a few herds of unruly white cattle in the north of England and Scotland allowed them the run of their large estates and fantasized that they were surviving remnants of the aboriginal aurochs (see Ritvo 1992). Similar dreams have inspired twentieth-century attempts, sometimes inflected politically and sometimes inflected environmentally, to resurrect the aurochs by breeding back from existing cattle perceived to be relatively primitive (see Lorimer and Driessen 2016).

Hybrid Issues

Analogous fantasies have subsequently become available to more modest (and less institutional) proprietors involving acknowledged cross species hybridization rather than wishful conflation of species. Received wisdom to the contrary notwithstanding, it is not unusual for fertile individual hybrids (and even hybrid populations) to result from mating across mammalian species and even genera; the deer, cattle, and dog families offer ready examples, as does the cat family. So ailurophiles wishing to inject some excitement into their home menageries can purchase cats who result from crosses between the domestic cat and various similarly sized wild relatives. Such animals are expensive even in comparison to other pedigreed cats, but they cost much less to acquire and to maintain than pedigreed cattle, whether ostensibly wild or otherwise. According to the website of the International Bengal Cat Society,[2] the breed is "a medium to large domestic feline that originates from crossings of the small Asian Leopard Cat (ALC) to the domestic cat in an attempt to create a companion with an 'exotic' look but a domestic temperament." (Nevertheless, even though the kittens that are offered as pets are at least four generations removed from a fully wild progenitor, prospective owners are warned that "the energetic Bengal is not for people who just want a leopard print cat for decoration.") Other feline hybrids designed to appeal to a similar market include the Savannah (domestic cat and African serval) and the Chausie (domestic cat and Asian jungle cat).

This is not to say that wildness has definitively triumphed in every context—and indeed one explanation for the difficulty of distinguishing between wild and domesticated species is that more or less identical animals can seem very different depending on their circumstances. The modern pit bull is the latest of a series of dogs (predecessors include the bulldog, the German shepherd, and the Doberman pinscher) that have been appreciated initially for their ferocity (or other qualities associated with their wild relatives) and subsequently for an appearance and a temperament that retains some of the cachet of toughness without any of its danger. Thus, a typical apologist locates them firmly within the realm of domestication, declaring that "pit bulls are not the stereotypical devil dog put forth in media myth. They are companion animals who have enhanced the lives of many through their devoted people-loving natures, [and their] positively channeled physical prowess, bravery, and intelligence."[3] Or, as Vicki Hearne, a much less typical apologist, put it with characteristic intensity, "many Americans believe that there is a breed of dog that is irredeemably, magically vicious. That is not the only reason the current era is going to go down in history as one of the most remarkably hysterical and

2. http://www.bengalcat.com/whatis.php.
3. http://www.paw-rescue.org/petbulls.php.

superstitious of all time, but it is a bigger reason than current speculation allows for" (Hearne 1991, 7).

Such dual significance can be conveyed by animals who begin as wild as well as by those that begin as domesticated. Among the principle attractions for visitors to southern Africa are the national parks and private game reserves, where many kinds of large wild animals can be viewed in habitats that appear natural and behaving in ways that also appear natural. But it is also possible to view their conspecifics in situations that give a very a different impression—for example, in roadside paddocks that implicitly present various antelope species as incipient food items for people (livestock rather than game) and in tourist attractions that implicitly present such animals as ostriches or elephants as pets. In a more generalized, less immediate way, most zoo animals also have similar functions—not just made harmless by captivity and enclosure and micromanaged according to the policies or whims of their guardians but available for metaphorical purchase as "adoptees" and as cuddly toys.

In particular, breeding offers a more abstract way to overlay wildness with the trappings of domestication. As the untrammeled reproductive options historically available to both house and barn cats have made them seem a little more wild (or feral), the application of the machinery of pedigree developed for elite domesticated breeds can make even tigers seem a little less so. Studbooks have controlled the mating of zoo animals, especially if they belong to species that have become scarce in the wild, for more than half a century (Olney 2001). The standard justification for this practice is to maintain genetic diversity and to avoid the inbreeding that may otherwise weaken small captive populations. But it has also frequently been used to reify the category of subspecies (i.e., to maintain racial purity). Both agendas mean that zoo animals whose parentage is unknown are precluded from breeding, and zoo animals whose parentage is deemed inappropriate may be precluded from breathing. A notorious episode at the Copenhagen Zoo provided an extreme (or at least spectacular) case of the possible consequences of such policies. A young giraffe named Marius (the very fact that he was named is another indication of his initial status as a notional pet), just past the stage of baby cuteness, was shot, then publicly dissected, then fed to the local lions. In language that resonates at least as much with economics and marketing as with zoology and conservation, he was declared surplus genetically (i.e., there were no suitable partners for him within the network of approved European zoos) and also surplus physically (i.e., he took up a lot of room, accommodation for large zoo animals is limited, and zoo accreditation rules precluded his adoption by institutions outside the network).[4]

The advent of DNA analysis in recent decades has made it both easier to distinguish wild species from domesticated ones and more difficult. For example, the Scottish Wildcat Association was established in 2007 to protect the small remaining British subpopulation of the very widely distributed species ancestral to domestic cats.[5] (Again, the fact that such creatures are considered worthy of protection signals a distinctively modern valuation of wild animals; Victorian gamekeepers hunted down the ancestors of these animals and nailed their skins to barn doors.) The targeted felines strongly resemble domesticated tabbies, although they tend to be larger and more irascible. Perhaps for this reason, the distinction between pure wild animals and those contaminated by miscegenation features prominently on the association's website: "In 2004 a team of scientists . . . estimated that 400 wildcats remained, the other 5000 or so being feral domestic cats or hybrid mixes of domestic and wildcat." It further advocated "improving legal protection, launching a public awareness campaign, supporting the captive breeding program and creating special reserves for wildcats which would in turn benefit many other species." As a result of these efforts, the Scottish wild cat has been declared a "priority species" (at least in Scotland). It has therefore become eligible to benefit from the establishment of a studbook, a captive breeding program, and other measures that blur the cultural boundary between the wild and the domesticated even as they attempt to reinforce the genetic boundary that separates them. The efficacy of these measures has been questionable, however; and the association currently supports an enterprise, unironically entitled "Wildcat Haven," devoted to "complete feral cat removal across a vast landscape."[6]

The case of the American bison is still more puzzling. Having teetered on the brink of extinction in the late nineteenth century, it has become one of the success stories of species preservation. Although their free-ranging populations remain far below their historical maximum (in the tens of thousands compared to estimates as high as fifty million or more[7]) bison

4. This episode received international media attention. For a sample account, see Eriksen and Kennedy (2014).

5. http://www.scottishwildcats.co.uk/conservation.html.

6. https://www.wildcathaven.com/about/actionplan/actionferalcats/.

7. http://www.bronxzoo.com/animals-and-exhibits/animals/mammals/bison.aspx; http://library.sandiegozoo.org/factsheets/bison/bison.htm.

are now sufficiently numerous to be eaten undiluted as buffalo burgers or in hybridized form as beefalo (the name itself indicates hybrid descent from the American bison (*Bison bison*) and the domestic cow (*Bos taurus*)). But the relation of contemporary bison to the noble former inhabitants of the Great Plains is far from straightforward. The animals who end up in fast-food restaurants and grocery stores come from domesticated stock, not from the wild herds that roam places like Yellowstone National Park. But it appears that beneath their reassuring demographic success, even the apparently wild bison populations may be similarly compromised. That is, the impressive herds of bison that wander around preserved and protected landscapes in the American West look and act like wild bison; they seem indistinguishable from the iconic beast who formerly adorned the American nickel. But a recent article in the Sierra Club's magazine pointedly celebrates the 3,700 Yellowstone bison as "free of cattle genes . . . our last wild bison"(Loomis 2013, 28). Despite their reassuring phenotype, most of the current American bison (in public herds as well as in private herds) include substantial genetic contributions from domesticated cattle— that is, among their ancestors lurk members not only of another species but also of another genus (Derr et al. 2012). At least in theory (and if it is assumed that genotype trumps phenotype), this raises substantial questions about exactly what has been saved and why.

Suggestions for Further Reading

Darwin, Charles. (1859) 1964. *On the Origin of Species*. Facsimile of the first edition. Cambridge, MA: Harvard University Press.

Dupré, John. 1993. *The Disorder of Things: Metaphysical Foundations of the Disunity of Science*. Cambridge, MA: Harvard University Press.

Friese, Carrie. 2013. *Cloning Wild Life: Zoos, Captivity, and the Future of Endangered Animals*. New York: New York University Press.

Kunz, Werner. 2012. *Do Species Exist? Principles of Taxonomic Classification*. Weinheim: Wiley-Blackwell.

Ritvo, Harriet. 1997. *The Platypus and the Mermaid, and Other Figments of the Classifying Imagination*. Cambridge, MA: Harvard University Press.

Sites, Jack W., and Jonathon C. Marshall. 2003. "Delimiting Species: A Renaissance Issue in Systematic Biology." *Trends in Ecology and Evolution* 18: 462–69.

Wilkins, John S. 2009. *Species: The History of an Idea*. Berkeley: University of California Press.

Winston, Judith E. 2009. *Describing Species: Practical Taxonomic Procedure for Biologists*. New York: Columbia University Press.

References

Allaby, Michael, ed. 2009. *Oxford Dictionary of Zoology*. Oxford: Oxford University Press.

Bewick, Thomas. 1824. *A General History of Quadrupeds*. Newcastle: T. Bewick and Son.

Darwin, Charles. (1859) 1964. *On the Origin of Species*. Facsimile of the first edition. Cambridge, MA: Harvard University Press.

Derr, James N., P. W. Hedrick, N. D. Halbert, L. Plough, L. K. Dobson, J. King, C. Duncan, D. L. Hunter, N. D. Cohen, and D. Hedgecock. 2012. "Phenotypic Effects of Cattle Mitochondrial DNA in American Bison," *Conservation Biology* 26: 1130–36.

Driscoll, Carlos A., Juliet Clutton-Brock, Andrew C. Kitchener, and Stephen J. O'Brien. 2009. "The Evolution of House Cats." *Scientific American* 300 (6): 68–75.

Eriksen, Lars, and Maev Kennedy. 2014. "Marius the Giraffe Killed at Copenhagen Zoo Despite Worldwide Protests." *Guardian*, February 9. https://www.theguardian.com /world/2014/feb/09/marius-giraffe-killed-copenhagen-zoo-protests.

Gentry, Anthea, Juliet Clutton-Brock, and Colin P. Groves. 2004. "The Naming of Wild Animal Species and Their Domestic Derivatives." *Journal of Archaeological Science* 31: 645–51.

Hearne, Vicki. 1991. *Bandit: Dossier of a Dangerous Dog*. New York: Harper Collins.

Kunz, Werner. 2012. *Do Species Exist? Principles of Taxonomic Classification*. Weinheim: Wiley-Blackwell.

Loomis, Molly. 2013. "Bison and Boundaries." *Sierra*, November/December.

Lorimar, Jamie, and Clemens Driessen. 2016. "From 'Nazi Cows' to Cosmopolitan 'Ecological Engineers': Specifying Rewilding through a History of Heck Cattle." *Annals of the American Association of Geographers* 106: 631–52.

Olney, Peter J. S. 2001. "Studbook." In *Encyclopedia of the World's Zoos*, vol. 3, edited by Catharine E. Bell, 1180. Chicago: Fitzroy Dearborn.

Ritvo, Harriet. 1992. "Race, Breed, and Myths of Origin: Chillingham Cattle as Ancient Britons." *Representations* 39: 1–22.

———. 1997. *The Platypus and the Mermaid*. Cambridge MA: Harvard University Press, 1997.

Stamos, David N. 2003. *The Species Problem: Biological Species, Ontology, and the Metaphysics of Biology*. Lanham, MD: Lexington Books.

Wilkins, John S. 2009. *Species: A History of the Idea*. Berkeley: University of California Press.

27 VEGAN

Annie Potts and Philip Armstrong

The term *vegan* is relatively new. It was coined in 1944 by Donald Watson, who (with Elsie Shrigley) cofounded the Vegan Society in the United Kingdom, via the amalgamation of the first three and last two letters from the word *vegetarian* (Spencer 2000). In the broadest and most commonly understood sense, veganism extends concern for animals beyond a meat-free diet to a purposeful way of life that opposes and avoids the exploitation of animals in any form (including the use of animals for clothing, entertainment, transport, sport, in vivisection, and so on).[1] While the word itself has emerged only recently, what we now understand as veganism has existed for thousands of years and across cultures, although a diet avoiding all animal products has never been the default in any culture at any time (Wright 2015). In Eastern cultures influenced by Buddhism or Hinduism, vegetarianism—including in some cases the most radical and consistent ("vegan-like") forms of vegetarianism—is prevalent and supported socially. In Western societies, where vegetarianism continues to be marginalized, the reasons people give for being vegan tend to fall into one of two categories: for some, adopting veganism is to promote personal health or to adopt a particular lifestyle (including spirituality); for others, veganism is associated with political or ethical concerns regarding animal

Thanks to Lori Gruen for detailed, conscientious editorial help with this chapter.

1. Watson argued at the time that veganism took vegetarianism to its logical conclusion (Puskar-Pasewicz 2010); however, the British Vegetarian Society declined to support veganism or even publicize the vegan view as it was considered "both extreme and antisocial" (Wright 2015, 4). The words *veg*n* and *veg*nism* connote, respectively, both vegetarians and vegans, and vegetarianism in its broadest sense.

suffering and slaughter and/or environmental/habitat destruction caused by the production of "animal products" (Cole and Stewart 2011). Whichever reasons motivate people to reject the consumption of animal products, veg*nism functions as a marker of difference in relation to dominant "meat culture" (Potts 2016). Thus, vegan practice is a form of resistant biopolitics (see chap. 5), an effective way of undermining carnism, the ideology that normalizes and naturalizes systems of animal exploitation and meat consumption in Western culture (Joy 2011).

With the emergence of Animal Studies, the word *vegan* has begun to function not only as the descriptors of a practice—a way of living and being in the world—but also as a critical term. In this regard we can say that *vegan* refers to a particular kind of critical and conceptual disposition, one characterized by an ethical and political commitment to uncover and reject the ideologies that justify and enable the consumption of nonhuman animals. *Vegan* may describe the presence of this disposition in the object(s) of study (persons, practices, social or conceptual systems), it may describe (in the form of self-positioning) the disposition of the researcher or scholar undertaking the study, or it may describe the ethical or political disposition that shapes and motivates the mode of analysis being undertaken. The dispositional challenge represented by the term *vegan* constitutes, among other things, one of the major ethical and political debates of Animal Studies as discipline: the question of whether Animal Studies should itself be vegan is inevitably generated by the field's foundational ethical and political commitment to addressing the interests of nonhuman animals. The term cannot be separated from practice—from the material decisions that human beings make, in their everyday lives, about their treatment of animals and their support for systems that make animals consumable. To understand the operation of these terms in a critical and conceptual context, then, we must also address their function in practice.

Theorizing Veganism

Why and how do people turn against prevalent beliefs that animals are food?

In a study involving interviews with vegans, social scientist Barbara McDonald (2000) identified a process of transformation associated with learning about and adopting veganism. This generally entailed some awareness of concern for animals before a life-changing event (e.g., visiting a slaughterhouse) that then propelled interviewees into a period of knowledge seeking that culminated in the decision to become vegan. McDonald speculated that for some vegans at least, this process is a manifestation of an inherent affinity for all life. Thus, becoming vegan for McDonald is the expression of a kind of preexisting "orientation" associated with empathy for others that is triggered by a "catalytic experience" or by contact with information or educative resources (6). An orientation in this sense is more deep rooted than a preference. While Animal Studies scholars are in the main critical of this kind of essentialism,[2] Laura Wright (2015) argues that for those who become vegans primarily because of a deeply felt connection to and empathy for other species, explanations of this kind may be very relevant:

> Considering veganism as an orientation allows for an understanding of that minority position as a delicate mixture of something both primal and social, a category—like sexual orientation or left- or right-handedness—that constitutes for some people, just perhaps, something somewhat beyond one's choosing. (Wright 2015, 7–8)

It is certainly the case that some vegans experience their refusal to exploit and consume animals on a profoundly embodied level; over time, living as an ethical vegan may even involve a major shift in sensory and perceptual responses. For example, as Stephanie Jenkins (2012) points out, "the smell of bacon may no longer recall childhood memories" (508) but instead come to represent suffering and slaughter against the pig's will. Some vegans may even reject sexual partners who eat animal products on the grounds of an "ethical embodied sexuality" associated with veganism, claiming that omnivores' bodies (and bodily fluids) are tainted by the slain animals who sustain them (see Potts and Parry 2010, 2014).

The notion of veganism as an orientation nevertheless remains controversial largely because of the implied essentialism at its heart. If "real" or "authentic" veganism is associated with an innate "calling" thought to be the outcome of an inherent greater empathetic "nature," then other contextual factors that limit knowledge of and options to practice veganism may be devalued or dismissed (Wright 2015). On the other hand, the recognition of radical *changes* in embodied reactions—the emergence of

---

2. *Essentialism* refers to the existence of fixed (i.e., ahistorical cross-cultural) characteristics or attributes. In biological essentialism, these attributes are fixed within genetics, physiology, etc.

disgust at the smell of meat in someone who once relished it—means that the study of vegan phenomenology opens up new ways of thinking about the cultural construction of even the most apparently immediate and intimate of experiences, including taste, sexual desire, and even hunger.

Philosophers Lori Gruen and Robert C. Jones (2015) theorize that people come to understand and practice ethical veganism in two distinct ways: one is based on lifestyle where being "a vegan" becomes part of one's identity; and the other views veganism as an aspiration. Identity veganism is associated with a sense of ethical "purity," since people subscribing to this version tend to believe their lifestyle is morally advanced with respect to nonvegan lifestyles. This perception of superiority may be reinforced by the solidarity found in identifying with other vegans. Identity veganism is asserted to different degrees, but the most dedicated proponents are liable to appear to others—nonvegan and vegan—as self-righteous; they may even, as Gruen and Jones (2015) put it, "[act] as the 'vegan police' who promulgate veganism as the universal, one-and-only way to fight systemic violence against animals" (155). Thus, identity veganism adheres to a universal human-animal politics demanding the complete cessation of all human practices in which matters of power and control over animals are evident. This "all or nothing" approach to veganism is idealistic and perhaps impossible, since being human involves killing on some level, however unintentionally: we might think, for example, of the insects squashed underfoot on a routine walk, the insects and small animals who are killed as part of veg*n food processing, and the medicinal and surgical developments that are accomplished using animals.

The second type of veganism Gruen and Jones examine they call aspirational veganism. This sort of veganism is not so much about a lifestyle but rather a practice or process of eliminating or minimizing exploitation of, violence against, and the killing of others. Aspirational veganism is about "*striving* for a moral goal; [it] is something that one works at rather than something one is" (156, emphasis added). Aspirational veganism acknowledges that veganism *is* an ideal that cannot be imposed as a universal moral imperative. Gruen and Jones suggest that aspirational veganism is (following ecofeminist Marti Kheel 2004) "*an invitation* in response to the violence, exploitation, domination, objectification, and commodification that sentient beings endure in modern industrialized food production processes, part of a larger resistance to such harm and destruction" (156, emphasis added). Aspirational veganism thus takes into account "the complexities of being a consumer in late capitalist industrialized countries," and, because it refrains from idealizing veganism, it facilitates a more "grounded form of individual political commitment, fostering a deeper understanding of intersecting oppressions" (169).

It is important to note that the distinction between identity and aspirational veganism is not complete or fixed. Some movement across categories may occur: those vegans who identify strongly with vegan practices and a cruelty-free lifestyle may not necessarily seek connections with other vegans, nor are they necessarily "vegangelical" in their relations with others; similarly, aspirational vegans may be deeply involved in vegan communities, animal activism, and some forms of abolitionist politics (see chap. 1). These distinctions provide a means to analyze and explain some of the ways in which vegans and veganism can be stereotyped and misrepresented, both in everyday culture and in academic and scholarly discourse.

More conceptual fine-tuning of this kind is provided by sociologist Richard Twine (2014b), whose extensive work on veganism shifts the focus from individual motivations and beliefs toward the actual "practices" of vegans. Here, following Shove et al.'s "practice theory" (Shrove, Pantzar, and Watson 2012), Twine proposes that vegan practice is comprised of three main elements: *competency* (skills and know-how), *materiality* (which relates to the body and its connection to objects and technologies), and *meanings* (ideas, norms, symbolism). The study of veganism thus undertakes what Twine calls an "ontological shift" so that these elements become "qualities of a practice in which the single individual participates, not qualities of the individual" (627).

Twine argues that the benefits of researching practice rather than identity reside in the posthumanist focus on people as "carriers" of practices, which allows a more nuanced way of understanding how practices manifest and change or might disappear as connections between elements (i.e., competency, materiality, and meanings) are made, modified, or broken. Twine's (2016) examination of vegan practice brings affect (emotions) and praxis into dialogue; specifically, he scrutinizes the interdependency between practice and relationships.

One of the important insights emerging from Twine's research stems from people's experiences of transitioning from omnivorous to vegan eating. Adapting Sara Ahmed's (2010) notion of "the feminist killjoy" (a figure who challenges the happiness of the normative affective community and social order by bringing to attention oppositional or conflictual belief systems and sets of social practices), Twine discusses the possibilities of "the vegan killjoy." While the feminist killjoy rejects the "happiness" and "security" associated with complying with conventional heterosexual gender performance, the vegan killjoy rejects the happiness and security related to convivial family and other social gatherings (such as family dinners, holidays, etc.). The vegan killjoy is an "affect alien" who struggles against the dominant affective order that is part of hegemonic "meat culture" and in doing so draws attention to the (usually) invisibilized and unmentioned violent origins of the food that marks the social celebration.[3]

Twine (2014b) discusses how "even at the point of being challenged, specific hegemonies and their practitioners try to consolidate and limit damage by making the killjoy herself the object of bad feeling" (625). Thus, vegans become scapegoated as the origin of the discomfort of nonvegans—not the meat, dairy products, eggs, or other animal-derived ingredients themselves. Such projection may also account for why few people take up veganism when exposed to education and other political insights—precisely because the practice of meat rejection carries with it an emotional and personal risk of alienation or exclusion by others.

One way that carnist vernacular tries to regenerate an ethical position and challenge veganism is through the use of repetitive scripts endorsing the notion of "the happy oppressed," aided by tropes such as "happy meat" and "humane slaughter" (626). Such concepts function by relieving the discomfort and guilt of those who position themselves as more caring about the ways that animals destined for the dinner table live and are killed.[4] However, vegans can still kill the joy that might result from having faith in "humane slaughter" and "happy meat." Their refusal to eat free-range foods, animals killed by Mobile Slaughter Units, or locally grown and "harvested" animal products questions the validity of such notions and draws attention to the paradox of "happy violence." Twine speculates that "[to] have happiness reunited with its inherent violence through the killjoy discourse or mere presence of the vegan invites a troubled self-conception for the omnivore" (628–29).[5]

Twine's work provides an excellent example of the way research on

3. See also Carol Adams's ([1990] 2010) theory on the "absent referent."

4. For a critique of so-called happy meat, see Gillespie (2011) and Cole (2011).

5. Note also that James Stanescu has suggested vegans can by "killjoys" to each other, too, particularly when highlighting intersecting forms of oppression and discrimination. For example, the vegan feminist may well challenge the "shared happiness" of an animal activist meeting (that neglects gender issues) or the anthropocentrism of a feminist gathering (see Stanescu 2013).

veganism as a practice requires and produces new "vegan-related" concepts, terminologies, categories, and methodologies that can in turn generate a whole range of productive epistemological tools. It also provides a way of critically distinguishing veganism from dietary change movements, such as "reducitarianism" or "flexitarianism" (see Kateman 2017 and Bittman 2013).

Intersectional Veganism

If veganism, both as a practice and a critical disposition, is examined from an intersectional theoretical perspective,[6] then the ways in which racialized, gendered, socioeconomic, and ableist contexts influence representations of meat and meat consumption (as well as representations of meat refusal and veg*nism) and produce differing incentives and opportunities for veg*n practice and critique become integral to the analysis. For instance, it is much easier for a wealthy white person living in a large city to practice vegan ethics than it is for an Inuit person surviving on animal-derived foods available in their specific geography and climate in the Arctic Circle. This is the point where universalist and contextual approaches to veganism come into conflict. Within scholarship and activism on ethical veganism, there has been heated debate about whether vegan critique and advocacy can be universalized (i.e., applied across all peoples and cultures) or whether contextual factors such as geography, ethnicity, gender, ability, and socioeconomic circumstances should be taken into account when advocating for veganism. In this way the vegan disposition demands and enables the creation of new critical thinking about such essential issues as nutrition, environmental impact, and cultural identity.

These questions have been most rigorously interrogated within ecofeminist theory and practice. Unlike some abolitionist approaches to animal advocacy in which veganism is represented as a moral imperative for those opposing violence (see, e.g., Francione 2008), ecofeminist theory scrutinizes—with the aim of dismantling—multiple interacting modes of oppression of both human and nonhuman animals. These interactions are viewed as vital to any understanding of and movement away from

---

6. Intersectionality here refers to the ways in which various forms of oppression, marginalization, exploitation, and discrimination are linked. The inclusion of speciesism as an intersectional form of oppression (alongside heterosexism, racism, ableism, classism) has not always been welcomed by more anthropocentric approaches to the study of power and oppression.

dominance over others. As Traci Warkentin (2012, 500–501) explains: "universal moral vegetarianism has long been a subject of deep critical debate among environmental/eco/feminist philosophers, particularly in terms of its dislocated, universal application and potential for a Western ethnocentric and androcentric bias."

Veganism has too often attracted criticism for its construction as a project practiced predominantly by white privileged people (Bailey 2007; see chap. 19). African American vegan-feminist activist A. Breeze Harper has, for example, pointed out that this representation of veganism has resulted in the critical omission of vegan experiences of people of color. The Sistah Vegan Project, initiated by Harper, highlights the ways in which veganism manifests and is variously practiced within white versus black communities in the United States; it demonstrates how factors such as socioeconomic class, race, gender, and geography affect the reasons for and ability to practice veganism. Harper (2010) notes, for example, that the uptake of plant-based diets by African American women may be less about animal rights and more likely part of a deliberate move to mitigate racial health inequalities and "decoloniz[e] their bodies from the legacy of racialized colonialism" (157).[7]

Other scholars, too, caution against the use of multicultural arguments against veganism that employ simplistic notions of "tradition" and "cultural difference" as a response to animal cruelty occurring within different cultural contexts (see Kim 2015). Cultures are not static or homogenous, nor are they able to be essentialized. Rather they encompass diverse perspectives, experiences, and practices—some more prominent and influential, others more subversive or alternative. Furthermore, veg*nism has never been exclusively a Western practice anyway (see indigenous scholar Kirsty Dunn's [2017] work on *kaimangatanga*—Māori plant-based food and ethics), nor are all animal advocates practicing vegan consumption. Twine (2014a) suggests that recognizing these issues can facilitate intersectional coalition; this is because "both animal advocates and discriminated communities have a shared interest in opposing a capitalism that instrumentalizes according to constructions of race and species via intertwined legacies of animalization and racialization" (199). Moreover, and importantly, as human exceptionalism is increasingly challenged by new ideas and research emerging from the humanities and the social and natural

7. See also Yarbrough and Thomas, "Women of Color in Critical Animal Studies," special issue, *Journal of Critical Animal Studies* 8, no. 3 (2010).

sciences, considerations of "culture" are more inclusive of other species, too. "Culture" is increasingly understood as a conglomeration of human and more than human interactions socially and corporeally.

Intersections between veganism, gender, and sexuality have also been taken up under the mantle of "queer veganism." Rasmus Simonsen (2012) advocates this veganism as a way of denaturalizing both heteronormativity and anthropocentrism. Much as queerness disturbs the ways in which heteronormativity is naturalized in masculinist cultures, veganism challenges the "preconceived notions of what a 'proper' diet consists of," the "long established traditions and conventions that govern how and what we eat" (52)—and especially the naturalization of carnism (or meat consumption). Rejecting any vision of veganism as "pure" or as some kind of utopian enterprise, Simonsen proposes that veganism eschews identity politics in favor of embracing the fluidity and productivity of practices that subvert both heteronormative and anthronormative ideologies. In this sense, Simonsen disputes the importance of a "norm" of veganism, emphasizing instead the queerness (nonnormativity) of veganism itself, as one of its appeals and objectives. "We should not refer to veganism as a lifestyle. Veganism shares with queer ethics a refusal to carry on, or reproduce, the social order of anthro/heteronormativity. Assimilation is not an option" (65). Importantly, in his connection of veganism and queerness, Simonsen argues that veganism challenges "the foundational character of how we 'act out' our selves," not least in the arena of gender, sexuality, and reproduction (ibid.). This intersectional approach to evaluating veganism as a mode of queerness highlights the ways in which *eating* (like gender) is performative. Simonsen also proposes that vegans are "despised" in mainstream carnist culture precisely because they do not shy away from, but in fact draw attention to, death—in particular, to the deaths of the animals who make "meat" possible. He speculates that it is this "'morbid' and 'stubborn' preoccupation with the death of nonhuman others that renders veganism so markedly queer" (71).

While Simonsen focuses on the intersection of queerness and veganism, ecofeminist Greta Gaard (1997) constructs a model for "queer vegan studies" that potentially reconciles the various perspectives and positions on veganism encountered so far in this essay. Her approach facilitates acceptance of both the view that human and animal rights are of equivalent value *and* the perspective that animal rights are unattainable until human rights, and especially decolonization, have been achieved. Gaard's queer vegan studies approach disrupts binaries such as natural/unnatural in order that those voices that have been more or less marginalized within ecofeminist discourse itself (namely, queer, nonwhite, and differently abled voices) have the space to bring diversity to the field (see also Taylor 2017). Laura Wright (2015) sees this inclusive approach providing "a space for veganism to exist at once as multiple things: as an orientation, as a socially conscious choice, and as a decision based on a politics of health-based racial decolonization" (10).

Should Animal Studies Be Vegan?

There is continuing debate about the place of veg*nism in Animal Studies. Because the field includes multiple disciplinary perspectives, scholars take various approaches to understanding and representing veg*nism. Arguably, the most vociferous objections to veg*nism as a concept and as a practice come from those whose scholarship is heavily invested in evolutionary theory. Claims that veg*nism is unnatural or "maladaptive" rely on the hypothesis that meat consumption played a crucial role in the evolution of the human brain.[8] In contrast, within philosophical discourse, veganism may be seen as a rational endpoint following consideration of the logic (or lack of logic) behind continued meat production and consumption, particularly in industrialized societies (see Steiner 2010).

Warkentin (2012), in an essay titled "Must Every Animal Studies Scholar Be Vegan?," approaches the question from the point of view of vegan-feminist critical theory precisely because of its analysis of links in representations in Western societies (via media, advertising, etc.) between women's bodies and the bodies of animals killed for food or exploited in other ways. She cites the trailblazing work of Carol J. Adams ([1990] 2010, 2003) showing how both women and animals are represented as "analogous, eroticized, consumable objects for a masculinized diet and gaze" and how meat consumption is associated with "violence against women, feminized men and animals" (500). Yet as Warkentin emphasizes, Adams examined these issues within a sociocultural and political framework rather than one that personalized blame on individual consumers. According to Warkentin, "such an opening also enabled other important critiques

---

8. See Bulliet (2005) and Herzog (2010) for useful overviews of these perspectives. For an incisive summary of the alternative view, which debunks the idea that any single diet, meat based or not, can be considered universal and therefore formative among ancestral humans, see Marlene Zuk's *Paleofantasy* (2013).

to emerge" (500). She suggests that contextual veganism will ensure that diverse viewpoints are acknowledged and that the complexity of any given situation can be taken into account, "thereby avoiding counterproductive allegations of hypocrisy based on an all-or-nothing type purity; it requires that each specific situation may require its own unique and partial resolution" (500). For Warkentin, then, it is these debates about the ethics of dietary choices and practices that should inform and provoke the wider Animal Studies community.

Taking a more distinctly pro-vegan position, Stephanie Jenkins (2012) argues for the return of the ethical and political to Animal Studies. She questions why, when the number of animals farmed, killed, and eaten has increased dramatically, vegan practice remains such a provocation to the field and calls for an "affective feminist practice" that positions nonhuman animals as "grievable, vulnerable, and valuable" (505).

> When built upon feminist ethics, vegan practice is not a universal obligation or a fantasy of purity but rather [a] "bodily imperative" to respond to another's suffering and to reject the everyday embodied practices that make certain animate others killable. (Jenkins 2012, 505)

Jenkins also maintains that many contemporary philosophical debates about animals and human-animal power relations are "hypocritical." Contending that "Animal Studies theorists often seek to bracket, postpone, or eradicate questions of ethics" (505), Jenkins cites conference presentations where speakers comment along the lines of "I'm not for animal rights" as contexts within which pressing ethical issues get sidelined. This is especially the case when posthumanist (relativist) Animal Studies scholars confront the ethical views of academics who are also animal activists: for example, following Donna Haraway's multispecies posthumanism, Kathy Rudy argues that it is possible to kill and consume other species without making them "killable" (Rudy 2012); that is, rather than avoiding violence in our relationships with animals, scholars advocate killing "thoughtfully."[9] For Jenkins, however, this kind of Animal Studies constitutes a nonresponse to animal suffering and slaughter. To counter Haraway's "nonresponse" to animal ethics, Jenkins (following Judith Butler) proposes instead that animal lives are "grievable." Based on her study of the so-called

---

9. See Jeffrey Williams's interview with Donna Haraway in *The Conversant* for more detailed explanations regarding the notions of "killability" and "non-killability."

War on Terror, Butler (2009) analyzed the ways in which nation politics and societal norms construct ideas about who is grievable and who is not: "[there are] subjects who are not quite recognizable as subjects, and there are lives that are not quite—or indeed, are never—recognized as lives" (4). Applying this notion to human-animal relations, Jenkins argues that, similarly, some animal lives are viewed as less worthy or valuable than others: specifically, those animals considered "livestock" are not deemed worthy on their own terms; they are valuable only as monetary units within the animal agriculture economy. Because they are "units" (objectified), they are not grievable. Jenkins advocates that we not just ask ourselves "What is a life?" but *also* "How can I prepare myself to be addressed by a life that lives beyond my ability to apprehend it?" Veganism, from this perspective, can be seen as "a practice of expanding the realm of grievable life," or as "a precautionary principle of moral standing in action" (Jenkins 2012, 509).

Perhaps rather than asking whether Animal Studies should be vegan, a more appropriate question to ask, and one in fact posed by Greta Gaard (2012), is *"Has the growth of animal studies been good for animals?"* (520, emphasis added). Any response to this query requires deliberation on the value of Animal Studies not only for endangered or charismatic species—or for beloved companion animals, or for nonhuman primates and other mammals deemed most like humans—but also for those animals systematically objectified and killed for food, those animals whose very being is trivialized through the name we have given them: "livestock."

Theorizing and practicing veganism aims to be "good for animals" by disturbing anthropocentrism insofar as the interests of all nonhuman animals become part of decisions and practices. Tensions arise because of the challenge to try and represent the interest of all species as well as those of humans. This is where intersectional politics are vital to a thorough understanding and practice of veganism. Intersectionality reveals, for example, that in today's capitalist consumerist Western societies, it is much easier to practice veganism if you are white and privileged educationally and financially. The interests of other animals are not so easily recognized, nor are they so easily honored, in other contexts and cultures where financial or bodily survival might be the main task of the day. Practicing vegans in Western cultures also have the privilege to be able to consider and debate other aspects of animal exploitation beyond food consumption, such as whether or not to take medicines that contain animal products or that have been tested on animals, and whether or not to wear clothing made from animal skins or other parts. The intersectional perspective always insists that veganism be understood as a diverse practice and that its goals are not helped by being judgmental of the paths followed by others in the direction of more compassionate living.

Such goals, of course, require thinking and behaving differently. As this chapter has sought to show, understanding veganism as a critical method—as a way of thinking differently—is crucial to the aim of veganism as an ethics, a politics, and a practice. The debates and theories surveyed above all demonstrate the ways in which the vegan critical disposition, as it interacts with other forms of social and cultural critique, generates new vocabularies and new conceptual perspectives to enhance our understanding of human-animal relations.

Ultimately, veganism is about helping to create and maintain a world where *less misery* is made and where *more joy* is possible, not just for humankind but for all species. This seems a very respectable goal for Animal Studies, too.

Suggestions for Further Reading

Castricano, Jodey, and Rasmus Simonsen. 2016. *Critical Perspectives on Veganism*. London: Palgrave.

Cole, Matthew, and Kate Stewart. 2011. "Veganism Contra Speciesism: Beyond Debate." *Brock Review* 12 (1): 144–63.

Harper, A. Breeze, ed. 2010. *Sistah Vegan: Black Female Vegans Speak on Food, Identity, Health, and Society*. New York: Lantern.

Potts, Annie, ed. 2016. *Meat Culture*. Leiden: Brill.

Twine, Richard. 2014. "Ecofeminism and Veganism: Revisiting the Question of Universalism." In *Ecofeminism: Feminist Intersections with Other Animals and the Earth*, edited by Carol J. Adams and Lori Gruen, 191–207. New York: Bloomsbury.

Wright, Laura. 2015. *The Vegan Studies Project: Food, Animals, and Gender in the Age of Terror*. Athens: University of Georgia Press.

References

Adams, Carol J. (1990) 2010. *The Sexual Politics of Meat*. New York: Continuum.

———. 2003. *The Pornography of Meat*. New York: Continuum.

Ahmed, Sara. 2010. "Killing Joy: Feminism and the History of Happiness." *Signs* 35 (3): 571–94.

Bailey, Cathryn. 2007. "We Are What We Eat: Feminist Vegetarianism and the Reproduction of Racial Identity." *Hypatia* 22 (2): 39–60.

Bittman, Mark. 2013. *VB6: Eat Vegan before 6:00*. New York: Random House.

Bulliet, Richard W. 2005. *Hunters, Herders, and Hamburgers: The Past and Future of Human-Animal Relationships*. New York: Columbia University Press.

Butler, Judith. 2009. *Frames of War: When Is Life Grievable?* Brooklyn: Verso.

Cole, Matthew. 2011. "From 'Animal Machines' to 'Happy Meat'? Foucault's Ideas of Disciplinary and Pastoral Power Applied to 'Animal-Centred' Welfare Discourse." *Animals* 1 (1): 83–101.

Cole, Matthew, and Kate Stewart. 2011. "Veganism Contra Speciesism: Beyond Debate." *Brock Review* 12 (1): 144–63.

Dunn, Kirsty. 2017. "Wharekai on Line: Maori Perspectives on Veganism and Dietary Ethics." Paper presented at Australasian Animal Studies Conference, Adelaide.

Francione, Gary L. 2008. *Animals as Persons: Essays on the Abolition of Animal Exploitation*. New York: Columbia University Press.

Gaard, Greta. 1997. "Toward a Queer Ecofeminism." *Hypatia* 12 (1): 137–56.

———. 2012. "Speaking of Animal Bodies." *Hypatia* 27 (3): 520–26.

Gillespie, Kathryn. 2011. "How Happy Is Your Meat? Confronting (Dis)connectedness in the 'Alternative' Meat Industry." *Brock Review* 12 (1): 100–128.

Gruen, Lori, and Robert C. Jones. 2015. "Veganism as an Aspiration." In *The Moral Complexities of Eating Meat*, edited by Ben Bramble and Bob Fischer, 153–71. Oxford: Oxford University Press.

Haraway, Donna. 2009. "Donna Haraway with Jeffrey Williams." Interview by Jeffrey Williams. *Conversant*, July 6. http://theconversant.org/?p=2522.

Harper, A. Breeze, ed. 2010. *Sistah Vegan: Black Female Vegans Speak on Food, Identity, Health, and Society*. New York: Lantern.

Herzog, Hal. 2010. *Some We Love, Some We Hate, Some We Eat: Why It's So Hard to Think Straight about Animals*. New York: Harper Collins.

Jenkins, Stephanie. 2012. "Returning the Ethical and Political to Animal Studies." *Hypatia* 27 (3): 504–9.

Joy, Melanie. 2011. *Why We Love Dogs, Eat Pigs, and Wear Cows: An Introduction to Carnism*. Newburyport, MA: Conari Press.

Kateman, Brian. 2017. *The Reducetarian Solution*. New York: Random House.

Kheel, Marti. 2004. "Vegetarianism and Ecofeminism: Toppling Patriarchy with a Fork." In *Ecofeminism: Women, Animals, Nature*, edited by Greta Gaard, 243–70. Philadelphia: Temple University Press, 2004.

Kim, Claire Jean. 2015. *Dangerous Crossings: Race, Species, and Nature in a Multicultural Age*. Cambridge: Cambridge University Press.

McDonald, Barbara. 2000. "'Once You Know Something, You Can't Not Know It': An Empirical Look at Becoming Vegan." *Society and Animals* 8: 1–23.

Potts, Annie, ed. 2016. *Meat Culture*. Leiden: Brill.

Potts, Annie, and Jovian Parry. 2010. "Vegan Sexuality: Challenging Heteronormative Masculinity through Meat-Free Sex." *Feminism and Psychology* 20 (1): 53–72.

———. 2014. "Too Sexy for Your Meat: Vegan Sexuality and the Intimate Rejection of Carnism." In *Critical Animal Studies*, edited by John Sorenson, 234–50. Toronto: Canadian Scholars' Press.

Puskar-Pasewicz, Margaret, ed. 2010. *Cultural Encyclopedia of Vegetarianism*. Santa Barbara, CA: Greenwood.

Rudy, Kathy. 2012. "Locavores, Feminism, and the Question of Meat." *Journal of American Culture* 35 (1): 26–36.

Shove, E., M. Pantzar, and M. Watson. 2012. *The Dynamics of Social Practice: Everyday Life and How it Changes*. London: Sage.

Simonsen, Rasmus R. 2012. "A Queer Vegan Manifesto." *Journal for Critical Animal Studies* 10 (3): 51–81.

Spencer, Colin. 2000. *Vegetarianism: A History*. New York: Four Walls Eight Windows.

Stanescu, James. 2013 "Vegan Feminist Killjoys (Another Wilful Subject)." *CriticalAnimal* (blog), September 23. http://www.criticalanimal.com/2013/09/vegan-feminist -killjoys-another-willful.html.

Steiner, Gary. 2010. *Anthropocentrism and Its Discontents: The Moral Status of Animals in the History of Western Philosophy*. Pittsburgh: University of Pittsburgh Press.

Taylor, Sunaura. 2017. *Beasts of Burden*. New York: New Press.

Twine, Richard. 2014a. "Ecofeminism and Veganism: Revisiting the Question of Universalism." In *Ecofeminism: Feminist Intersections with Other Animals and the Earth*, edited by Carol J. Adams and Lori Gruen, 191–207. New York: Bloomsbury.

———. 2014b. "Vegan Killjoys at the Table: Contesting Happiness and Negotiating Relationships with Food Practices." *Societies* 4: 623–39.

———. 2016. "Negotiating Social Relationships in the Transition to Vegan Eating Practices." In *Meat Culture*, edited by Annie Potts, 243–63. Leiden: Brill.

Warkentin, Traci. 2012. "Must Every Animal Studies Scholar Be Vegan?" *Hypatia* 27 (3): 499–504.

Wright, Laura. 2015. *The Vegan Studies Project: Food, Animals, and Gender in the Age of Terror*. Athens: University of Georgia Press.

Yarbrough, Anastasia, and Susan Thomas, eds. 2010. "Women of Color in Critical Animal Studies." Special issue, *Journal of Critical Animal Studies* 8 (3).

Zuk, Marlene. 2013. *Paleofantasy: What Evolution Really Tells Us about Sex, Diet, and How We Live*. New York: W. W. Norton.

28 VULNERABILITY

Anat Pick

Santa Cruz Biotechnology, dubbed, "the Walmart of biotech," is one of the largest purveyors of antibodies in the US biotech market. Until 2013, the company held tens of thousands of animals, including goats, rabbits, rats, mice, and birds for the in vivo "harvesting" of antibodies. To produce antibodies, animals are injected with antigens that trigger the immune system. The animals' blood is then drawn, and the antibodies are isolated and sold to researchers.[1] In addition to the large-scale confinement and harming of animals, the operation relies on animals' natural resistance as a source of profit. Animals are thus at once vulnerable and resistant, a fact that shapes the violent mechanism used against them.

Like all living beings, animals are temporal and finite: they are born, they live, and they die. This fact is mundane. In their peculiar relations to human beings, however—as agricultural, medical, symbolic, and emotional fodder—animals are uniquely vulnerable. This, as the example of Santa Cruz Biotechnology shows, is a mundane if highly lucrative fact.

The reality of animals' lives and deaths concerns the duality of vulnerability as the condition of fragility and finitude shared by everything that lives and as susceptibility and exposure to orchestrated violence that affects some lives more than others. In other words, vulnerability is

---

1. Santa Cruz Biotechnology has a checkered history of grave violations of the federal Animal Welfare Act. Currently before the California Court of Appeal is a lawsuit brought by the Animal Legal Defense Fund for violations of the California Cruelty Code. See Cat Ferguson's "Valuable Antibodies at a High Cost," *New Yorker*, February 12, 2014, and "Animal Advocates Appeal Case against Santa Cruz Biotechnology," Animal Legal Defense Fund, June 13, 2013, http://aldf.org/press-room/press-releases/animal-advocates-appeal-case-against-santa-cruz -biotechnology-2/.

universal but unequally distributed. This duality is not easily parsed. The tension inherent in the concept of vulnerability as something shared yet disproportionately endured animates the field of Animal Studies. To speak of the vulnerability of animals is not, then, to resort to convenient jargon. Animal vulnerability asks us to confront the host of phenomena, ideas, and sensations that define our living alongside animals, yet separately from them, in the midst of so great a violence that it has no clear beginning and end.

As universal and shared, vulnerability blurs species distinctions because humans and nonhumans alike are subject to natural law, to injury, and to death. As purposeful and targeted, vulnerability singles out animals as vulnerable outliers at the mercy of mechanisms that serve and stimulate human desires and needs. Industrialized farming and medical research, for example, painstakingly refine specific physical and mental attributes in animals that render them more pliable and profitable commodities. If vulnerability connects humans and animals via our shared corporeality, it also sets animals apart from most humans most of the time in the scale and reach of the infliction of harm. This makes vulnerability an urgent critical site in which questions of cross species ontology, ethics, and, policy intersect.

The aim of this essay is twofold: to define and contextualize vulnerability and to explore the ramifications and tensions of vulnerability as a focal point of proanimal thought. I begin by pointing out the explicit and implicit place of vulnerability in theories of animal liberation and in recent ethical philosophy that does not pertain directly to animals. My argument is, first, that despite their proximity, vulnerability and ethics are, in a sense, incompatible. The two are what Lissa McCullough (2014) calls "correlative oppositions" (15). For not only does the state of being vulnerable *not* call forth any given ethical response, it may incite, indeed entrench, a recoiling from, even retaliation against the weak. Nevertheless, it is precisely this strange twinning, the correlation of the opposites of vulnerability and violence, that illuminates the lived intricacies of our dealings with other creatures. To think of vulnerability as a complicated site allows for a clearer understanding of the dynamics of power that brutalize animals. I will therefore insist throughout on the inextricability of vulnerability and violence, a connection that is reluctantly pondered by theorists and activists because it is too painful, seemingly hopeless, or inordinately dark.

Emerging in the past two decades in fields as diverse as human rights, bioethics, philosophy, and the social sciences, the discourse of vulnerability encompasses two main ideas: the state of exposure to injury and risk, and the understanding that in being exposed, individuals are mutually dependent. My discussion is restricted to those strands of theory and philosophy that have been most influential for Animal Studies, in particular, those areas of inquiry that at the turn of the twenty-first century underwent a so-called ethical turn inspired by the philosophy of Emmanuel Levinas in response to the modern perpetration of mass violence.

Vulnerability and Animal Ethics

It is possible to trace the notion of vulnerability across the field of pro-animal theory. In his *Introduction to the Principles of Morals and Legislation* (1780–1823), Jeremy Bentham, philosopher, social reformer, and the founder of utilitarianism, declared pleasure and pain (not right and wrong) to be the guiding moral principles. In his often-cited passage on animals, Bentham argued that feeling pain and the ability to suffer, not cognitive or discursive abilities, make animals morally significant. Following Bentham, Peter Singer argues in the utilitarian tradition that the morality of an action is determined by the balance of suffering and pleasure produced by the action regardless of the species experiencing the pain or pleasure. But vulnerability is not reducible to the capacity to feel pain. Its remit is far wider.

For Jacques Derrida (2002), Bentham's question marks a significant break in moral thinking: "'Can they suffer?' asks Bentham simply yet so profoundly. . . . The form of this question changes everything" (396). The question, Derrida says, is "disturbed by a certain *passivity*."

> It bears witness, manifesting already, as question, the response that testifies to a sufferance, a passion, a not-being-able. . . . "Can they suffer?" amounts to asking "can they *not be able*?" And what of this inability [*impouvoir*]? What of the vulnerability felt on the basis of this inability? What is this nonpower at the heart of power? (Derrida 2002, 396)

Susceptibility to suffering "changes everything" insofar as it does not provide new answers to old questions. The neural or mental fact of sentience is hardly the point. Vulnerability shifts the terms of the debate; it disrupts the familiar logic that grants animals moral consideration in accordance with certain innate capacities (to speak, to reason, even to suffer). In place of capacity, ability, capability—euphemisms of power—Bentham's question introduces the notion of powerlessness, what Derrida enigmatically calls "nonpower." Vulnerability as nonpower is not the absence of force but its suspension. In the apparent paradox of a power without power lies the true radicalism of an ethics of vulnerability.

Attention to vulnerability is also central to feminist care ethics and ecofeminist thought. Care ethicists and ecofeminists such as Josephine Donovan, Carol J. Adams, Val Plumwood, and Marti Kheel avoid analytical quarrels "over what should constitute the basis of an appropriate ethics for the natural world" (Kheel 1993, 243). In place of general moral principles, these feminists propose a relational approach to ethics as "a natural outgrowth of how one views the self, including one's relation to the rest of the world" (244). Not all animals are vulnerable in the same way all the time, and different cases require different responses, decisions taken from the ground up, as it were.

In the tradition of care ethics, Lori Gruen's approach is rooted not in abstract principles (the calculation of suffering or weighing of interests) but in the specificities of context and the cultivation of "caring perception," which Gruen calls "entangled empathy" (Gruen 2015, 28). Here, the perception of vulnerability is part of one's "moral experience" rather than an ethical rule that applies across the board of different situations. "Empathetic attunement or perception is directed toward the wellbeing of another" (45) in a particular situation and context. Empathy acknowledges the dependence and fragility of others, and by extension, our own.

If entangled empathy is the concrete response to another's vulnerability, how far could, or should, it be stretched? And what of the relation between vulnerability and violence? Might intimately linking vulnerability and violence itself betray a patriarchal worldview that overemphasizes aggression and domination and ignores the realities of empathy and cooperation?

Precarious Creatureliness

"To be a created thing," writes the philosopher and mystic Simone Weil, "is not necessarily to be afflicted, but it is necessarily to be exposed to affliction" (Weil 1998, 66). Creatureliness is an iteration of vulnerability as "a mark of existence" (Weil 2004, 108). Weil's work is devoted to the elucidation of vulnerability as a state of creaturely exposure and a gauge of reality. When someone is utterly vulnerable, they are stripped of the privileges of species and social rank. Although such stripping is often the result of injustice, it invariably reveals the contingency of species identity (as well as the identities of race, gender, and class).

Creaturely vulnerability opens up zones of "indistinction" (Calarco 2015), where species identities blur and where different beings, or creatures, are perceived as corporeal and vulnerable. More radically still, creaturely vulnerability, as I understand it, calls for the *contraction* of humanity rather than its benevolent extension to nonhumans. Creatureliness—the state of being exposed to natural necessity and the ravages of power—does not call for the alleviation of vulnerability via gestures of "humanization" but for more profound forms of "dehumanization."[2] The creaturely, then, is focused on unseating the structures of human exceptionalism (less on the generation of empathy). By imbuing materialism with a sense of reverence for everything that *is*, creatureliness encompasses all life, from animals to plants.

Creatureliness also informs the ways we think about art. In refusing the cliché of art as an expression of "the human condition," that ill-begotten, passive-aggressive idiom desperate to shore up the human as a unique ontology, creaturely vulnerability sets into motion different modes of artistic expression and, crucially, an alternative poetics and critical practice rooted in what I have described as the contraction of humanism and an exploration of affliction (Pick 2011).

No such contraction is found in Judith Butler's influential work on "precariousness," which, despite being anonymous and universal, remains the exclusive marker of human existence. For vulnerability to become such a marker is already to challenge deeply held beliefs on human prowess, and while Butler's "new corporal humanism" (Murphy 2011, 589) is neither forthright nor sure footed, it is expansively, not contractedly, humanistic. Though she does not specifically speak about animals, Butler's telling omission can help to unsettle "the human condition" and draw attention to the mechanisms that brutally subject animal life.

All lives are vulnerable, but the loss of some lives goes unmourned.

---

2. The problem of violence cannot be understood solely as the result of a so-called humanist deficit, and the remedy to violence cannot, therefore, be more humanity. For a more sustained critique of the accepted wisdom that violence depends primarily on processes of "othering" and dehumanization, see my essay "Turning to Animals between Love and Law," *New Formations* 76 (2012): 68–85.

Butler's *Precarious Life* (2004) and its companion piece *Frames of War* (2009) explore the ideological apportioning of value that makes some lives matter and others not:

> The differential allocation of grievability that decides what kind of subject is and must be grieved, and which kind of subject must not, operates to produce and maintain certain exclusionary conceptions of who is normatively human: what counts as a liveable life and a grievable death? (Butler 2004, xiv–xv)

Butler (2009) is careful to avoid appealing to universal humanity as the source of value. "The point will be to ask how such norms [of what or who is human] operate to produce certain subjects as 'recognizable' persons and to make others decidedly more difficult to recognize" (6). Recognition is an act of framing by which those in the frame assume reality as living beings while others are edited out: "the frame tends to function . . . as an editorial embellishment of the image" and "implicitly guides the interpretation" of what we see (8).

Using Butler's approach, it is clear how commonplace violence against animals remains invisible *as* violence. By framing animals as "food," for example, we can go on killing and eating them and not see such killing as violence. The editorializing that removes animals from the epistemic, legal, and emotional frameworks that would make their lives matter ensures that violence continues and animals go ungrieved. Consequently, "those of us who value the lives of other animals," writes James Stanescu,

> live in a strange, parallel world to that of other people. Every day we are reminded of the fact that we care for the existence of beings whom other people manage to ignore, to unsee and unhear as if the only traces of the beings' lives are the parts of their bodies rendered into food: flesh trans- formed into meat. To tear up, or to have trouble functioning, to feel that moment of utter suffocation of being in a hall of death is something rendered completely socially unintelligible. (Stanescu 2012, 568)

Vulnerability, then, does much more than argue for animal rights or the reduction of suffering. It brings another world into view in which animal are not food. As the frame shifts and perception transforms, different moral arguments are possible.

Though Butler (2009) herself stops short of claiming that animal life, too, is grievable (16), her work on precariousness and framing is central to the formulation of an inclusive ethics of vulnerability.[3] Butler's residual anthropocentrism aside, I want to reflect on her difficult pairing of vulnerability and violence.

As a counter to acts of exclusionary framing that make some lives (white, male, human) matter, Butler offers an ethics of precariousness inspired by Emmanuel Levinas. For Levinas, the other person, in its very fragility, calls the self into being. In the sway of responsiveness to someone else, subjectivity forms. In this way, Levinas argues, the other precedes the self, and *ethics*—the primordial encounter with another—precedes *ontology* (my existence as an autonomous subject). Neither Levinas nor Butler envisages the self-other encounter as naturally harmonious. On the contrary, the threat of violence hangs over the encounter with alterity. Vulnerability is central to the encounter with the other, which, though wordless, Levinas subtitles with the words "Thou shalt not kill." To encounter another is to come into being via the threat of violence and the possibility of care. Like talk of "inalienable rights," the prohibition on killing would be unnecessary if at the outset life were not already exposed to violence (and rights were precisely alienable). And so, for Levinas and Butler, vulnerability and violence are copresent.

In *Precarious Life*, Butler (2004) spends some time on this problem. Why should the other's vulnerability "prompt in anyone a lust for violence?" (136). Various explanations come to mind (self-preservation; fear of the other), but Butler concedes that Levinas "presumes that the desire to kill is primary in human beings" (137). In this view, vulnerability functions as a *provocation* and an *invitation*, and the relation between vulnerability and violence is tautological, for where else would violence turn if not toward vulnerability? This, it seems, is something that the political left (and some in the animal movement), forever aghast at the crushing of the weak, has yet to fully grasp. Vulnerability offers violence the path of least resistance. To imagine the flow of violence in the other direction, toward power, is to imagine politics under the conditions of zero gravity.

Animals are the clearest case of the doubling of vulnerability and violence. A nonviolent ethics of vulnerability that recognizes and grieves not

---

3. On Butler's anthropocentrism, see Chloe Taylor's "The Precarious Lives of Animals: Butler, Coetzee, and Animal Ethics" *Philosophy Today* 52 (1: 60–72), and my "Animal Rights Documentaries, Organized Violence, and the Politics of Sight," *Routledge Companion to Cinema and Politics*, edited by Yannis Tzioumakis and Claire Molloy, 91–102 (New York: Routledge, 2016). Stanescu offers a more favorable assessment of Butler's contribution to animal ethics.

just human but *all* life is one that recognizes the threat of violence wherever vulnerable life presents itself.

Horrorism and Violence

The philosopher Adriana Cavarero (2009) coined the neologism *horrorism* to name forms of contemporary political violence that she perceives as new: "a certain model of horror," she writes, "is indispensable for understanding our present" (29).[4] Twentieth- and twenty-first-century political violence targets the most vulnerable, mainly unarmed civilians, whose destruction is not merely physical but ontological: an uprooting of the victims' very humanity.

"If we observe the scene of massacre from the point of view of the helpless victims rather than that of the warriors . . . the picture changes. . . . More than terror, what stands out is horror" (1). While terror is dynamic, prompting a flight for survival, horror is passive and petrifying, like the look of the mythical Medusa. Here, again, vulnerability and violence are strangely entwined: "irremediably open to wounding and caring, the vulnerable one exists totally in the tension generated by this alterative" (30). Wherever there is need for care, there is already the potential for violence. The vulnerable body is the site of this tension.

Vulnerability is an optic that conveys violence from the point of view of the victim:

> Today it is particularly senseless that the meaning of war and its horror . . . should still be entrusted to the perspective of the warrior. . . . The civilian victims, of whom the numbers of dead have soared from the Second World War on, do not share the desire to kill, much less the desire to get killed. (Cavarero 2009, 65)

Although Cavarero does not see animals as particularly vulnerable ("a child is the vulnerable being par excellence" [30]), horrorism aptly captures the

---

4. Cavarero's *Horrorism* and Butler's work on precarity and war can be compared to Steven Pinker's *The Better Angels of Our Nature: A History of Humanity and Violence* (2012). All are attempts to come to terms with forms of contemporary violence in an era that seems, superficially at least, particularly volatile. In a section titled "Animal Rights and the Decline of Cruelty to Animals," Pinker marshals familiar arguments against animal liberation and in favor of vague notions of animal welfare to support his thesis of modernity's gradual reduction of violence (548–72). Although they differ markedly in their assessment of contemporary violence, Cavarero, Butler, and Pinker are also invested in narratives of violence as a gateway to the human condition.

ferocity and unilateralism of violence against animals as well as the ambivalence of killing and care.

Yet to what extent does horrorism reveal something new about contemporary violence, especially once animals are brought into the mix? While horrorism offers a way of dealing not only with mass violence but with responses to it (Stanescu's grocery store scenario is one example), other terms do similar work. Weil's "affliction" describes states of profound vulnerability and violation. Like horrorism, the destruction wreaked in affliction is total: the afflicted person is not only physically damaged but "loses half his soul" (Weil 1998, 41). For Cavarero, terrorism and war fail to properly address the nature of contemporary violence. While in the case of human victims, a military focus risks obscuring the significance of violations incurred by civilians, in the case of animals, whose victimhood is profound but remains largely invisible, the frame of war has the opposite effect of brining violations into view. Derrida and J. M. Coetzee, among others, have described violence against animals as a war. But in *The War against Animals* (2015), Dinesh Wadiwel provides a comprehensive exploration of the ramifications of animal vulnerability, intelligible precisely as war (see chap. 5).

Vulnerability and Resistance

In Elizabeth Bishop's frequently anthologized "The Fish," vulnerability arises in a state of conflict and designates both violence and resistance. This tough-minded, deeply moving poem about a fish caught, observed, then returned to the water pays close attention to bodily detail, its ethical coda a startling response to the facticity, or reality, of the titular fish.

As the speaker pulls out the "tremendous fish" and holds it "beside the boat half out of water, with my hook fast in the corner of his mouth," she takes her time, sadistically we might say, to peruse the physical minutia of the dying animal. There are the "the frightening gills, fresh and crisp with blood," breathing in "the terrible oxygen." Though initially the fish "didn't fight. He hadn't fought at all," later the speaker notices the marks of past struggles:

> I admired his sullen face,
> the mechanism of his jaw,
> and then I saw

that from his lower lip
—if you could call it a lip
grim, wet, and weaponlike,
hung five old pieces of fish-line,
or four and a wire leader
with the swivel still attached,
with all their five big hooks
grown firmly in his mouth.
A green line, frayed at the end
where he broke it, two heavier lines,
and a fine black thread
still crimped from the strain and snap
when it broke and he got away.
Like medals with their ribbons
frayed and wavering,
a five-haired beard of wisdom
trailing from his aching jaw.[5]

Breaking the fishing line must have been excruciating, but the poem makes scant mention of suffering. Instead, what matters is the constellation of forces: the apparatus of fishing, the "battered" body of the fish, and the speaker-fisherwoman. Fish vulnerability and resistance go hand in hand. Although at certain moments the fish is personalized ("his sullen face") the bulk of the poem is descriptive and the situation that of war: the old fish lines are "weaponlike," worn as military decorations. The fish's past triumphs make his capture even more exhilarating: "victory filled up the little rented boat . . . until everything was rainbow, rainbow, rainbow!"

Vulnerability in this case is not a prompt for humane treatment due to fish sentience. Instead, the activity of fishing itself confirms animal resistance. As Wadiwel (2016) points out, "it is precisely because fish resist . . . that recreational fishing becomes a 'sport'; since the supposed pleasure and art of these fishing practices relies upon the capture of an animal who eludes the recreational fisher, and will struggle against the line when hooked" (208).

The time that Bishop's detailed descriptions require (in the real time of fishing and reading) makes the poem a ticking clock. By its end, we, too,

5. https://www.poets.org/poetsorg/poem/fish-2.

are out of breath. The closing line—"And I let the fish go"—directly follows the speaker's celebration of victory in the battle of fishing and is a visceral relief. If it is possible to speak of empathy here, it's of the respiratory kind. The poem's surgical precision is the source of its ethical charge. When imagining the inside of the fish, "shiny entrails, and the pink swim-bladder like a big peony," we are still on the outside, at the level of vulnerable but resilient flesh. The fish is let go. The poem ends. But compassion, too, is a winner's whim.

The most intriguing and important animal artworks are those, like Bishop's poem, that capture something of the collision between human might and animal life, an encounter in which human dominance exercises its prerogative semiautomatically yet at the same time stands to discover, in the midst of power, its own contingency and automation, its own afflicted animality.

J. M. Coetzee's *The Lives of Animals* (1999) is a foundational text on vulnerability and affliction. In it, the elderly writer Elizabeth Costello grapples with the horrors of the reality of animals' lives and deaths. Costello does not lay out a systematic argument about the rights of animals. Instead, she invokes animals' aliveness, pitilessly extinguished on farms and biomedical facilities. Costello does not believe that violence is primarily a philosophical matter, but she seems on shaky ground. Her comparison between factory farming and the Nazi death camps meets with derision. Her belief in poetry's capacity for "sympathetic imagination" (reminiscent of Gruen's entangled empathy) is thought to lack analytical rigor.

Part of what is dramatized in *The Lives of Animals* is the vulnerability of Costello's stance on animal ethics: the somewhat vague appeal to animals' lives over rights, the absence of clear concepts in support of animal protection, and Costello's own inconsistencies (a vegetarian with leather shoes). As a text, *The Lives of Animals* is also vulnerable. Delivered as lectures by the real Coetzee at Princeton University, then published as a novella about lectures given by the fictional Costello at the made up Appleton College, we are never sure whether the views expressed in the piece are Costello's or those of the author.

Interpretive instability, the questioning of authority and mastery, and the anti-Platonic preference for poetic modes of address over philosophical argument is all cleverly metafictional, of course, but it is much more than that. For Cora Diamond, writing in *Philosophy and Animal Life* (2008), the novella is concerned with the state of "a profound disturbance of soul"

(54). This is the real theme of the piece, and it takes precedence over the question of the moral status of animals and how we should treat them.

Costello is a wounded creature. "What wounds this woman, what haunts her mind, is what we do to animals. This, in all its horror, is there, in our world. How is it possible to live in the face of it? And in the face of the fact that, for nearly everyone, it is as nothing, as the mere accepted background of life?" (Diamond 2008, 47). This woundedness is also the weakness of Coetzee's text. But the privation is really the heart of the work and a demonstration of the creative force of vulnerability.

In arguing from a place of vulnerability, and in making a vulnerable argument, *The Lives of Animals* avoids what Diamond calls "deflection": philosophy's fortifications against potentially unbearable realities. Deflection is "what happens when we are moved from the appreciation, or attempt at appreciation, of a difficulty of reality to a philosophical or moral problem apparently in the vicinity" (57). In the translation from concrete difficulty to abstract problem, thought does not merely shield us from pain but contributes to the sense of impregnability that has a hand in the creation of difficulty in the first place. Formulating a strong argument for a general case is satisfying (and elegant). But abstractions can harbor deflection. And where deflection lurks, we are also likely to find injustice.

For Diamond (as for Gruen, Butler, and Cavarero) the work of philosophy and the work of critique must connect to the lived experience and embodied knowledge that guard against deflection:

> The awareness we each have of being a living body, being "alive to the world," carries with it exposure to the bodily sense of vulnerability to death, sheer animal vulnerability, the vulnerability we share with them. This vulnerability is capable of panicking us. To be able to acknowledge it at all, let alone as shared, is wounding; but acknowledging it as shared with other animals, in the presence of what we do to them, is capable not only of panicking one but also of isolating one. . . . Is there any difficulty in seeing why we should not prefer to return to moral debate, in which the livingness and death of animals enters as facts that we treat as relevant in this or that way, not as presences that may unseat reason? (Diamond 2008, 74)

The Lives of Animals illustrates the meaning of vulnerability for thought as it struggles to take in reality. Coetzee's text is the incarnation of vulnerability as critical practice.

Beyond Vulnerability

I have argued that in virtually all of their dealings with humans, animals are extraordinarily (yet mundanely) vulnerable. Animals are on the receiving end of violence exercised with the fewest moral, legal, and technical restrictions. As such, animals are outliers in the operation of power and its most commonplace case. The few controls placed on inflicting "needless suffering" on animals are little more than window dressing, and this, too, in order to render animal use even more total and secure.

There is a risk of becoming too attached to vulnerability as the sole prism for viewing animal life, reducing animals to the status of victims, adopting a paternalistic attitude with regard to their protection and welfare, and failing to offer workable alternatives to our current treatment of animals. To avoid these pitfalls, I have suggested that vulnerability be seen in the wider context of critiques of power and in relation to the perpetration of violence. Vulnerability is not the absence of power but the product of relations of power. To speak of an ethics of vulnerability is to apprehend the ubiquity of power *and* imagine its suspension (as, in different ways, do Levinas, Derrida, and Weil).

But most importantly, perhaps, vulnerability is the tug of reality, an attunement to "the difficulty of staying turned . . . toward flesh and blood" (Diamond 2008, 77). Acknowledging vulnerability in its three manifestations—a living body, a biopolitical resource, and a critical practice—chips away at established ways of thinking and acting that continue to deny animals their place in the sun.

Suggestions for Further Reading

Butler, Judith. 2004. *Precarious Life: The Power of Mourning and Violence*. London: Verso.

Coetzee, J. M. 1999. *The Lives of Animals*. Edited by Amy Gutmann. Princeton, NJ: Princeton University Press.

Diamond, Cora. 2008. "The Difficulty of Reality and the Difficulty of Philosophy." In *Philosophy and Animal Life*. 43–89. New York: Columbia University Press.

Pick, Anat. 2011. *Creaturely Poetics: Animality and Vulnerability in Literature and Film*. New York: Columbia University Press.

References

Bentham, Jeremy. 1907. *An Introduction to the Principles of Morals and Legislation.* Oxford: Clarendon.

Butler, Judith. 2004. *Precarious Life: The Power of Mourning and Violence.* London: Verso.

———. 2009. *Frames of War: When Is Life Grievable?* London: Verso.

Calarco, Matthew. 2015. *Thinking Through Animals.* Stanford, CA: Stanford University Press.

Cavarero, Adriana. 2009. *Horrorism: Naming Contemporary Violence.* New York: Columbia University Press.

Coetzee, J. M. 1999. *The Lives of Animals.* Edited by Amy Gutmann. Princeton, NJ: Princeton University Press.

Derrida, Jacques. 2002. "The Animal That Therefore I Am (More to Follow)." Translated by David Wills. *Critical Inquiry* 28 (2): 369–418.

Diamond, Cora. 2008. "The Difficulty of Reality and the Difficulty of Philosophy." In *Philosophy and Animal Life*, 43–89. New York: Columbia University Press.

Gruen, Lori. 2015. *Entangled Empathy: An Alternative Ethic for Our Relationships with Animals.* New York: Lantern.

Kheel, Marti. 1993. "From Heroic to Holistic Ethics: The Ecofeminist Challenge." in *Ecofeminism: Women, Animals, Nature*, edited by Greta Gaard, 243–71. Philadelphia: Temple University Press.

McCullough, Lissa. 2014. *The Religious Philosophy of Simone Weil.* London: I. B. Tauris.

Murphy, Ann V. 2011. "Corporeal Vulnerability and the New Humanism" *Hypatia* 26 (3): 575–90.

Pick, Anat. 2011. *Creaturely Poetics: Animality and Vulnerability in Literature and Film.* New York: Columbia University Press.

Stanescu, James. 2012. "Species Trouble: Judith Butler, Mourning, and the Precarious Lives of Animals." *Hypatia* 27 (3): 567–82.

Wadiwel, Dinesh Joseph. 2015. *The War against Animals.* Amsterdam: Rodopi.

———. 2016. "Do Fish Resist?" *Cultural Studies Review* 22 (1): 196–242.

Weil, Simone. 1998. "The Love of God and Affliction." In *Simone Weil*, edited by Eric O. Springsted, 41–71. New York: Orbis Books.

———. 2004. *Gravity and Grace.* Translated by Emma Crawford and Mario von der Ruhr. London: Routledge.

29 WELFARE

Clare Palmer and Peter Sandøe

The basic idea underlying the term *welfare* is that things can go better or worse for the individual concerned; what is done or happens can be good or bad for them. Nonliving things, like cars or stones, don't have a welfare; while they can break down, or fall apart, there aren't states that are better or worse *for* them.

The meaning of the term *welfare* may be extended to apply to all living things. A drought, for example, can be bad for an oak tree, and sunlight will typically be good for it. However, most discussions of welfare focus on sentient beings, that is, humans and some nonhuman animals, who are aware of and care about what happens to them. Welfare understood as being not only good or bad *for* an individual but also as belonging to an individual who is an experiencing subject may be viewed as the "core" sense of welfare (Sumner 1996, 14). This is the sense of welfare with which we will work here.

Concern for animal welfare, then, is a concern for how the lives of sentient nonhuman animals are going for the sake of and from the perspective of the animals themselves. In most developed countries, this concern is officially recognized through so-called animal welfare legislation, which requires people to look after the welfare of animals in their care.

At the same time, animal welfare is highly controversial and divisive largely because of disagreement about how to balance concerns for animal welfare against concerns for human interests. For example, it's often maintained that if scientific experiments can help cure human disease, it's acceptable to experiment on animals even if this compromises their welfare. This frames the animal welfare goal as being about minimizing the compromise of animal welfare needed to achieve human benefits. Similarly, efforts to improve farm animal welfare tend to assume that it's acceptable for humans to raise animals for meat, milk, and eggs; the welfare goal is to ensure that animals do not suffer unnecessarily in the process.

Because animal welfare is often framed in this way, some people who don't accept the use of animals for human benefit reject the whole idea of animal welfare (see chaps. 1, 22). But this throws the baby out with the bathwater. Animal welfare, understood as what matters from animals' own perspectives, should be of relevance to all who care about animals. In this chapter, we will try to develop an idea of what matters—of what "good welfare" is—for sentient animals. We will begin with the emergence of modern ideas of animal welfare in the United Kingdom in the 1960s, though we will argue that early ideas of welfare inadequately capture what matters to animals themselves. We will then begin to develop a fuller concept of animal welfare focused on "animal autonomy."

The Emergence of the Idea of Animal Welfare

Concern about animal suffering has, of course, a long heritage. In Victorian Britain, and shortly afterward in the United States, there was an upsurge of interest in the prevention of *cruelty* to animals. The United Kingdom enacted the world's first anticruelty law in 1822; other European countries soon followed. The idea that animals should be treated *humanely* also grew in popularity. Concern about cruelty (understood as causing gratuitous suffering to animals or being grossly negligent) and humane treatment (understood as something like compassionately reducing harms) both have a relationship to the idea of welfare. But neither of these ideas gets close to the modern idea of animal welfare that emerged in the 1960s.

Prompted by the publication of Ruth Harrison's book *Animal Machines* in 1964, increasing public misgivings were expressed about conditions in intensive farm animal production. This led the UK government to form the Brambell Committee to investigate and report on welfare conditions in British livestock farming. In 1965, the committee produced a *Report of the Technical Committee to Enquire into the Welfare of Animals Kept under Intensive Livestock Husbandry Systems*. The Brambell report, as it became known, was *theoretically* important for the way in which it conceptualized animal welfare—both in terms of the ideas it made prominent and those it omitted. It was also *practically* important because its recommendations formed the basis of subsequent British and European animal welfare legislation.

A major conceptual shift took place in the Brambell report and the animal welfare legislation that followed it. Whereas previous anticruelty legislation had focused on preventing what was seen as pointless or "wanton" suffering without human benefits, this new development in animal welfare involved protecting animals against the adverse consequences of practices useful to efficient food production. For example, although keeping sows or calves confined by chains or crates could be seen as integral to the efficient production of pork or veal, these methods were still criticized for denying animals the fulfillment of their basic needs. (However, as we shall see, this change was not as radical as it may first appear, since the focus remained on solving welfare problems within the context of efficient, large-scale animal production.)

The Brambell report understood animals' needs as requirements that would cause suffering if they were not met. And the report came with a new and wider understanding of suffering. Suffering had previously been conceptualized in terms of persistent and significant pain. However, the report introduced the idea that suffering could also follow from the frustration of "behavioral urges" in the form of discomfort, stress, and other negative mental states. This sense of suffering made it possible to criticize the confinement of sows and calves not based on the claim that confinement causes pain but rather that animals were prevented from engaging in behaviors they were highly motivated to perform.

Based on these ideas the Brambell report formulated the requirements that farm animals should be free to "to stand up, lie down, turn around, groom themselves and stretch their limbs" (13). These requirements became called the "Brambell Freedoms" (later developed into the Farm Animal Welfare Council's "Five Freedoms"; Farm Animal Welfare Council 2009). Viewed in the context of current critical debates about farm animal production, these requirements do not go very far. In particular, the report focuses entirely on reducing what is now called "negative welfare"—bad experiences and outcomes, such as pain and frustration. The report also understood behavioral needs narrowly—for instance, not including the need to engage in social and exploratory behaviors.

Nonetheless, this conceptual change, in particular the extension of the idea of suffering to include mental states other than pain, was both new and highly influential. Together with similar initiatives in other European countries, the Brambell report set off a process of legal reform taken up internationally, first by the Council of Europe and later by the EU. These reforms led to improvements in the conditions of farm animals across Western Europe. However, despite intending to ensure that farm animals' needs were met, early formulations of animal welfare, beginning with the Brambell report, remained loyal to the basic idea that intensive farm animal production is a legitimate human endeavor. This limited the scope of possible critique in several key ways.

First, the focus was mainly on the *absence* of suffering; there was no mention of positive welfare—that is, the presence of positive states, such as pleasure. Even the central principle of the Brambell report, that animals require certain basic freedoms, was understood to be violated only when the infringement of a freedom led to suffering. Second, the protection from suffering secured in most animal welfare legislation applied only to so-called unnecessary suffering—understood as suffering that could be avoided without giving up animal production. In practice, this allowed some welfare problems to be accepted as necessary, for example, those related to keeping laying hens in battery cages. Third, the emphasis was solely on a limited range of animals' subjective experiences without consideration of a wider range of natural behaviors.

So, a fuller account of animal welfare needs to move beyond the limitations of the Brambell report (see Sandøe and Jensen 2013). We will consider some important ways of doing this below.

Animal Welfare beyond the Brambell Report

Almost everyone agrees that animals should not be subjected to the worst instances of negative animal welfare, such as suffering from strong, protracted hunger, frustration, or pain. But can bad states always be avoided? Is avoiding bad states all that matters for good welfare? Is animal subjective experience all that matters, or does natural behavior matter, too?

The Role of Negative and Positive Welfare

While intense and lasting negative states are clearly bad for welfare, Mellor (2016) argues that it's impossible to achieve complete freedom from hunger, thirst, discomfort, pain, and injury across an animal's whole life. All that one can achieve is a minimization of the frequency and intensity of such negative states. And as Mellor also points out, it's not desirable to eliminate these states altogether; to some degree they are necessary for survival. The experience of pain is, for example, an important mechanism that warns animals away from potential harms.

Second, when we think about human welfare, we're not only concerned with the absence of negative states. We don't normally think someone has good welfare just because she or he is not suffering. Positive states, such as happiness or excitement, are also part of what makes for good welfare. The same is surely true for animals. Some positive states may also help animals to deal with unavoidable negative states—for instance, positive social engagement may help animals to cope with injuries or illnesses—just as it helps people to deal with these states. In addition, important positive things in life may only be achieved through a certain amount of suffering—grief can be seen as an unavoidable component of highly rewarding, close human relations, and similar relations may hold true in the case of nonhuman animals.

The Role of Subjective Experience in Animal Welfare

The Brambell report, as we've noted, understood animal welfare in terms of avoiding negative subjective experiences of strong and/or lasting pain and other forms of suffering. But it isn't only the focus on *negative* experience about which questions can be raised; it's the focus on *subjective experience in itself*. Is all that matters only what the animal actually experiences?

The idea of welfare understood solely in terms of subjective experience has a long history in the human case, where it's usually called the *hedonistic* view. The best life, on this view, is one with as many stimulating, comforting, and pleasurable experiences and as few frustrating, unpleasant, or painful experiences as possible. The more positive experiences relative to negative ones there are (i.e., the higher the net level of positives), the better the level of welfare.

Where humans are trying to promote good animal welfare, then, this approach requires the weighing of different possible lives an animal might have and choosing the one that's likely to bring the best balance of positive experiences over negative experiences for the animal concerned. Doing this is not always easy. For instance, if we have to choose whether to keep a cat indoors or not, we may have to try to weigh the potential pleasure to the cat of being allowed outdoors against (say) a 5 percent risk of the cat catching a fatal disease and a 5 percent risk of being run over by a car (Palmer and Sandøe 2014) It's very unclear how to make such decisions. It's also difficult to know what animals are actually experiencing. Some species of animals, including cats and donkeys, are more "stoical" than others, giving few indications when they are suffering.

In addition, subjective welfare may not simply involve adding and subtracting positive and negative states. While an experience of mild pain may not significantly affect an animal's welfare, at a certain threshold the animal may no longer be able to cope and may start to suffer. We can't assume that either pains or pleasures change in linear ways; there may be, as it were, relatively abrupt "tipping points." This also complicates decision making about subjective welfare.

Despite these difficulties, the idea that animal welfare should be understood in terms of animal subjectivity—how animals feel—is widely accepted. The biologist Marian Dawkins (1980), for instance, maintains that "to be concerned about animal welfare is to be concerned with the affective feelings of animals, particularly the unpleasant subjective feelings of suffering and pain." Mellor, Hunt, and Gusset (2015) in *Caring for Wildlife: The World Zoo and Aquarium Animal Welfare Strategy*, develop this view by advocating a version of what's called the "Five Domains" model of animal welfare (see fig. 1). While animal welfare, they suggest, should be understood along multiple domains, the four physical/functional domains (physical health, nutrition, environment, and behavior) are important inasmuch as they contribute to animals' subjective experiences in the

FIGURE 1. The "Five Domains" model of animal welfare

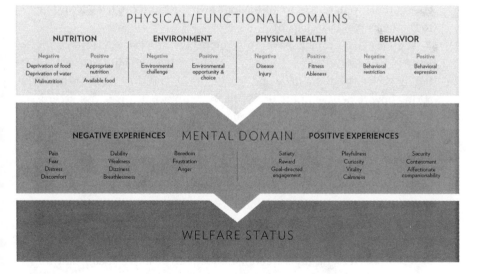

Reproduced by permission from D. J. Mellor, S. Hunt, and M. Gusset, eds., *Caring for Wildlife: The World Zoo and Aquarium Animal Welfare Strategy* (Gland: WAZA Executive Office, 2015). Figure modified from original in D. J. Mellor and N. J. Beausoleil, "Extending the 'Five Domains' Model for Animal Welfare Assessment to Include Positive Welfare States," *Animal Welfare* 24 (2015): 241–53.

mental domain. It's these subjective experiences that make for good or poor welfare.

Yet while having a good life in terms of experiences is clearly important, experiences may not be all that matters for welfare. In a famous thought experiment, the philosopher Nozick (1974) asks whether we would welcome the idea of being plugged into an experience machine for the rest of our lives if that machine gave us a constant stream of the most pleasurable experiences we could imagine and we believed them to be real. Most people don't regard this as a remotely tempting prospect, which suggests that it's not generally accepted that all that's good in our lives flows from our experiences.

Of course, people may be wrong to think this, and even if this were the case for humans, it may not be the case for other animals. But this does lead us to ask whether there's more to welfare than subjective experience alone. One of the most important ideas here concerns living according to one's nature.

The Role of Natural Behavior in Animal Welfare

Alongside subjective experience, animal welfare can also be understood in terms of *naturalness*: that is, the idea that at least part of what makes for good animal welfare is animals having the opportunity to live "according to their natures" (see chap. 6).

The idea that animals have, to some degree at least, species-specific natures is widely accepted. For instance, wolves live and hunt in extended family packs, the alpha male and female breed, and they teach their offspring how to behave in the pack. That's part of what it is to live according to "wolf nature." As Bernard Rollin (1993) puts it, "animals, too, have natures—the pigness of the pig, the cowness of the cow, 'fish gotta swim, birds gotta fly'—which are as essential to their well-being as speech and assembly are to us" (11). Here, Rollin suggests that good welfare (he uses the term *well-being*) is achieved when animals are free to live according to their natures.

This can be interpreted in two ways, however. It might mean that if animals live according to their natures, they are likely to have good experiences—essentially a hedonistic view. Or it might mean that for animals, to live according to their natures constitutes good welfare independently of their resulting experiences. Both approaches are interesting. Even an approach that does not go beyond a hedonistic view may, in practice, serve to make people more aware of how to improve animal welfare by enriching animal lives.

In the human case, this idea of "naturalness" as an approach to welfare is standardly called *perfectionism*: a good life for a human depends on the development of certain capacities central to human nature. On a "naturalness" view of animal welfare, being able to realize significant natural, species-specific potentials is the precondition of a good life. So, for instance, even if a wolf in a zoo has never hunted with his pack and isn't in any conscious sense aware of missing anything, nonetheless, something is lacking. The wolf has been denied the opportunity to fulfill his species-specific nature—which may be of value in its own right (from a strict perfectionist view) or may be valuable as a source of creating rewarding experiences (from a broad hedonist view).

Philosopher Martha Nussbaum has proposed a related view of a "good life" for an animal in her "capabilities" approach. Nussbaum (2006) maintains that "animals are entitled to a wide range of capabilities to function, those that are most essential to a flourishing life, a life worthy of the dignity of each creature" (392). According to Nussbaum, to ensure that an animal has a good life, we need to identify which capabilities are most important to animals in enabling their flourishing based on what's normal for the species. Once we have identified these core capabilities, we should recognize that animals can be harmed if prevented from fulfilling their core capabilities even if they are not aware of the harm.

Nussbaum does not just focus on *anything* animals naturally do but on *core* capabilities that are particularly important. Animal welfare scientists, too, often focus on "behavioral needs"—behaviors that are particularly important to animals because they are *internally motivated* to carry them out. Suckling, for instance, is important to calves, and if udders are not available, calves will suck on whatever they can, including other calves (Keeling, Rushen, and Duncan 2011, 23). So, suckling behavior is internally motivated, carried out even without environmental cues, and is important on this account of calf welfare (as it also is on a broad hedonistic account, since calves that cannot suckle are likely to feel frustration).

This understanding of animal welfare, particularly in its pure perfectionist form, is not without its critics. It can be argued that a species-specific approach to animal natures is far too general, just as "human nature" is too general not least because humans vary widely. Arguments of this kind have been made particularly strongly in the context of animals altered by genetic modification or conventional breeding, where the animals are clearly still members of a particular species but where what's "in their nature" is different (by design) from what's "in the nature" of unmodified species members. Judging good welfare for these modified animals by what's species specific could be highly problematic for the modified animals concerned.[1] This objection could be expanded to a more general friendly amendment: that animals' natures are not entirely determined by their species.

Natural behavior is also controversial when it creates negative subjective feelings. For instance, while fighting between male members of a species may be natural, it is also stressful, can cause injury, and gives rise to pain, fear, and other negative experiences. So in terms of subjective experience, it may seem better to prevent stressful—though natural—fights that result in injuries, for instance by castrating males to reduce aggression.

This all points toward a bigger problem. If we adopt a wide and multifaceted concept of welfare, including positive and negative welfare, and subjective experience as well as the freedom to express natural behaviors, decision making becomes extremely complex and risks being piecemeal. How can we identify the "animal's perspective"? Is there a more unified way to think about animal welfare?

Making Decisions about Animal Welfare: Measurement and Aggregation

Animal welfare science has, until recently, had to deal with exactly this problem of a piecemeal approach to animal welfare. Scientists typically investigated the effects of different conditions and treatments on animals and measured outcomes in terms of factors such as levels of stress hormone (cortisol), pathologies, mortality levels, stereotypies, and other forms of abnormal behavior. This work was carried out in artificial experimental settings with one group of animals being studied and another serving as a control.

However, these experimental studies caused frustration because they seemed ineffective at identifying the real problems for animal welfare and because such piecemeal welfare measures are extremely difficult to unite into an idea of overall welfare. Therefore, animal welfare scientists were increasingly led both to focus on tools that aggregate welfare and to develop on-farm animal welfare assessment. Arguably, this development

---

1. See Sandøe et. al 2014 for a discussion about whether welfare is all that matters when animals are modified in these ways.

culminated in the recently concluded large-scale EU project Welfare Quality®.

Welfare Quality® ultimately aimed to collect welfare scores that allowed for comparisons across farms and for individual farms to document progress over time. Since the scores were expressed in single numbers, derived from many measurements, the need to aggregate is very obvious—but the process of aggregation generates perplexing questions.

Difficult questions began to arise when different values were added together to arrive at an overall score. For instance: to what extent should a low value in one criterion be compensated for by a high value in another criterion? We can see how this plays out in practice, taking the three criteria that feed into the principle "Good Health" for fattening pigs as an example.

Here are examples of five sets of criteria scores feeding into the principle "good health" (table 1). Low scores (on this chart, scores of 25) mean negative welfare and animal suffering. High scores (on this chart, scores of 75) mean reasonably good but not perfect welfare. Each combination in this chart has the same average (50). But the aggregated "good health" score is not the same across all five cases. That's partly because the higher values (of 75) are not permitted to fully compensate for the lower values (25) when criteria scores are added. This is explained by an underlying ethical principle giving greater weight to the prevention of suffering (low values) than it does to the promotion of good welfare (high values). In addition, a different relative weight is given to the three criteria (otherwise the first two and the last two principle scores would have been the same).

TABLE 1. Example of scores for the overall principle "good health" in fattening pigs given by scores obtained for the criteria "absence of injuries," "absence of diseases," and "absence of pain due to management procedures"

Criteria			Principle
Absence of injuries	Absence of diseases	Absence of pain	Good health
25	50	75	32
25	75	50	35
50	50	50	50
75	25	50	28
75	50	25	34

Source: I. Veissier, K. K. Jensen, R. Botreau, and P. Sandøe, "Highlighting Ethical Decisions Underlying the Scoring of Animal Welfare in the Welfare Quality® Scheme," *Animal Welfare* 20 (2011): 89–101.

This gives us the outcome that row three has the best health. But it's very unclear on what grounds these weightings are made (see more in Veissier et al. 2011).

This brief discussion of the Welfare Quality® project suggests that there are significant problems involved in unified approaches to welfare that involve the aggregation of scores along different dimensions. The key problem is that to deal with the many potentially conflicting elements of an animal's welfare, human experts are called in to do multiple balancing and weighting exercises. It is almost as if the animals themselves drop out of the picture and animal welfare becomes dependent on human judgment despite the fact that Welfare Quality® was devised as a way of building animal welfare assessment purely on "animal-based" measures.

An alternative approach, which we consider more successful, tries to emphasize animals' own preferences about their welfare. To this we will now turn.

Making Decisions about Animal Welfare: Animal Autonomy

A different way of approaching overall animal welfare is to focus on what we might call, "animal autonomy"—where animals have more control over their own lives and are able to make their own choices. This hasn't featured as strongly in discussions of animal welfare as it should—perhaps because the idea of animal welfare emerged in the context of farm animal welfare, where animals are largely passive recipients of human provision. In this more autonomous sense of welfare, wild animals are likely to have the most autonomy (despite the fact that they may also undergo considerable suffering). Of course, wild animal autonomy is also limited in various ways—for instance by predators, food supply, or environmental conditions—but wild animals are, nonetheless, largely in charge of their own lives.

Terry Maple (2015) has developed something like this idea in the context of zoo animals, arguing that the goal for zoo animals should be "wellness" (or "well-being") rather than "welfare." Applying ideas from humanistic psychology, Maple argues that animals need the opportunity for self-actualization by facing challenges and overcoming them, thus becoming more resilient. Optimal wellness for zoo animals, Maple argues, requires a "stimulating environment" that encourages the animal "to express its full repertoire of behavior and reach a state of thriving, rather than mere coping with the constraints and demands of a zoo." This relates both to the idea of allowing animals the opportunity to perform natural behavior and to an idea of animal autonomy where animals make self-directed choices and have as much control as possible over their own lives.

However, this idea of animal autonomy raises two obvious difficulties. First is the problem of "bad choices," when animals under human care opt to do things that are obviously self-destructive. The second concerns animals kept under close human management—especially in farms and laboratories. Can these animals approach anything like autonomy? Or is an emphasis on animal autonomy as key to overall welfare inconsistent with keeping animals in farms and laboratories at all?

Of course, humans as well as animals can make self-destructive choices. In the human case, it's usually argued that people should be free to act on even self-destructive choices if they understand the consequences and others are not harmed. However, some animals can't reflect on their choices as most adult humans do, and they may well not understand the consequences of their actions. For instance, some Labrador retrievers left to feed autonomously would eat themselves to death.

The situation with animals in human care may more closely parallel that of children. When children make obviously self-destructive choices, it's expected that those who care for them will intervene. People can make choices on an animal's behalf as they would for children, reflecting what one thinks an animal would choose were he or she able to make decisions in an informed and reflective way. That interventions to protect an animal (or a child) from making self-destructive choices are sometimes justified does not imply that there is no value in the idea that in normal situations animals (and children) should be allowed to make their own choices about what matters to them.

The second concern was about animals under close human management, largely living as passive recipients of what humans offer them. Can these animals have good welfare in the suggested autonomous sense?

To some degree, even in controlled systems, it's possible to get a sense of what animals choose by preference, motivation, and aversion tests (Widowski 2015). Preference tests were developed by Marian Dawkins in the 1980s and have since been pursued by a number of prominent animal welfare scientists. These tests may present animals with different forms of bedding or different foodstuffs to choose between, or animals may have to work in order to access something. This provides some indication of what animals want and how motivated they are to get it. However, while these tests work within narrow parameters, such choices are obviously limited by what the animals are actually offered.

Work in the area of animal decision making has recently been reinvigorated and promises to become much broader and more sophisticated. For instance, at the opening lecture of the 2016 conference of the International Society for Applied Ethology, Christine Nicol (2016) argued, "As we look towards the future we can see that work on animal decision-making will be a vital adjunct of studies of animal emotion, since the only way of defining the valence of a stimulus or a situation in operational terms is via an animal's decision to approach or avoid." This commitment to animal decision making could usher in a general approach to animal welfare—applicable to wild animals as well as farm animals—based on some degree, at least, of animal autonomy.

However, while the recognition of autonomy as important to welfare in the case of farm and laboratory animals is growing, for obvious reasons, it's unlikely that these animals' autonomy will ever approach that of most wild animals. And even where animals are given more opportunities to take control of their own lives, we need to be aware that the way in which they have previously been under human control might shape what they do. For instance, adult cats that have never been outdoors may, given the opportunity, be too afraid to go outside. But this choice is shaped by the fact that they have always been confined. In the human case, preferences shaped by restricted or oppressive circumstances are sometimes described as "deformed" or "adaptive." Perhaps indoor cats who prefer to stay indoors (or at least do not prefer to go outdoors) might also have "adaptive preferences."

Still, in our view, autonomous choices made by animals should be considered the primary way of trying to find out what's best from an animal's perspective and for helping us to choose between potentially conflicting situations or treatments in welfare terms.

Ideas about understanding animal welfare continue to change, from an early focus on avoiding significant suffering to more recent approaches that emphasize animal thriving and autonomy. These recent approaches help us to understand what matters about the lives of sentient nonhuman animals for the sake of and from the perspective of the animals themselves, that is, what matters about animal welfare.

Good animal welfare as we see it should mean a clear balance of positive over negative states, the opportunity for animals to carry out natural behaviors, and animals' ability to choose what matters most to them—provided that natural behaviors and choices are not in an obvious sense harmful to the animals. This notion of animal welfare can both be construed as a form of perfectionism or as a version of what we call "broad hedonism."

We acknowledge that the goal of "good animal welfare" may still be rejected from abolitionist and animal rights perspectives that see it as a gesture that legitimates animals' continued exploitation (see, e.g., Wadiwal 2016). And even for those who do not, in principle, reject every form of animal use, there's an important discussion to be had about how to balance human interests with animal welfare. Our concern here has been, rather, to try to ensure that discussion about this human/animal balance does not disadvantage animals from the outset by adopting an overly narrow definition of what counts as good animal welfare.

Suggestions for Further Reading

Dawkins, M. 1980. *Animal Suffering: The Science of Animal Welfare*. London: Chapman and Hall.

Keeling, L. J., J. Rushen, and I. J. H. Duncan. 2011. "Understanding Animal Welfare." In *Animal Welfare*, 2nd ed., edited by Michael Appleby, Barry Hughes, Joy Mench, and Anna Ollson, 13–26. Wallingford: CABI International.

Maple, T. L. 2015. "Four Decades of Psychological Research on Zoo Animal Welfare." *WAZA Magazine* 16 (1): 41–44.

Mellor, D. 2016. "Updating Animal Welfare Thinking: Moving beyond the 'Five Freedoms' towards 'A Life Worth Living.'" *Animals* 6 (3): 21.

Nussbaum, M. 2006. *Frontiers of Justice*. Cambridge, MA: Harvard University Press.

References

Brambell, F. W. R. 1965. *Report of the Technical Committee to Enquire into the Welfare of Animals Kept under Intensive Livestock Husbandry Systems*. Command Rep. 2836. London: Her Majesty's Stationery Office.

Dawkins, M. 1980. *Animal Suffering: The Science of Animal Welfare*. London: Chapman and Hall.

Farm Animal Welfare Council. 2009. *Farm Animal Welfare in Great Britain: Past, Present and Future*. London: Farm Animal Welfare Council.

Houe, H., P. Sandøe, and P. T. Thomsen. 2011. "Welfare Assessments Based on Lifetime Health and Production Data in Danish Dairy Cows." *Journal of Applied Animal Welfare Science* 14: 255–64.

Keeling, L. J., J. Rushen, and I. J. H. Duncan. 2011. "Understanding Animal Welfare." In *Animal Welfare*, 2nd ed., edited by Michael Appleby, Barry Hughes, Joy Mench, and Anna Ollson, 13–26. Wallingford: CABI International.

Maple, T. L. 2015. "Four Decades of Psychological Research on Zoo Animal Welfare." *WAZA Magazine* 16 (1): 41–44.

Mellor, D. 2016. "Updating Animal Welfare Thinking: Moving beyond the 'Five Freedoms' towards 'A Life Worth Living.'" *Animals* 6 (3): 21.

Mellor, D. J., and N. J. Beausoleil. 2015. "Extending the 'Five Domains' Model for Animal Welfare Assessment to Include Positive Welfare States." *Animal Welfare* 24: 241–53.

Mellor, D. J., S. Hunt, and M. Gusset, eds. 2015. *Caring for Wildlife: The World Zoo and Aquarium Animal Welfare Strategy.* Gland: WAZA Executive Office.

Nicol, C. 2016. "Decisions, Decisions: Animals, Scientists and the Time That Is Given Us." In *Proceedings of the 50th Congress of the International Society for Applied Ethology 12–15th July, 2016, Edinburgh, United Kingdom*, edited by C. Dwyer, M. Haskell, and V. Sandilands, 69. Wageningen, the Netherlands: Wageningen Academic.

Nozick, R. 1974. *Anarchy, State and Utopia.* New York: Basic Books.

Nussbaum, M. 2006. *Frontiers of Justice.* Cambridge, MA: Harvard University Press.

Palmer, C., and P. Sandøe. 2014. "For Their Own Good: Captive Cats and Routine Confinement." In *The Ethics of Captivity*, edited by Lori Gruen, 135–55. New York: Oxford University Press.

Rollin, B. E. 1993. "Animal Production and the New Social Ethic for Animals." In *Food Animal Well-Being: Conference Proceedings and Deliberations*, 3–13. West Fayette, IN: Purdue University, Office of Agricultural Research Programs.

Sandøe, P., P.-M. Hocking, B. Förkman, K. Haldane, H. H. Kristensen, and C. Palmer. 2014. "The Blind Hens' Challenge: Does It Undermine the View That Only Welfare Matters in Our Dealings with Animals?" *Environmental Values* 23 (6): 727–42.

Sandøe, P., and K. K. Jensen. 2013. "The Idea of Animal Welfare: Developments and Tensions." In *Veterinary and Animal Ethics: Proceedings of the First International Conference on Veterinary and Animal Ethics, September 2011*, edited by C. M. Wathes, S. A. Corr, S. A. May, S. P. McCulloch, and M. C. Whiting, 19–31. UFAW Animal Welfare Series. Oxford: Wiley-Blackwell.

Sumner, L. W. 1996. *Welfare, Happiness, and Ethics.* Oxford: Clarendon.

Veissier I., K. K. Jensen, R. Botreau, and P. Sandøe. 2011. "Highlighting Ethical Decisions Underlying the Scoring of Animal Welfare in the Welfare Quality® Scheme." *Animal Welfare* 20: 89–101.

Wadiwal, D. 2016. *The War against Animals.* Leiden: Brill.

Widowski, T. 2015. "Why Are Behavioral Needs Important?" In *Improving Animal Welfare: A Practical Approach.* 2nd ed. Edited by T. Grandin, 247–66. Wallingford: CABI.

Acknowledgments

I am deeply grateful to all of the authors included here. Their brilliance and good humor made the work of editing this volume a joy. I'd like to thank Susan Basow, pattrice jones, Claire Jean Kim, Fiona Probyn-Rapsey, and especially Margot Weiss for good advice and welcome support.

Christie Henry was an incredibly skilled editor to work with from the conception of the volume until the final manuscript was approved. I am so glad we had the opportunity to work closely together; her wisdom is sprinkled throughout the preceding pages. I'm grateful also to Alan Thomas, Miranda Martin, and Susan Zakin at the University of Chicago Press for bringing the book through production.

I am deeply grateful to the University Center for Human Values at Princeton University for supporting me during the final stages of proofreading and indexing, while I served as a Laurance S. Rockefeller Visiting Professor for Distinguished Teaching.

My heartfelt thanks to Zinnie for keeping me company, under my feet, while I was at the computer, and to Taz, who seems to always know just when I need to get up from the computer to spend time with animals.

Contributors

KRISTIN ANDREWS is York Research Chair in the Philosophy of Animal Minds at York University in Toronto, Canada, and the author of two books: *Do Apes Read Minds? Toward a New Folk Psychology* (MIT Press 2012), and *The Animal Mind* (Routledge 2015). She is also the coeditor of *The Routledge Handbook of the Philosophy of Animal Minds* (2017).

PHILIP ARMSTRONG is the codirector (with Annie Potts) of the New Zealand Centre for Human-Animal Studies (www.nzchas.canterbury.ac.nz). He is the author of *Sheep* (Reaktion Books, 2016) and *What Animals Mean in the Fiction of Modernity* (Routledge, 2008), the coauthor (with Annie Potts and Deidre Brown) of *A New Zealand Book of Beasts: Animals in Our History, Culture and Everyday Life* (Auckland University Press, 2013), and the coeditor (with Laurence Simmons) of *Knowing Animals* (Brill, 2007).

VICTORIA A. BRAITHWAITE was a faculty member at Edinburgh University for twelve years before becoming professor of fisheries and biology at Pennsylvania State University. She is currently the codirector of the Center for Brain, Behavior and Cognition. She works on animal cognition, including sensory perception and pain. She is the author of a popular science book, *Do Fish Feel Pain?* (Oxford 2010).

ALICE CRARY is professor of philosophy and cofounder of the Gender and Sexuality Studies Program at the New School for Social Research. She is the author of *Beyond Moral Judgment* (Harvard University Press 2007) and *Inside Ethics* (Harvard University Press 2016), and she is the editor of three additional books and the author of numerous articles.

COLIN DAYAN is professor of English, Robert Penn Warren Professor in the Humanities, and professor of law at Vanderbilt University. Her books include *The Law Is a White Dog*; *Haiti, History, and the Gods*; and, most recently, *With Dogs at the Edge of Life*. A recipient of National Endowment for the Humanities and Guggenheim fellowships, she was elected to the American Academy of Arts and Sciences in 2012.

MANEESHA DECKHA is professor and Lansdowne Chair at the Faculty of Law, University of Victoria in British Columbia, Canada. Her research and teaching interests include critical animal law, feminist analysis of law, post-colonial legal studies, reproductive rights, health law, and bioethics. She has published in Canada and internationally in sociolegal and interdisciplinary venues and is the recipient of grants from the Canadian Institutes of Health Research, the Social Sciences and Humanities Research Council, and the Canada-U.S. Fulbright Program. Professor Deckha has held the Fulbright Visiting Chair in Law and Society at New York University.

SUE DONALDSON is an associate researcher in the Department of Philosophy, Queen's University, Canada. She is the coauthor (with Will Kymlicka) of *Zoopolis: A Political Theory of Animal Rights* and of numerous articles exploring conceptions of multispecies society, politics, and democracy.

THOM VAN DOOREN is associate professor of environmental humanities at the University of New South Wales, Australia, and coeditor of the journal *Environmental Humanities*. His research and writing focus on some of the many philosophical, ethical, cultural, and political issues that arise in the context of species extinctions and human-wildlife entanglements. These themes are explored in a sustained manner in his second book, *Flight Ways: Life and Loss at the Edge of Extinction* (Columbia University Press, 2014). Van Dooren has held visiting positions at the University of California, Santa Cruz (2005, 2010) and the KTH Environmental Humanities Laboratory in Stockholm (2014) and has been a Humboldt Research Fellow at the Rachel Carson Center, Munich (2014–2016).

AGUSTÍN FUENTES is a professor and chair of the anthropology department at the University of Notre Dame. His current research focuses on cooperation, community, and semiosis in human evolution; ethnoprimatology and multispecies anthropology; evolutionary theory; and interdisciplinary approaches to human nature(s). Fuentes's recent books include *Race, Monogamy, and Other Lies They Told You: Busting Myths about Human Nature* (University of California Press), *Conversations on Human Nature(s)* (Routledge), and *The Creative Spark: How Imagination Made Humans Exceptional* (Dutton).

LORI GRUEN is the William Griffin Professor of Philosophy at Wesleyan University. She is also a professor of feminist, gender, and sexuality studies and coordinator of Wesleyan Animal Studies. She is the author and editor of nine books, including *Ethics and Animals: An Introduction* (Cambridge, 2011), *The Ethics of Captivity* (Oxford, 2014), and *Entangled Empathy* (Lantern, 2015). She is a fellow of the Hastings Center for Bioethics, a faculty fellow at Tufts's Cummings School of Veterinary Medicine's Center for Animals and Public Policy, and was the first chair of the faculty advisory committee of the Center for Prison Education at Wesleyan.

ALEXANDRA HOROWITZ studies dog cognition and teaches psychology at Barnard College, Columbia University. The Horowitz Dog Cognition Lab at Barnard conducts research on a wide range of topics, including dog olfaction, play behavior, and attributions of secondary emotions to dogs. In addition to many scholarly articles relating to dog behavior and cognition, she is author of *Inside of a Dog: What Dogs See, Smell, and Know* (2009), *On Looking: Eleven Walks with Expert Eyes* (2013), *Being a Dog: Following the Dog into a World of Smell* (2016), and editor of *Domestic Dog Cognition and Behavior* (Springer, 2014).

CLAIRE JEAN KIM is professor of political science and Asian American studies at the University of California, Irvine, Her book *Bitter Fruit: The Politics of Black-Korean Conflict in New York City* (Yale University Press, 2000) won two awards from the American Political Science Association. Her book *Dangerous Crossings: Race, Species, and Nature in a Multicultural Age* (Cambridge University Press, 2015), is also a recipient of a Best Book Award from the American Political Science Association. She was co–guest editor of a special issue of *American Quarterly*, "Species/Race/Sex" (September 2013). She is the recipient of a grant from the University of California Center for New Racial Studies and has been a fellow at the University of California Humanities Research Institute and a visiting fellow at the Institute for Advanced Study in Princeton, New Jersey.

BARBARA J. KING is emerita professor of anthropology at the College of William and Mary in Williamsburg, Virginia, and now is a full-time freelance science writer. Her latest book, published in 2017, is *Personalities on the Plate: The Lives and Minds of Animals We Eat*. Her *How Animals Grieve* from 2013 has been translated into Japanese, Portuguese, French (winning a book prize), and Hebrew. She blogs each week for National Public Radio's 13.7 *Cosmos and Culture* blog.

EDUARDO KOHN is the author of the book *How Forests Think: Toward an Anthropology beyond the Human*, which won the 2014 Gregory Bateson Award and has been translated into several languages. His ongoing ethnographic work, in Amazonia and elsewhere, on the relations we humans have to the many other kinds of beings (animals, plants, and also spirits) with whom we share life on this planet aims to contribute to a new kind of ethical orientation for the so-called Anthropocene. He teaches anthropology at McGill University in Montreal and is research associate at FLACSO-Ecuador in Quito.

CHRISTINE M. KORSGAARD is Arthur Kingsley Porter Professor of Philosophy at Harvard University. She was elected to the American Academy of Arts and Sciences in 2001, was a corresponding fellow of the British Academy in 2015, held a Mellon Distinguished Achievement Award from 2006 to 2009, was president of the Eastern Division of the American Philosophical Association in 2008–2009, and holds honorary degrees from the University of Illinois at Urbana-Champaign (2004) and the University of Groningen (2014). She works on moral philosophy and its history, practical rationality, the nature of agency, personal identity, and the ethics of our treatment of animals. She is the author of five books: *Creating the Kingdom of Ends* (Cambridge, 1996), *The Sources of Normativity* (Cambridge, 1996), *The Constitution of Agency* (Oxford, 2008), and *Self-Constitution: Agency, Identity, and Integrity* (Oxford, 2009), and *Fellow Creatures* (Oxford, forthcoming), and is working on *The Natural History of the Good*, a book about the place of value in nature.

WILL KYMLICKA is the Canada Research Chair in Political Philosophy at Queen's University. He has published eight books and over two hundred articles, which have been translated into thirty-two languages. His books include *Contemporary Political Philosophy* (1990; 2nd ed. 2002), *Multicultural*

Citizenship (1995), which was awarded the Macpherson Prize by the Canadian Political Science Association and the Bunche Award by the American Political Science Association, *Multicultural Odysseys: Navigating the New International Politics of Diversity*(2007), which was awarded the North American Society for Social Philosophy's 2007 book award, and *Zoopolis: A Political Theory of Animal Rights* (2011), coauthor with Sue Donaldson.

LORI MARINO is a neuroscientist and expert in animal behavior and intelligence who formerly taught at Emory University in the neuroscience and behavioral biology program. She was also a faculty affiliate at the Emory Center for Ethics. She is currently president of *The Whale Sanctuary Project*, whose mission is to create the first seaside sanctuaries for orcas (killer whales) and beluga whales in North America. She is also founder and executive director of The Kimmela Center for Animal Advocacy, a science-based nonprofit organization focused on bringing academic scholarship to animal protection efforts.

ROBERT R. MCKAY teaches literature, film, and critical theory at the University of Sheffield, where he is codirector of the Sheffield Animal Studies Research Centre. His writing focuses on the politics of species in modern and contemporary literature and film. He is the coauthor (with the Animal Studies Group) of *Killing Animals* (Illinois University Press, 2006) and coeditor of *Against Value in the Arts and Education* (Rowman and Littlefield, 2016) and of *Werewolves, Wolves and the Gothic* (Wales University Press, 2017). He is also series coeditor for *Palgrave Studies in Animals and Literature* and associate editor (literature) for *Society and Animals*.

TIMOTHY PACHIRAT is assistant professor of political science at the University of Massachusetts, Amherst and author of *Every Twelve Seconds: Industrialized Slaughter and the Politics of Sight* (Yale University Press, 2011) and *Among Wolves: Ethnography and the Immersive Study of Power* (Routledge, 2018).

CLARE PALMER is professor of philosophy and Cornerstone Fellow in the Liberal Arts at Texas A&M University. She works in environmental ethics, animal ethics, and climate ethics. She is the author of *Animal Ethics in Context* (Columbia University Press, 2010), and she coauthored *Companion Animal Ethics* with Peter Sandøe and Sandra Corr (Wiley Blackwell, 2015).

She has edited or coedited a number of volumes in environmental and animal ethics, including *Animal Rights* (Ashgate, 2008), and, with J. Baird Callicott, the five-volume set *Environmental Philosophy* (Routledge, 2005)

ANAT PICK lectures in film studies at Queen Mary, University of London. She is the author of *Creaturely Poetics: Animality and Vulnerability in Literature and Film* (2011), and *Maureen* (2016), and she is coeditor of *Screening Nature: Cinema beyond the Human*. She has published widely on animals, ethics, and the visual and is working on a new book titled *Vegan Cinema: Looking, Eating, Letting Be.*

NATALIE PORTER is an assistant professor of anthropology at the University of Notre Dame, where her research explores zoonoses and multispecies relations, the intersections of global health and livestock economies, and emerging regimes of property in biomedical research and development. Natalie is currently completing a book based on long-term ethnographic research on avian flu in Vietnam.

ANNIE POTTS is associate professor and the codirector of the New Zealand Centre for Human-Animal Studies at Canterbury University. She is the author of *Chicken* (Reaktion, 2012), coauthor (with Philip Armstrong and Deidre Brown) of *A New Zealand Book of Beasts: Animals in our Culture, History and Everyday Life* (Auckland University Press, 2013) and (with Donelle Gadenne) of *Animals in Emergencies* (Canterbury University Press, 2014), and editor of *Meat Culture* (Brill, 2016). Along with Jovian Parry, Annie has also coauthored several articles on vegan sexuality and the gendered construction of meat consumption and plant-based diets.

FIONA PROBYN-RAPSEY is professor in the School of Humanities and Social Inquiry at the University of Wollongong, Australia. Her research connects feminist critical race studies and Animal Studies, examining where, when, and how gender, race, and species intersect. She is the author of *Made to Matter: White Fathers, Stolen Generations* (2013) and the coeditor of *Animal Death* (2013) and *Animals in the Anthropocene: Critical Perspectives on Non-human futures* (2015). Fiona is also series editor (with Melissa Boyde) of the Animal Publics book series through Sydney University Press.

HARRIET RITVO is the Arthur J. Conner Professor of History at the Massachusetts Institute of Technology. She is the author of *The Dawn of Green:*

Manchester, Thirlmere, and Modern Environmentalism (University of Chicago Press, 2009), *The Platypus and the Mermaid, and Other Figments of the Classifying Imagination* (Harvard University Press, 1997), *The Animal Estate: The English and Other Creatures in the Victorian Age* (Harvard University Press, 1987), and *Noble Cows and Hybrid Zebras: Essays on Animals and History* (University of Virginia Press, 2010). Her articles and reviews on British cultural history, environmental history, and the history of human-animal relations have appeared in a wide range of periodicals, including the *London Review of Books, Science, Daedalus, American Scholar, Technology Review*, and the *New York Review of Books*,\ as well as scholarly journals in several fields. Her current research concerns wildness and domestication.

PETER SANDØE is professor of bioethics at the University of Copenhagen. The major part of his research has focused on ethical issues related to animals, biotechnology, and food production. He is committed to interdisciplinary work combining perspectives from natural science, social sciences, and philosophy and publishes in a wide range of scholarly journals. His books include *Ethics of Animal Use* (coauthored with Stine B. Christiansen), published by Blackwell (2008), and *Companion Animal Ethics* (coauthored with Sandra Corr and Clare Palmer), published by Wiley/Blackwell (2016).

JEFF SEBO is clinical assistant professor of environmental studies and director of the Animal Studies MA program at New York University. He works primarily in bioethics, animal ethics, and environmental ethics. His book *Food, Animals, and the Environment: An Ethical Approach* (coauthored with Christopher Schlottmann) is forthcoming from Routledge, and his book *Why Animals Matter for Climate Change* is forthcoming from Oxford. In addition to his academic work, Jeff sits on the board of directors at Animal Charity Evaluators, the board of directors at Minding Animals International, and the executive committee at the Animals and Society Institute.

PETER SINGER is Ira W. DeCamp Professor of Bioethics in the University Center for Human Values at Princeton University, a position that he now combines with the position of Laureate Professor at the University of Melbourne. His books include *Animal Liberation, Practical Ethics, The Ethics of What We Eat* (with Jim Mason), *The Life You Can Save*, and *The Most Good You Can Do*. He also has edited *In Defense of Animals* and its successor, *In Defense of Animals: The Second Wave*. Together with Paola Cavalieri, he cofounded The Great Ape Project, an attempt to gain basic rights for great apes, and coedited an anthology, also called *The Great Ape Project*.

JAMES K. STANESCU is assistant professor of communication studies and director of oral communication at Mercer University. He is the editor, with Kevin Cummings, of *The Ethics and Rhetoric of Invasion Ecology* (Lexington, 2017). He publishes and lecturers widely in the fields of Animal Studies and environmental studies, and is currently working on a monograph about meliorism in animal ethics.

KRISTEN STILT is professor of law at Harvard Law School. She also serves as faculty director of the Animal Law and Policy Program, director of the Islamic Legal Studies Program: Law and Social Change, and is the deputy dean. Stilt was named a Carnegie Scholar for her work on Constitutional Islam, and in 2013 she was awarded a John Simon Guggenheim Memorial Foundation Fellowship. She received a JD from the University of Texas School of Law and a PhD in history and Middle Eastern studies from Harvard University. Her research focuses on animal law and in particular the intersection of animal law and religious law; Islamic law and society; and comparative constitutional law. Publications include *Islamic Law in Action* (Oxford University Press, 2011), "Constitutional Innovation and Animal Protection in Egypt," *Law and Social Inquiry* (forthcoming), and others. She is currently working on a new book project titled *Halal Animals*.

MALINI SUCHAK is assistant professor in the undergraduate animal behavior, ecology and conservation program and the anthrozoology graduate program at Canisius College. Her research explores social cognition in nonhuman animals and the intersection of sociality and welfare. She is interested in better understanding how our companion animals navigate the multispecies social worlds that they live in. Previously, she explored the influence of social relationships on cooperation in chimpanzees and capuchin monkeys while she was at Emory University.

GARY VARNER is professor of philosophy at Texas A&M University. He has published extensively on topics in environmental ethics and animal ethics. His 2012 book, *Personhood, Ethics, and Animal Cognition: Situating Animals in Hare's Two-Level Utilitarianism* provides a detailed discussion of how the two-level utilitarianism of R. M. Hare would apply to animal ethics, something that Hare never did during his lifetime. Hare was Peter Singer's dissertation advisor, and Varner compares and contrasts the results to Singer's work.

DINESH JOSEPH WADIWEL is senior lecturer in human rights and sociolegal studies in the School of Social and Political Sciences, University of Sydney. Dinesh has background in social and political theory with research interests in theories of violence, critical Animal Studies, and disability rights. He is author of the monograph *The War against Animals* (Rodopi / Brill 2015) and coeditor (with Matthew Chrulew) of the collection *Foucault and Animals* (Brill 2016).

KARI WEIL is university professor and director of the College of Letters at Wesleyan University. She is the author of *Androgyny and the Denial of Difference* (University Press of Virginia, 1992), *Thinking Animals: Why Animal Studies Now* (Columbia University Press, 2012), and has published numerous essays on literary representations of gender, feminist theory, and, more recently, on theories and representations of animal otherness and human-animal relations. With Lori Gruen she coedited of a special issue of *Hypatia* on *Animal Others* 2012. Her current book project, *Meat, Motion, Magnetism: Horses and Their Humans in Nineteenth-Century France*, is forthcoming from the University of Chicago Press.

CYNTHIA WILLETT is the Samuel Candler Dobbs Professor of Philosophy at Emory University. Her authored books include *Interspecies Ethics* (2014), *Irony in the Age of Empire: Comic Perspectives on Freedom and Democracy* (2008), *The Soul of Justice: Racial Hubris and Social Bonds* (2001), and *Maternal Ethics and Other Slave Moralities* (1995). She has edited *Theorizing Multiculturalism* (1998) and is a coeditor for the Symposia on Race, Gender, and Philosophy. She is currently completing a book on humor and is beginning two new projects, the musicality of ethics and criticality theory and the Anthropocene.

Index

of, 47–48, 54–55; cows and, 59; Darwin and, 59; difference and, 115, 117n6; dogs and, 47–48, 58; emotion and, 125, 135–36; ethics and, 49–52, 56; extinction and, 170; farm animals and, 51, 60; feminists and, 48, 54–55; fish and, 58; Gruen and, 53; human chauvinism and, 48–50, 54; hunting and, 57–58; inevitable, 48–51; kinship and, 56, 58; language and, 59; law and, 204; life and, 48, 51–59; matter and, 222–26, 231; mind and, 47, 56; morals and, 50n1, 53, 55–56, 58; pain and, 58–59; personhood and, 277; philosophy and, 48, 56, 59; pigs and, 59; postcolonial analysis and, 54–55, 280, 282, 284, 287–89; racism and, 52–55, 58, 60; rats and, 58; representation and, 315, 317; resisting, 55–58; sentience and, 56, 59; sexism and, 48, 52–54, 58, 60; slaughter and, 51; slavery and, 29; sociality and, 50; suffering and, 60; veganism and, 400n5, 401n6, 403, 406; vulnerability and, 416; welfare and, 52

anthropodenial, 115
anthropological machine, 50, 83, 92, 119
anthropomorphism, 75, 115, 130, 135–36, 240, 311n3, 312
anthrozoology, 11, 450
anticruelty laws, 200, 206, 281–82, 425–26
ants, 68
apes. *See specific kinds*
Aphro-ism, 351
Arendt, Hannah, 82, 274
Are We Smart Enough to Know How Smart Animals Are? (de Waal), 135
Aristotle, 113, 191, 223, 294, 301
Armstrong, Philip, 395–409, 443
Arnhem Zoo, 142
Asian Leopard Cat (ALC), 390
Austin, J. L., 226
automatons, 222
autonomy: captivity and, 99–100; kinship and, 186; law and, 206; mind and, 246; moral, 328n12; personhood and, 273–74; sanctuary and, 346–48; sentience and, 356–59, 362–64; sexual, 206;

vulnerability and, 416; welfare and, 425, 434–37

baboons, 127, 134, 323, 372–76
bacteria, 184–85, 226
Bailey v. Poindexter's Executor, 270
Baker, Steve, 317–18
Balcombe, Jonathan, 137
Barad, Karen, 120n7, 223–24, 227, 229
Baron-Cohen, Simon, 143n3
Baur, Gene, 347
Bayesian reasoning, 239
bears, 101, 374, 376–79
beef, 20, 393
bees, 58, 68, 245
behavior, 13; abnormal, 103–6, 432; activism and, 33, 38; anthropocentrism and, 48, 75; birds and, 65, 67; captivity and, 103–8; cats and, 68; chickens and, 65; chimpanzees and, 67, 70; concept of, 64–65; Darwin and, 65; difference and, 114; dogs and, 67–69, 71–73, 75; dyadic play and, 71–73; emotion and, 65, 68, 72, 75, 126, 128, 133–37; empathy and, 141–43, 149; ethology and, 64, 66, 70, 74–75; evolution and, 67; extinction and, 175; fear grimace and, 72; fish and, 66n2, 72; hunting and, 71; infants and, 66–67; intelligence and, 65; kinship and, 187–88, 193; law and, 197–98, 201, 208; levels of, 70–71; life and, 66, 71, 74; matter and, 229; mind and, 65–66, 237–38, 241–42, 244–48; monkeys and, 75; naming, 71–73; neurons and, 71; nondefinitive definitions and, 67–69; observation of, 73–75; pain and, 258–61; pant threat and, 72; physiology and, 67–68; plants and, 247–48; prosocial, 141, 374, 377–78; psychology and, 64–66, 69, 75; rationality and, 295–99; rats and, 68–69; response and, 64–66, 68–69; sanctuary and, 248, 346n20, 351; science of, 75–76; sociality and, 374–79; species and, 383, 386; stimuli and, 66, 69; stress and, 67; studying, 65–67; welfare and, 426–37
behaviorism, 65–66, 75